Plants, Biotechnology
and Agriculture

Plants, Biotechnology and Agriculture

Professor Denis Murphy

University of Glamorgan, UK

www.cabi.org

CABI is a trading name of CAB International

CABI Head Office
Nosworthy Way
Wallingford
Oxfordshire OX10 8DE
UK

Tel: +44 (0)1491 832111
Fax: +44 (0)1491 833508
E-mail: cabi@cabi.org
Website: www.cabi.org

Learning Resources
Centre
13856952

CABI North American Office
875 Massachusetts Avenue
7th Floor
Cambridge, MA 02139
USA

Tel: +1 617 395 4056
Fax: +1 617 354 6875
E-mail: cabi-nao@cabi.org

A catalogue record for this book is available from the British Library, London, UK.

Library of Congress Cataloging-in-Publication Data

Murphy, Denis J.
  Plants, biotechnology and agriculture / Denis Murphy.
    p. cm.
  Includes bibliographical references and index.
    ISBN 978-1-84593-688-4 (pbk. : alk. paper) -- ISBN 978-1-84593-871-0 ('cabi south asia' edition : alk. paper)
1. Plants, Cultivated. 2. Agriculture. 3. Botany, Economic. 4. Plant biotechnology. I. Title.

SB91.M87 2011
631.5'233--dc22

                                        2011011316

ISBN-13: 978 1 84593 688 4 (paperback)
ISBN-13: 978 1 84593 913 7 (hardback)
CABI South Asia Edition ISBN: 978 1 84593 871 0

Commissioning editor: Nigel Farrar
Editorial assistant: Alexandra Lainsbury
Production editor: Shankari Wilford

Typeset by SPi, Pondicherry, India.
Printed and bound in the UK by Cambridge University Press, Cambridge, UK.

Contents

Preface

The aim of this book is to provide readers with a modern perspective on plants both as biological organisms and useful resources for people to exploit. The early chapters cover plant evolution, genomics, metabolism, organization, development and responses to the environment. In each case there is an emphasis on how biotechnology can be used to manipulate such processes for the benefit of humanity. During the past decade, many of our notions about plants and their manipulation have been challenged by the application of modern biotechnological methods. As well as contributing to our understanding of plant function, biotechnology has been an invaluable tool for crop improvement in agriculture. More recently it has also been used in bio-pharming, which is the production of pharmaceuticals in plants, and for the bioengineering of microalgae to provide renewable resources such as a new generation of carbon-neutral fuels and recyclable industrial raw materials.

Although most bio-based technologies for crop improvement, such as hybrid production or induced muta-genesis, attracted little public attention, one method, namely genetic engineering or GM, has become highly controversial in many parts of the world. As we pass through the second decade of the 21st century, the cultivation of genetically manipulated (GM) crops is still severely restricted in a few regions, particularly in Europe. In contrast, GM crops are grown widely across the Americas and in Asian countries such as India and China. Although the great majority of plant scientists are supportive of using the full range of modern biotechnologies to assist crop improvement, the general public remains sceptical about the role of GM technologies in particular. What are the scientific and social factors behind this phenomenon and what is the future for GM and other plant biotechnologies?

The scope of the book ranges from the earliest evidence of pre-agricultural plant manipulation over 30,000 years ago to the latest recombinant DNA methods used in 21st-century agricultural systems. Unlike most textbooks where there is a focus on the technical aspects of genetic engineering, this book will take a wider view of what constitutes modern plant biotechnology. In addition, the broader social, economic, commercial, legal and ethical contexts of all forms of crop-related technology will also be examined. These include an analysis of the immense contributions of chemical and mechanical technologies.

As well as a retrospective look at the interaction between biotechnology and society, we will consider the role of plant biotechnology as we look forward to an increasingly uncertain future. Here, its tools will be required to tackle enormous challenges to the welfare of human populations around the world. Such challenges include the predicted massive population increases over the next few decades, irreversible depletion of non-renewable resources of all types, and the spectre of climate change that might have unpredictable effects on crop growth, e.g. by reducing rainfall, altering temperatures, or leading to the emergence of new pests and diseases. These developments have the potential to seriously impact on crop productivity in some of the most densely populated and vulnerable regions of the world.

This book is aimed principally at senior level undergraduates and postgraduates studying for degrees including biology, plant science, agriculture, biotechnology, crop science and breeding. The book may also be useful to a wider readership including researchers in agbiotech, pharmaceutical, biofuel or breeding companies and other professionals, such as patent attorneys, environmentalists, journalists etc., seeking a broader insight into how contemporary plant science is being harnessed for crop improvement to address the many challenges of the 21st century.

Units

All units are metric and prices are in inflation-adjusted US dollars as of 2011. Ancient dates are listed as BP (before present) while modern dates are written in the normal common-era form.

Some common units/abbreviations

cM, centimorgan; BP, before present ('present' is 1950); ha, hectare; Gb, giga-base pairs of DNA; kb, kilo-base pairs of DNA; Mb, mega-base pairs of DNA; Mha, million hectares; $, US dollar; t, tonne (1000 kg); Mt, million tonnes.

Genes and proteins

Formal names for genes are *italicized* and their protein products are CAPITALIZED. Hence, one of the floral identity genes in angiosperms is *apetala3*, while its protein product is the transcriptional activator, APETALA 3. All genome sizes in the text refer to the 'haploid' state of the genome after meiosis.

1 Plants and their Exploitation by People

1.1 Chapter Overview

This chapter is an introduction to the nature and exploitation of plants. We begin by defining what exactly we mean by a plant and then examine the various ways in which human societies have interacted with plants during the course of our evolution. This leads on to a discussion of some of the technologies that have assisted with the management and improvement of crops, especially within the context of agriculture. The final part of the chapter is concerned with the various forms of biotechnology, ranging from ancient crafts such as brewing to the latest high-tech methods of genome manipulation being applied to microbes, animals and plants.

1.2 What is a Plant?

The definition of 'what constitutes a plant' has exercised the minds of people since ancient times. An obvious way of categorizing living organisms is to divide them into things that move (animals) and things that do not (plants). This was the system originally adopted by the ancient Greeks and was commonly used until modern times. However, according to this definition fungi would be included in the same group as plants simply because both were sessile (non-moving) organisms. Thanks to DNA sequence analysis, we now know that fungi are phylogenetically much closer to animals than to plants. Therefore, in modern classification systems plants, animals and fungi are placed in separate groups or taxa.

Unfortunately, the convenient division between plants, animals and fungi breaks down when we consider some of the many microorganisms that are clearly plant-like (they can photosynthesize) but also have animal characteristics (they are motile and can even ingest smaller cells as prey). For example, some unicellular green algae, such as *Euglena*, can use light to photosynthesize just like higher plants. However, in reduced light these algae can become heterotrophic and use their flagella to move around and hunt for prey similarly to animals. To complicate matters further, although the capacity for photosynthesis is generally regarded as the defining feature of a plant-like organism, there are numerous non-photosynthetic plants. For example, several groups of angiosperms, such as some forms of dodder (e.g. *Cuscuta* spp.) and some orchids (e.g. the bird's nest orchid, *Neottia nidus-avis*), have completely lost the ability to photosynthesize as a result of their parasitic lifestyles. These organisms are obviously plants but, because they cannot photosynthesize, they now lack one of the fundamental characteristics that would normally define them as plants.

In reality there is no simple way to define a plant and there will always be exceptions to any definition that we can come up with. So-called 'true plants' are usually defined in textbooks as members of the phylum Plantae. These organisms are normally multicellular, sessile, eukaryotic autotrophs that derive their main nourishment from oxygenic (oxygen evolving) photosynthesis, using water as an electron donor and atmospheric CO_2 as a carbon source. The vast majority of 'true plants' are terrestrial, although several have secondarily become aquatic. The 'lower plants' range from relatively simple non-vascular species such as liverworts and mosses to seedless vascular plants like the ferns. The seed-bearing 'higher plants' include the gymnosperms and angiosperms and are the most widespread, complex, diverse and successful groups of true plants. Gymnosperms and angiosperms are also the most thoroughly studied and economically useful group of plants and include all the major crop and forestry species. However, although higher plants are by far the most important group, several other types of plant-like organisms are also useful, especially in a biotechnological context.

The two main groups of oxygenic photosynthetic organisms that are not defined as 'true plants' are the algae and cyanobacteria. The algae, all of which are aquatic eukaryotes, range from simple phytoplankton species to more complex multicellular organisms such as the green, brown and red sea-weeds often seen on beaches at low tide. One particular group of relatively complex multicellular green algae, the Streptophytes, gave rise to the earliest land plants about 500 million years ago.

The algae also include many unicellular species, known as microalgae, all of which are eukaryotes. The final group of plant-like organisms is the cyanobacteria, which are the only group of prokaryotes capable of oxygenic photosynthesis. As shown in Fig. 1.1,

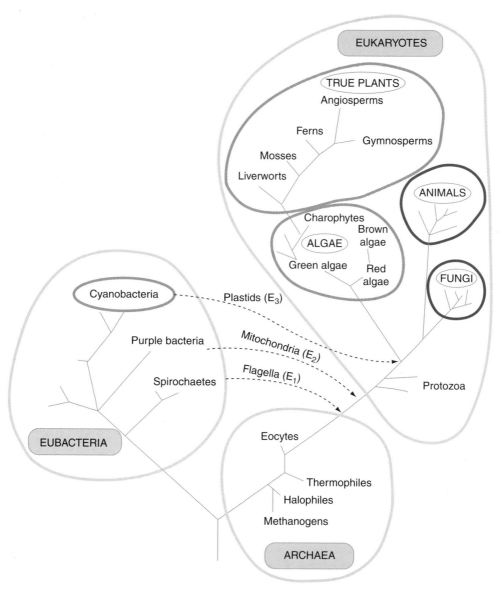

Fig. 1.1. The major groups of plants and plant-like organisms. There are three major groups of plant-like organisms. The largest group is the so-called 'true plants' that dominate most terrestrial ecosystems and provide the vast majority of crops and other useful plants that are exploited by people. The second major group is the algae, a diverse collection of photosynthetic eukaryotes, some of which have been long exploited by coastal communities. The third group is the cyanobacteria, which are the only prokaryotes capable of oxygenic photosynthesis. Several microalgae and cyanobacteria are currently being biotechnologically manipulated to act as bioreactors or sources of renewable biofuels. E1, E2 and E3 refer to the three endosymbiotic events leading to flagella, mitochondria and plastids respectively. See also Fig. 2.5.

the cyanobacteria are one of the evolutionary precursors of both algae and land plants. Marine microalgae and cyanobacteria together make up the phytoplankton, which in terms of its overall global biomass and photosynthetic productivity is more significant than all of the land plants combined.

With the increasing recognition of the key roles played by these microorganisms in marine food chains and in maintaining the global carbon cycle, greater attention is now being paid to analysing and exploiting these plant-like organisms. For example, one of the major biotechnological efforts of the 21st century is the engineering of some microalgae and cyanobacteria to produce renewable carbon-based fuels as alternatives to non-renewable fossil fuels. Such organisms could also provide renewable sources of valuable complex hydrocarbon-based industrial materials, such as plastics and lubricants. Other microalgae are being cultivated on a large scale *in vitro* to produce high-value protein- and vitamin-rich foods, for both human and animal consumption. Some of the practical uses of plants and plant-like organisms are depicted in Fig. 1.2.

While cyanobacteria and algae are not defined as 'true plants', they are obviously closely related to plants, especially in their ability to use photosynthesis as a renewable source of a wide range of potentially useful products for humankind. It is also the case that many of the modern biotechnological methods, especially DNA-based technologies, are now being used to exploit these diverse groups of organisms. For the purposes of this book, therefore, we will define plants, together with the plant-like organisms, as *organisms capable of oxygenic photosynthesis*. This definition includes all of the photosynthetic bacteria, the algae, and of course the 'true plants' themselves. We will return to a more detailed analysis of the evolution of photosynthetic organisms in Chapter 2.

1.3 Plant–Human Interactions

Having defined what we mean by a plant, we will now briefly examine the changing nature of plant–human interactions during human evolutionary development (see Table 1.1). Modern humans, or *Homo sapiens*, are descended from a branch of anthropoid apes that left their forest habitats to become bipedal and ground dwelling at least 4 million years ago. These bipedal apes were omnivores capable of eating both plant- and

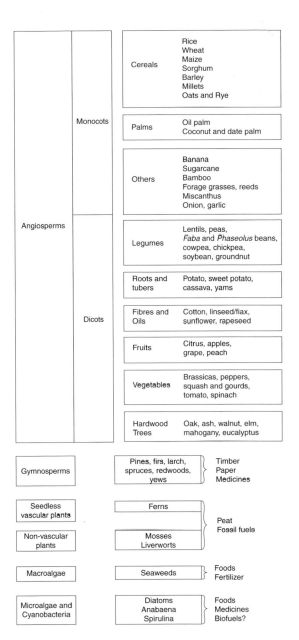

Fig. 1.2. Practical uses of various plant-like organisms. By far the most important group of plants for human use is the angiosperms, which includes all the starch-rich cereal crops and the nitrogen-fixing, protein-rich legumes. The other plant groups have much more minor roles, although many gymnosperms are important sources of wood for timber or paper production. Plant-based compounds make up about one-third of prescribed medicines and have been sourced from all groups of plants from cyanobacteria to dicots.

Table 1.1. Timeline of plant exploitation by people in Eurasia, 100,000–6,000 BP. Before the Younger Dryas, optimal food production strategies varied greatly as the climate changed, and with it the availability of such resources as megafauna, small game, fish and edible plants. The transition to agriculture started during the Younger Dryas cold period, about 12,000 BP.

Years BP	Climate	Human food resources[‡]	Technologies	Lifestyles
100,000–80,000	Cool and arid	Prolific fauna: aurochs, fallow deer, boar, lion and mountain gazelle	Stone and bone tools, some composite	Semi-nomadic, highly dispersed
75,000–60,000	Full Glacial Period Very cold and dry	Reduction in faunal diversity, especially larger and woodland species	More specialized stone and bone tools	Semi-nomadic, highly dispersed
45,000–25,000	Very unstable Generally cooler and drier than present	Decrease in megafauna, more small game, wild plants including grasses	Hunting tools Cereal grinding	Nomadic, highly dispersed
23,000–21,000	Cold and severe aridity	Small game, wild fruits, large-seeded grasses, e.g. barley and emmer	Cereal grinding Baking	Nomadic and semi-nomadic, dispersed, small huts
20,000–15,000	Last Glacial Maximum	Small game, wild fruits, large-seeded grasses, e.g. barley and emmer	Cereal grinding Baking	Nomadic and semi-nomadic, dispersed, small huts
15,000–13,000	Rapid warming, increase of 40% in atmospheric CO_2	Prolific fauna, less use of food plants	Geometric microlithic tools	Nomadic and semi-nomadic, dispersed, pit-houses
13,000–12,000	YOUNGER DRYAS Cold and dry	Much reduced fauna and flora, more use of wild einkorn barley and **RYE**	Loss of game and many wild plants, earliest tillage	Nomadic and semi-nomadic, dispersed villages, pit-houses
12,000–11,000	YOUNGER DRYAS Cold and dry	Wild einkorn, RYE Stock manipulation Wild sheep herding	Grinding querns Sickles adapted for harvesting	Multi-family small villages of 100–200 people
11,000	YOUNGER DRYAS ends, rapid rewarming	EMMER, EINKORN, RYE, 2-ROWED BARLEY, livestock	Grinding tools, flint-bladed sickles, some pastoralism	Small–medium villages of 100–400 people
10,500	Warm and moist	EMMER, EINKORN, livestock	More intensive tillage, manure, brick-making	Formalized multi-room brick houses, small towns
10,000	Warm and moist	EMMER, PULSES, EINKORN, RYE, 2-ROWED BARLEY, livestock	First use of clay tokens as precursor to writing	Spread of trade and village/town-based agro-urban culture
9,500–9,000	Warm and moist	EMMER, PULSES, EINKORN, RYE, 2-ROWED BARLEY, livestock	Agro-pastoral with seasonal game hunting	Modest towns 2,500–3,000 people, central planning
9,000–7,000	Warm and moist	(EPEB) EMMER, PULSES, EINKORN and 2-ROWED BARLEY	First pottery	Modest towns 5,000–6,000 people, central planning
7,000	Warm and moist	Higher yield EPEB crops, livestock	Improved irrigation, canals, copper working	Irrigation farming, increasing bureaucracy
6,000	Warm and moist	High yield EPEB crops, livestock	Potter's wheel, ploughs, mass-production	First cities, centralization of production, writing

[‡]Wild plants shown as lower case, domesticated crops are in CAPITALS; **RYE**, first definite evidence of a domesticated plant; EPEB, emmer, pulses, einkorn and barley. Adapted from Murphy (2007).

animal-derived foods. To achieve this, they developed thickened dental enamel and jawbones, and larger, flatter teeth that allowed them to cope with a more varied diet than other apes. The dietary range of members of the genus *Homo* was further enhanced by cultural innovations that favoured hunting, such as the use of fire, tools, weapons and other technologies, plus the development of complex social networks. Early *H. sapiens* tended to have a flexible diet and could exploit a wide range of food sources, including plants. This dietary flexibility was especially useful during the highly unstable climatic conditions that characterized the Pleistocene Epoch, which, as shown in Fig. 1.3, made up about 95% of the history of our species.

From hunter-gathering to agriculture

Homo sapiens first arose in Eastern Africa over 200,000 years ago and subsisted as fire- and tool-using hunter-gatherers exploiting a wide variety of

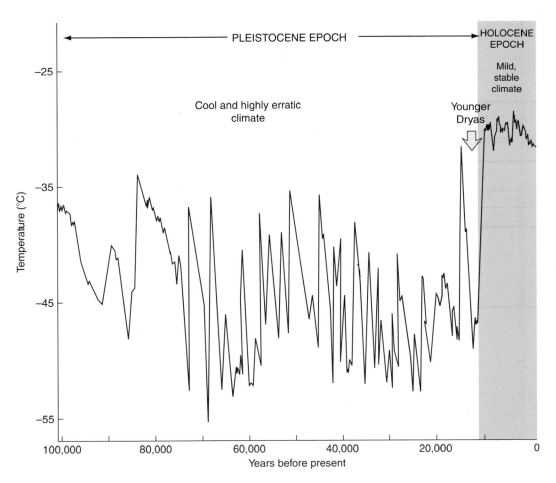

Fig. 1.3. In pre-agricultural times the climate was very variable. The climate of the past 100,000 years has been largely characterized by relatively cool temperatures with frequent and rapid fluctuations between extreme cold and milder conditions. These erratic and comparatively dry and cool conditions persisted during the entire Upper Pleistocene epoch as modern humans evolved and would have made it virtually impossible for agriculture to develop as a viable alternative food acquisition strategy to hunter-gathering. Vertical axis shows modern ice core temperatures, which are used as a proxy measurement for previous climatic conditions.

plant and animal foods. Some time after about 70,000 years ago small groups of people gradually spread from Africa to populate the rest of the world, from Australia and Asia to the Americas. These people were highly adaptable in their use of technologies for food acquisition. Where game was abundant, they focused on the development of tools, such as knives, spears, bows and arrows and slings for the hunting and processing of animals. In other cases where wild plant foods were more plentiful, people developed special tools for harvesting and processing edible plants. By 50,000 years ago, people in Africa and Eurasia were already using stone scythes and grinding tools to process the hard starchy seeds found in some grass species, especially the larger grained cereals. Grinding these hard seeds converted them into more easily digestible forms of powdery flour that could be baked or made into a porridge-like paste.

By this time, it is likely that many cultures around the world had also discovered that certain starchy or sugar-rich plant extracts could be left to ferment in a container where they were transformed into various alcoholic beverages. These processes were the first examples of biotechnology. Although our ancestors did not know the biochemical details of microbial fermentation, they soon learned how to prime their plant extracts with wild yeasts in order to achieve the desired conversion of the sugars and starches to ethanol. Like all pre-scientific technologies, this form of microbial biotechnology was based on empirical, 'trial-and-error' processes.

As early as 22,000 years ago people in the Near East had developed other plant processing technologies that are still in use today. One of the most important of these technologies was the baking oven. Baking allowed people to process hard, wild cereal grains into more easily digestible seed cakes, versions of which are still eaten in the region today. During this pre-farming era there was a particular abundance of starch-rich seeds of several wild pooid and panicoid grasses in many parts of Asia and Africa. The most useful pooid species included the wheats, barley and rye, while exploitable panicoid species included many of the millet group of plants. These wild grasses went on to become our major staple crops and are now key targets for manipulation via modern science-based biotechnologies.

Well before the start of organized agriculture, there is evidence that people were informally managing some wild plants, e.g. sowing their seeds into prepared soil. As a result of this new type of plant manipulation by humans, the plants would have experienced a subtly different environment from their previous 'wild' condition. Some of the plants adapted well and flourished in the new human-imposed conditions, while others did not. Human gatherers favoured the food plants that grew well and produced high yields under such circumstances. This led to the gradual, unconscious selection of a number of genetic attributes in some food plants, which modified their genetic profiles and initiated domestication. The process of unintentional, pre-agricultural domestication would have genetically altered some species more quickly than others. Those plants that became altered in favourable ways for the human gatherers eventually evolved into our main crop species. In different parts of the world, different plants became the favoured partners of humans although, where they were available, starch-rich cereals were normally selected as major staple crops.

One remarkable aspect of the development of agriculture is that, out of over 7000 wild plant species around the world regularly used for food by hunter-gatherers, only a tiny number of mainly grassy, tuberous or leguminous species were eventually selected and domesticated to serve as the primary dietary staples (see Table 1.2). Although about 300 plant species are cultivated around the world today, almost all of our food comes from only 24 species. A mere eight species supply about 85% of global food calories, while just three cereal grains (rice, wheat and maize) provide over half our food. Annual cereal production is about 1530 Mt compared with about 400 Mt of all other crops combined, including tubers, pulses, sugarcane and fruits. Indeed, despite all the impressive achievements of modern scientific breeding, the dozen or so plant species that were accidentally chosen by early Neolithic farmers are still our most important dietary items today.

The major food crops in order of commercial production are listed in Tables 1.3 and 1.4. The ancient crops from the grass family, including the cereals, wheat, rice, maize, barley, sorghum, millets, oats and rye, are still dominant in modern agriculture. The 'big three' cereal crops are wheat (28%), maize (27%) and rice (25%). Barley accounts for another 10% and all the remaining cereals together account for the final 10% of world production. Obviously the proportion of cereal-derived nutrients in the diet will be lower in richer

Table 1.2. Some important staple crops. Over 90% of calories consumed by people around the world today are provided by three groups of crop plants, namely cereals, pulses (legumes) and tubers/roots.

Crop plant	Domestication site	Domestication date, years BP	Type of crop	Food products of crop
Cereals				
Einkorn wheat	Near East	11,000	Starchy cereal	Rough bread, cakes, biscuits
Durum wheat	Near East	9,000	Starchy cereal	Pasta, couscous
Breadwheat	Central Asia	8,000+	Starchy cereal	Fine bread, cakes, biscuits
Barley	Near East	11,000	Starchy cereal	Rough bread, cakes, biscuits
Rye	Near East	12,000	Starchy cereal	Rough bread, cakes
Rice	China/India	10,000+	Starchy cereal	Boiled or fried rice
Maize	Mexico	8,000+	Starchy and oil-rich cereal	Tortillas, sweetcorn, cornbread, popcorn
Millets	Eurasia and Africa	10,500	Starchy cereal	Breads
Pulses				
Common bean	Mexico	10,000	Protein- and starch-rich legume	Cooked beans
Lentil	Near East	10,000	Protein- and starch-rich legume	Cooked beans
Pea	Near East	9,000	Protein- and starch-rich legume	Cooked peas
Soybean	Far East	4,000+	Protein-, oil- and starch-rich legume	Cooked beans, soy milk, soy sauce
Tubers/roots				
Potato	Andes	11,000+	Starchy tuber	Boiled or fried potatoes
Yams	West Africa	6,000+	Starchy tuber	Boiled yams
Cassava	Amazon Basin	5,000+	Starchy root	Boiled cassava

countries where meat is more readily available and affordable to most people. However, across much of the developing world, plant-derived foodstuffs still overwhelmingly predominate in the human diet. This is especially true for the most populous and fastest growing regions of Asia, Africa and South America, where the 'big three' cereal crops are especially dominant.

Why do we not have a wider range of staple crop species? As we will see in Chapter 7, the answer lies in the nature of plant genomes. Throughout the ages, people have repeatedly tried, and usually failed, to domesticate hundreds of other edible plants. For example, Manchurian wild rice, *Zizania latifolia*, was initially the preferred form of edible rice grown in parts of China. This wild rice is both more nutritious and tastier than the more common form of Asian rice, *Oryza sativa*, that is grown today. However, despite many centuries of effort by farmers, it proved to be impossible to domesticate plants from the *Zizania* genus. Although they gave a better food product, these wild rices tend to shed

their seed before they can be harvested and displayed other weedy traits that made them difficult to cultivate. As the human population increased, farmers increasingly switched to growing the higher yielding, but less tasty, Asian rice. With modern knowledge of the genetic basis of domestication traits in plants and access to new biotechnological methods, it may now be possible to modify some of these hitherto recalcitrant but potentially useful species, like *Zizania*, perhaps enabling them to be cultivated as fully domesticated food crops.

Agriculture, technology and civilization

Agriculture probably started just over 11,000 years ago when several groups of people in different parts of the world independently began to exploit plants in a totally new way. But why did people begin farming only about 11,000 years ago when they had been successfully manipulating and exploiting wild plants for over 100,000 years before that time? The most likely answer to this

Table 1.3. Major food crops in order of commercial production. Note the dominant position of the 'big three' cereals, wheat, maize and rice, which have been the major human crop staples since the dawn of agriculture, well over 10 millennia ago. This list does not include crops grown for subsistence only (otherwise rice would far out-yield the other crops).

Crop	Annual yield (Mt)	Annual or perennial	Climatic zone‡	Reproductive method†	Ploidy§
Wheat	468	A	M	S	2,4,6
Maize	429	A	S	C	2
Rice	330	A	S	S	2
Barley	160	A	M	S	2
Soybean	88	A	W	S	2
Sugarcane	67	P	T	V (C)	many
Sorghum	60	A	S	S	2
Potato	54	A	H	V (C)	2,4,6
Oat	43	A	M	S	2,4,6
Cassava	41	P	S	V (C)	4
Sweet potato	35	A	S	V (C)	6
Sugarbeet	34	A	C	C	2,3,4
Rye	29	A	M	C	2
Millets	26	A	S	C, S	2,4
Rapeseed	19	A	M	C	4,6
Common bean	14	A	S	S	2

‡M, Mediterranean; S, Savannah; W, Woodland; T, Tropical forest; H, Highland; C, Coastal.
†S, self-fertile; C, cross-fertilized; V, vegetative.
§2, diploid; 3, triploid; 4, tetraploid; 6, hexaploid; many, higher ploidy levels. Adapted from Harlan (1992).

Table 1.4. Annual production (edible dry matter) of major commercial crop/food groups. The starchy cereal, tuber and pulse crops together account for almost 90% of commercial food production, with cereals alone making up almost 75% of the total.

Crop/food group	Annual production (Mt)	Percentage
Cereals	1545	74.9
Tubers	136	6.6
Pulses	127	6.2
Animals (meat, dairy, egg products)	119	5.8
Sugar	101	4.9
Fruits	34	1.6
Total	2062	100.0

Adapted from Harlan (1992).

question is that the abrupt change to a less variable climate made agriculture a viable strategy for exploiting plants that was potentially more productive than hunter-gathering.

For most crops, their initial domestication occurred independently of other crops, often on different continents. In some cases, a crop might be domesticated at several different times in widely separated localities. Hence, common beans were domesticated at least twice, 2 millennia apart, first in Mexico and then in Peru. Rice was probably domesticated many times in several regions of eastern and southern Asia. Oats originated in the Near East where they were originally regarded as undesirable weedy contaminants of wheat and barley crops. But once wheat and barley cultivation spread to the cooler regions of Northern Europe, farmers found that the 'contaminating' oat grains often grew better than other cereals. As a result, oats were soon domesticated as useful crops in their own right.

As far as we can tell, some of the earliest domestication processes may have been directly linked with a sudden climatic change during the Younger Dryas interval (see Fig. 1.4 and Box 1.1). Compared with the previous 3 million years, the climate of the last 10 millennia has been relatively stable with generally mild and moist conditions in temperate regions. This stable climate meant that farming could be a more productive food acquisition strategy than hunter-gathering. The uptake of farming was also triggered by other factors including sedentism, cultural developments within societies, population increases and technological advances. The relative importance of these factors varied from place to place and from crop to crop, but one of the most important factors for successful domestication was the genetic constitution of the plants themselves. The shift from exclusive hunter-gathering to farming probably occurred in a series of transitional stages that lasted for several millennia. These stages would have established the necessary conditions for agriculture but would not have made it inevitable.

The kinds of conditions needed for farming include the availability of the 'right sort' of plants, i.e. plants that lent themselves to domestication due to their genetic make-up. People would also have needed to be very familiar with such plants; for example what they looked like, where they grew, when they set seed, what else ate them (animals) or competed with them (other plants), and so on. They would have needed the right technologies for harvesting and processing the edible parts of the plants into easily digestible food, such as sickles,

grinding stones and baking ovens. Some hunter-gatherer groups may have maintained a series of small gardens, which they periodically visited for tending and harvesting. This would have enabled people to familiarize themselves with the basics of plant cultivation and to experiment with strategies, such as tilling, sowing and weeding, that would encourage better growth of their favoured plants. Such activities could occur as part of a mobile hunter-gatherer lifestyle without involving an irrevocable commitment to full-time agriculture.

Farmers obtain far greater productivity than hunter-gatherers in terms of the food calories that can be gained from a particular area of land. This means that farming can sustain much larger populations, many of whom do not need to be involved in the process of food production itself. Non-farmers in agrarian societies were free to specialize in areas such as technology development. Therefore farming/sedentism tended to foster technological

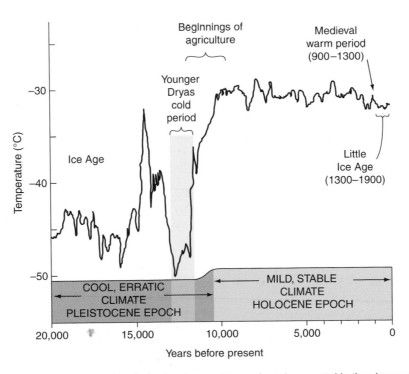

Fig. 1.4. During the past 10,000 years the climate has been milder and much more stable than in any period of the last million years. The hypervariable conditions of the Upper Pleistocene came to an abrupt end at the Younger Dryas Interval of 12,800–11,600 years ago. Since that time, the Holocene epoch has been dominated by a much more stable climate with milder and wetter conditions. To a great extent, the comparatively restricted group of crops still grown today depends on the continuation of this relatively stable climatic regime. Vertical axis shows modern ice core temperatures, which are used as a proxy measurement for previous climatic conditions.

Box 1.1. Climate change and agriculture

Agriculture involves the exploitation of a much smaller range of food crops than is used by hunter-gatherers. In general, farmers stay in a particular area and only grow a few staple crops plus a few minor species. These crops tend to be adapted for a relatively limited range of climatic conditions. In contrast, many hundreds of wild food plants growing in widely different climates can be utilized by highly mobile nomadic hunter-gatherers.

In crop-growing areas, the original farmers developed crops that suited the local climate. For example, rice was developed for the warm, moist climate of Southern Asia, maize was cultivated in the more seasonally arid conditions of Central America and oats were grown in the cool, damp, heavy soils of Northern Europe. If the climate were to change suddenly in a particular region, local farmers might not have access to any crops that could grow in the new conditions and they would be forced to abandon agriculture or starve.

As shown in Fig. 1.3, most of the past 100,000 years have been characterized by very unstable climatic conditions. In general, the weather was much cooler than today but there were also frequent and rapid fluctuations between cool, dry conditions and warm, moist conditions. These sudden changes would have had huge impacts on plant life. During most of recent human evolution conditions were especially variable and would have prevented the development of agriculture as a successful long-term strategy for producing food. Even if a local population had managed to cultivate a few food plants for several centuries, the next climatic shift would make this impossible and force them to resume hunter-gathering.

As shown in Fig. 1.4, the hyper-variable climate of the past few million years ended abruptly after the Younger Dryas cold period about 11,000 years ago. Since then, the earth has enjoyed an unusually warm and stable climate that we still live in today.

The exceptionally stable Holocene climate has enabled people to develop and grow relatively specialized crops, such as wheat and rice. Although the Holocene period so far only lasted for about ten millennia, this has been long enough to enable people around the world to domesticate several types of high-yielding crops. Once they had started growing such crops, it only took a few millennia for farming societies to develop complex cultures, new technologies, and advanced civilizations. Therefore, the farming-friendly climate of the Holocene is probably largely responsible for the technologically advanced and globalized societies that now dominate today's world. In the future, however, we might need all of our technological skills to ensure the continuation of agriculture, especially if the global climate begins to resume its former hyper-variable character.

---

innovation. Urban/agrarian societies were also able to operate on a much larger scale than the relatively small groups of clan-based hunter-gatherers. The greater numbers of people in farming-based societies and their better technologies gave them an advantage in the case of conflict with groups of hunter-gatherers. However, farming did not succeed overnight and it took several millennia for large-scale agro-urban cultures to emerge in several regions of the world.

The earliest domesticated crops in the Near East date from about 12,000 BP (before present). By about 9000 BP, there were many villages and small towns across the region and the first recorded cities date from about 5800 BP. In Mesopotamia, these new cities soon absorbed most of the population, which became progressively ever more urbanized until, by 4000 BP, an astonishing 90% of Sumerians were city dwellers. Many of these city dwellers would have 'commuted' to the nearby countryside to tend the intensively managed croplands while others were involved in early forms of manufacture, such as mass-produced pottery or in civic breweries in what is probably the earliest example of industrial-scale biotechnology. For example, by 5500 BP the Egyptian city of Hierakonpolis had a large commercial brewery capable of processing barley grains into 1400 l of beer per day.

Non-agricultural plant exploitation

Although agriculture is the predominant form of plant exploitation in use today, people have used non-agricultural methods of plant manipulation for over 95% of their history. Many forms of non-agricultural plant management were still in common use as late as the 20th century. We can define non-agricultural plant management as: 'the manipulation of plant development and distribution for human exploitation without using formal cultivation'. Non-agricultural plant management can involve techniques such as transplantation, controlled burning, and sowing of gathered wild seed. However, in the case of these particular plant species, such management practices have not given rise to the cascade of genetic changes that have produced domesticated varieties. Hence, the managed plants remained as wild forms that were encouraged to grow by humans, but these plants never became as completely dependent on people as did the fully domesticated crop species (see Table 1.5).

Across the world, various forms of non-agricultural plant management were used by people living as hunter-gatherers in the climatically variable period before 12,000 BP. After the development of agriculture, such practices continued in marginal climatic regions where crop cultivation was not feasible.

These people were still hunter-gatherers, but they also actively managed and exploited plant resources on a wide scale over many millennia, without ever adopting formal agriculture. In some cases, non-agricultural exploitation of plant resources reached a very high level of sophistication with an impressive degree of botanical knowledge. One example is the Kumeyaay people of southern California, which has a Mediterranean climate, with relatively sparse summer rainfall and a short, erratic wet season in winter. Farming in this region was not possible due to the periodic severe droughts that still affect the area.

The Kumeyaay used intensive plant management strategies to manipulate their floral landscape and successfully fed and clothed themselves for many

Table 1.5. Examples of non-agricultural plant exploitation. There are many well-documented cases of non-agricultural plant management, some of which have only died out in the last few decades. Examples include the burning, sowing and harrowing of grassland to increase production of favoured seed plants, the digging up and collection of edible wild roots and tubers, protection of plants from threats such as herbivores, and their dissemination by deliberate transplantation.

Common name	Systematic name	Location	Exploiters	Uses
Wild rice	Oryza rufipogon	Many regions of Australia	Many tribes	Starchy grains
Mitchell grass	Astrelba pectinata	South-east Australia	Bagundji	Starchy grains
Wild yams	Ipomoea costata	Many regions of Australia	Many tribes	Starchy tubers
Panic grass	Panicum spp.	Lower Colorado River Basin	Cocopa and others	Starchy grains
Wild rice	Zizania aquatica	North America	Many groups	Starchy grains
Indian rice grass	Oryzopsis hymenoides	North America	Many groups	Starchy grains
Blue grama	Bouteloua gracilis	North America	Many groups	Starchy grains
Wild agave	Agave seemanniana	Mesoamerica	Many groups	Succulent stem buds
Prickly pear cactus	Opuntia pilifera	Mesoamerica	Many groups	Succulent stems and fruits, medicinal
	Ferocactus latispinus	Mesoamerica	Many groups	Succulent stems and fruits
Reed grass	Phragmites australis	North America	Chipewyan	Stem bases
Black potato	Solanum nigrescens	Mesoamerica	Many groups	Leaves
Common cattail	Typha latifolia	North America	Many groups	Starchy rhizomes, shoots and flower spikes
Sugar maple	Acer saccharum	North America	Many groups	Sugary sap, bark
Smooth sumac and Staghorn sumac	Rhus glabra and Rhus typhina	North and Mesoamerica	Many groups	Shoots, fruits and seeds
Cow-parsnip or Indian celery	Heracleum lanatum	North America	Many groups	Leaf stalks and flower bud stalks
Wild carrot or Desert parsley	Lomatium macrocarpum	North America	Many groups	Fleshy taproots
Balsamroot	Babamorhiza sagittata	North America	Many groups	Taproots, shoots, leaves, grains
Wild sunflower	Helianthus annuus	North America	Many groups	Oil-rich seeds

hundreds of years without farming. For example, they relocated and replanted dozens of useful plant species. They created groves of wild oak and pine in higher altitude areas of their territory in order to harvest their edible nuts. They established desert palm and mesquite along the coast and planted agave, yucca and wild grapes in appropriate micro-habitats in various parts of their range. They planted cacti for use as emergency sources of water as close as possible to their villages, trails and campsites. In addition to transplantation, the Kumeyaay managed their floral environment by the systematic burning of tree groves to increase fruit yield; they used the controlled burning of chaparral grassland to improve forage for the (non-domesticated) deer that they hunted; and they re-sowed a proportion of the edible grain from wild grasses that they had harvested.

The major difference between agricultural and non-agricultural methods of plant exploitation is that agriculture depends on the intensive use of a few domesticated species, whereas non-agricultural methods typically involve use of large numbers of wild species. Non-agricultural methods of plant exploitation are probably more useful for food production in erratic and unpredictable climates. This makes them interesting to study today because they could highlight novel ways of coping with the possibility of drastic climate change in the future.

1.4 Agricultural Technologies

The exploitation of plants via agriculture involves biological and non-biological technologies. As discussed above, many non-biological technologies, such as the harvesting, grinding and baking of cereal grains, were developed many millennia before the first crops were grown. However, the increasingly intensive cultivation of crops after 10,000 BP gave the impetus for the invention of new technologies, such as seed sowers and ploughs, as well as new ways of managing the crops during their growing season. Although the climate of the last 10 millennia has been much more stable than any comparable period of human existence, there have been numerous climatic episodes such as periodic droughts or cool periods that have impacted on crop production. Also, climatically responsive crop pests and diseases have always been a threat, as have herbivores that might graze the crop before harvesting.

These factors mean that farmers constantly need to adapt methods of crop production to suit changed circumstances. The two main ways that a farmer can improve crop production are: (i) better management of crop cultivation, harvesting and processing; and (ii) the selection of better plant varieties. Crop management involves the use of non-biological methods, such as tools, fertilizers, weeding and pest control to facilitate production. In contrast, crop improvement involves the manipulation of the biology of a plant so that it gives better performance in the field. The management and improvement of crops normally involve different forms of technology. In historical terms, while technologies for crop management date back many thousands of years, crop improvement biotechnologies are barely more than a century old (see Chapter 8). Prior to this, crop improvement was based on empirical methods that were very limited in their scope compared with modern breeding. Nevertheless, despite their limitations, these empirical methods were remarkably successful in enabling pre-scientific agriculture to feed millions of people around the world (see Chapter 7).

Crop management

The major agricultural innovations in the ancient world were related to technology and agronomy, rather than crop genetics and breeding. Thanks to their previous use of non-agricultural forms of plant management, the earliest farmers already had access to crucial technologies, such as scythes and grinding tools. However, the expansion of agriculture from a subsistence activity to a level that could provide sufficient surplus to support larger communities required new forms of technology. One of the most effective of these was the plough, which was first invented in the Near East about 5000 BP. Ploughs facilitate crop production by making it possible for each worker to cultivate much larger areas, although the actual yield per hectare is generally lower than with manual methods. Therefore ploughing was more suitable for the larger scale, state-controlled systems of irrigation farming that were evolving at this time, where field size was less of a limitation than the supply of manpower. For example, the use of ox-drawn wooden ploughs slightly reduced the yield of cereals per hectare but enabled much larger areas to be worked by fewer people.

In the area of crop management, the ancient Babylonians introduced some significant innovations, including the seed plough and the use of draught animals such as oxen and horses, to increase crop productivity per worker. Early farmers improved other aspects of crop management such as systematic irrigation, invented over 7000 BP, and the establishment of communal grain stores, which were kept dry and vermin-free, enabling harvested grains to be stored safely for years at a time. The use of mechanical seeders for ploughing was another important advance, ensuring the more efficient and even broadcasting of seed. One major difference between the ancient and modern worlds was the speed of technology dissemination. It took almost 1000 years before seeder technology spread a few hundred kilometres across the Near East from Mesopotamia to Egypt. Further refinement of the Babylonian seeder into an automatic drill did not occur for another 2 millennia with Jethro Tull's famous invention in 18th-century England. In contrast, modern inventions such as DNA-based biotechnologies can spread around the world in just a few years.

The earliest recorded chemical inputs in agriculture include sulfur, chalk and alum, which were already being used as pesticides by 2500 BP. Other ancient crop-control chemicals include arsenical sulfides to control pests, potash as an anti-worm agent, and copper sulfate as a fungicide. In many parts of Europe, heavy acid soils were treated with lime (produced from limestone in small kilns) to reduce the pH and hence enable them to support crops. By the Middle Ages, use of heavier ploughs allowed farmers to cultivate a wider range of soils, especially in the damp climates of Western Europe. Farmers also started to adopt two- or three-field rotation systems that allowed a more intensive use of the land while maintaining crop yield and soil fertility. This was supplemented by the cultivation of legume crops that regenerated soil nutrients and extended an otherwise monotonous cereal-based diet.

However, these innovations were only patchily applied and for much of the medieval period agriculture and innovation progressed slowly or not at all. Indeed, in some areas agriculture actually went backwards in terms of the variety of crops and their productivity. For example, the Muslim rulers of early medieval Spain introduced many new crops and advanced irrigation methods that made it the most agriculturally diverse and

productive region of Europe. However, most of their innovations fell into disuse after the Christian reconquest and Spanish agriculture reverted to a relatively basic set of crops with much reduced yields. As we will see in Chapter 8, the dawn of the scientific era after the 17th century led to a new series of technological and management innovations in European agriculture. This agricultural revolution was especially effective in Britain where the resulting increased food production enabled the population to grow and helped lay the foundations for the industrial revolution of the 18th and 19th centuries. Some early achievements in crop management and improvement are listed in Table 1.6.

Crop improvement

From the beginnings of agriculture, farmers needed to adapt to changes in climate or new soil types as cultivation spread to new areas, and as new pests, diseases or other threats arose. One way of adapting was to select crop varieties that grew best in a particular climate and/or were more resistant to the prevalent diseases in that locality. This meant that farmers also needed to be breeders. They consciously looked for and selected superior variants within domesticated populations and, if they were available, wild relatives of their crops. Farmers then deliberately propagated new forms from such variants so that the genetics of the local crop population was gradually modified to suit the prevailing environmental conditions.

One of the best examples of this empirical breeding process is the domestication and selection of the various crops of the wheat genus, *Triticum*, where a few observant farmers with no knowledge of breeding were able to select naturally occurring hybrids or mutations to generate new varieties with much-improved yield and/or quality traits (see Chapter 3). However, this sort of empirical approach relies on comparatively rare biological events and the new crops were only very slowly spread from their point of discovery. Nevertheless, the breadwheat that was empirically selected and first grown by Neolithic farmers is now the second most important crop in the world (after rice) in terms of its yield and the numbers of people fed. Similar random and unplanned processes underpinned the selection of the other major crop groups, such as rice, maize and potatoes.

Table 1.6. Some biological and technological landmarks in early agriculture.

Date, BCE (before common era)	Location	Achievement
11,000	Peruvian Andes	Cultivation of potatoes
10,500	Near East	First good evidence for 'domesticated' forms of a crop, a large-seeded version of rye
9,400	Near East	Cultivation of 2-row barley
9,000	Near East	Cultivation of emmer and einkorn wheat
9,000–8,000	Near East	Cultivation of pulses, such as lentils, vetch and peas
7,000	Saharan Africa	Possible cultivation of sorghum, pearl millet and finger millet (this was before the desertification of the Sahara region)
7,000	Northern China	Cultivation of broomcorn and foxtail millet
6,000	Ethiopia	Cultivation of tef, noog and ensete
6,000	Near East	Water management, early irrigation
		Cultivation of 6-row barley
5,000	Northern China	Intensive rain-fed millet cultivation
	Southern China	Cultivation of rice
5,000	Nile Valley	Highly intensive irrigation agriculture
5,000	Mesoamerica	Cultivation of maize
4,500	Near East and Nile Valley	Use of ploughs
3,500	Near East and Nile Valley	Use of animal traction and large-scale brewing of beer from barley
3,000	Indus Valley	Complex, locally organized irrigation networks, introduction of sorghum. Domestication of water buffalo
3,000	Central and North-west Europe	Development of cold-tolerant cultigens of barley and wheat allow northerly expansion of farming
2,500	Near East	Imperial agriculture, widespread barley cultivation, virtually as a monoculture in the south after 2200 BCE
2,400	Peruvian Andes	Cultivation of common bean
2,000	Mesoamerica	Development of enlarged maize cobs of <4 cm
1,790	Near East	Laws of King Hammurabi with much detail on regulation of irrigation and cropping systems
ca. 1,700	Near East	Earliest known agricultural manual on clay tablet
1,200	Near East	Iron ploughs
1,100	Near East	King Tiglath-pileser I founded noted botanical gardens at Nineveh
800	Mesoamerica	Invention of highly intensive *chinampa* cropping system
669–623	Near East	King Ashurbanipal established the first systematic library at Nineveh, including a herbal listing 1000 useful plants
ca. 300	Greece	*Historiae Plantarum*, a comprehensive treatise on botanical and medicinal matters

Biologically speaking, the fundamental aim of crop improvement is to create new genetic variants with advantageous traits. In contrast, crop management is used to cultivate such varieties under the best possible conditions in order to optimize crop yield and quality. Biotechnologies can potentially play crucial roles in both of these processes. For example, plants can be biologically manipulated via breeding to produce higher yields or better quality, but they can also be bred to respond better to particular management or

input regimes. Breeding involves the selection by humans of genetic variants of a few chosen plant species according to their suitability for exploitation, for example as edible or non-edible resources. The two key prerequisites to both breeding and evolution are *variation* and *selection* (see Chapter 8).

Novel genetic variations in wild populations arise from a relatively slow process of naturally occurring mutation, plus the mixing of genomes that occurs with sexual reproduction. In contrast, science-based breeding, as practised since the early 20th century, is based on the creation of genetic variation via processes such as induced mutagenesis, hybridization, controlled introgression of traits from diverse populations of the same or different species, and transgenesis. The best variants are then selectively propagated to minimize genetic variation in their progeny. This inbreeding process creates a relatively uniform population that is then *managed* (e.g. cultivated, harvested and processed) for human exploitation. While so-called 'traditional' methods of variation enhancement, e.g. use of crop landraces, still have great and often untapped potential, the use of newer biotechnologies can create even wider genetic diversity. This gives modern breeders unprecedented opportunities for further crop improvement.

As we will explore in later chapters, our greatly increased ability to create new genetic variation has been matched in recent years by a revolution in the screening, identification and selection of potentially useful variants, using methods such as biochemical and genomic screening, plus molecular marker-assisted selection. Thanks to continued advances in basic plant research and in genomic and other -omic methods, there is great scope for the practical application of biotechnologies in plant breeding. During the coming years, this will apply especially to developing countries where population and environmental pressures are greatest (see Chapter 12). The major impacts of plant biotechnologies will be seen both in breeding improved crop varieties and in providing more sustainable methods for crop cultivation and management. Some targets for improved crop cultivation and management include:

- Improved production of disease-free, genetically normal propagation materials, especially in vegetatively propagated crops;
- Aspects of plant nutrition, such as production and use of biofertilizers;
- Use of symbiotic nitrogen-fixing bacteria and mycorrhizal fungi;
- Aspects of plant protection, including diagnostics and biopesticides;
- Conservation and management of crop genetic resources, both *in situ* and *ex situ*.

1.5 Biotechnologies, Ancient and Modern

Modern biotechnology is one of the most exciting, cutting-edge applications of contemporary biological science. However, in historical terms biotechnology can also be regarded as one of the most ancient traditional crafts practised by human cultures around the world with examples that date back many thousands of years. Some examples of traditional and modern biotechnologies are listed in Table 1.7.

Traditional biotechnologies

Perhaps the best-known example of traditional biotechnology is the transformation of starch- or sugar-rich plant extracts by aerobic fermentation using the brewer's yeast, *Saccharomyces cerevisiae*. This form of microbial fermentation allowed people to transform plant extracts into the vast range of alcoholic liquors used around the world. Hence we can change barley to beer, grapes to wine, or rice to saki. The earliest evidence of organized large-scale brewing of beer dates from the early city-states of Mesopotamia and Egypt, over 5500 years ago. In these densely populated urban societies, even weak beer was generally much safer to drink than the often-polluted water supply; a situation that continued until the introduction of chlorinated drinking water in the late 19th century. Another ancient use of fermentation is the incubation of bread dough with baker's yeast in a hot oven. Here, carbon dioxide produced by yeast fermentation causes the dough to rise, resulting in a lighter or 'leavened' form of bread that some people regard as more pleasant to eat and easier to digest than unleavened bread.

Modern biotechnologies

Some of the definitions of biotechnology are discussed in Box 1.2. Modern biotechnology can be separated into several different categories depending on the type of organism used and/or the market sector of the end product. One important point to make here is that by no means do all modern biotechnologies involve transgenic methods or the creation of GM (genetically modified) organisms. Although much of the general public might think that the terms 'plant biotechnology' and 'GM crops' are virtually synonymous, this is a misconception and in reality transgenesis is just one of several sophisticated biotechnological tools available to modern scientific plant breeders (see Box 1.3).

Industrial (white or grey) biotechnology

As we saw above, ancient biotechnologies were based on microbial fermentation, mostly involving

Table 1.7. Examples of traditional and modern biotechnologies.

Category	Period	Process	Raw material	Product
Empirical	Pre-agricultural	Fermentation	Sugary or starchy plant extracts	Alcoholic liquids: wine, beer, saki
Empirical	Pre-agricultural	Fermentation	Flour from cereal grains	Leavened breads
Empirical	From 6000 BP	Fermentation or oxidation	Milk from livestock	Alcoholic liquids, yoghurt, cheese, butter
Empirical	From 5000 BP	Clonal propagation	Explants and cuttings	Clonal crop plants, e.g. olives, dates
Early industrial	19th century	Non-sterile fermentation	Plant and animal extracts	Acetic acid, butanol, acetone
Early industrial	Early 20th century	Chemical methylation	Plant and animal fats or oils	Methyl esters for biofuel
Early industrial	Early 20th century	Non-sterile fermentation	Plant and animal extracts	Ethanol for biofuel
Modern industrial (white)	Mid-20th century	Sterile fermentation	Cultured microbes	Antibiotics, amino acids, organic acids, steroids
Modern industrial (white)	Mid-20th century	Sterile fermentation	Cultured microbes	Enzymes, e.g. protease, amylase, rennet, pectinase
Modern industrial (white)	1960s	Sterile fermentation	Pure strains of cultured microbes	Vaccines, monoclonal antibodies
Modern industrial (white)	1960s	Sterile fermentation	Pure strains of cultured microbes	Novel foods, e.g. quorn from *Fusarium* fungus
Medical (red)	1980s	Transgenic microbes	Bacteria or yeast cells	Therapeutics, diagnostics
Medical (red)	2000s	Transgenic eukaryotes	Eukaryotic cells, including human	Human therapeutic proteins
Marine (blue)	2010s	Transgenic fish	Farmed fish, e.g. salmonids	Larger farmed fish
Plant (green)	1910s	Tissue culture, e.g. embryo rescue	Seeds, plants or explants	Thousands of novel crop varieties
Plant (green)	1930s	Radiation or chemical mutagenesis	Seeds, plants or explants	>2000 new crop varieties
Plant (green)	2000s	Molecular mutagenesis, e.g. TILLING	Seeds, plants or explants	An increasing number of crop varieties
Plant (green)	1990s	DNA marker-assisted selection	Plant populations	An increasing number of crop varieties
Plant (green)	1990s	First-generation transgenic plants	Transformable plant varieties	Two modified input traits in four commercial crops
Plant (green)	2000s	Second-generation transgenic plants	Transformable plant varieties	Wider range of traits in more crop varieties
Plant (green)	2000s	Transgenic plant cell cultures	Transgenic plant cell cultures	High value products, e.g. therapeutic proteins

various forms of yeast. During the early 20th century, microbial-based biotechnology evolved into a much more systematic and scientifically informed process with the development of the classical industrial biotechnologies. Instead of taking place in poorly controlled open vats, fermentation or other useful biochemical processes now took place within precisely defined and regulated media housed inside large enclosed steel containers. And, instead of relying on undefined strains of microorganisms, pure clonal lines of yeast, fungi or bacteria were specially bred for each particular process.

Industrial biotechnology, also known as white biotechnology, can be defined as 'the modern use and application of biotechnology for the sustainable production of biochemicals, biomaterials and biofuels from renewable resources, using living cells and/or their enzymes'. It has made possible the large-scale manufacture of valuable complex molecules that cannot be synthesized by purely chemical processes. For example, in the 1920s Fleming discovered the well-known antibiotic, penicillin, in the fungus *Penicillium notatum*, but this organism is very difficult to culture. It was soon found that another species, *P. chrysogenum*, was easier to grow in large-scale culture and it proved to be a much better source of the antibiotic. Subsequently, new strains with even higher yields were developed using UV-mutagenesis and modern strains can produce 90 g/l of pure penicillin on an industrial scale that is sufficient to satisfy the huge global demand for this and many other medically important antibiotics.

Modern bioreactors or fermenters can hold in excess of 200,000 l of microbial cultures. The cultures are precisely maintained, using automated computer-controlled systems, to provide optimum growth conditions by regulating factors such as nutrient supply, temperature, pH, aeration and stirring. As well as important drugs like penicillin and cortisone, white biotechnology is widely used to produce industrial enzymes such as amylases, lipases and proteases that are in great demand for food processing and other industries. The use of microbial systems as cellular factories to manufacture useful compounds was greatly expanded following the development of recombinant DNA technologies in the late 20th century. Microbial

Box 1.3. Biotechnology is not just genetic engineering

Plant biotechnology is the application of scientific discoveries for the elucidation and manipulation of the genetics and development of plants. In addition to their practical uses in agriculture, many plant biotechnology techniques, such as recombinant DNA manipulation, are also powerful research tools for generating additional scientific discoveries. In turn, these discoveries have led to further improvements in our ability to manipulate plants. Although the terms 'GM (genetically modified) crops' and 'plant biotechnology' have become interchangeable for much of the general public, and even for some scientists, biotechnology involves much more than genetic engineering. One of the major themes of this book is to examine plant biotechnology in all of its forms, of which genetic engineering is just one example.

As we will see in subsequent chapters, over the past century numerous non-transgenic biotechnologies have been developed, ranging from F_1 hybrids to induced mutagenesis and micropropagation. Such technologies enable breeders to increase genetic variation in crop lines that are often relatively inbred and lacking in variation for important agronomic characters such as disease resistance. In recent years the immensely powerful genomic and other -omic technologies have revolutionized many of the earlier biotechnologies to create advanced molecular-based approaches to future crop manipulation. As well as enhancing genetic variation, biotechnology can contribute directly to crop improvement by enabling breeders to select superior varieties. Here, technologies such as DNA marker-assisted selection, supplemented by automated analytical devices, are being applied to an increasing number of crop species. Such technologies enable favourable genetic variants created by methods such as mutagenesis, wide crossing, or transgenesis, to be rapidly and easily selected from huge populations.

Biotechnology also contributes indirectly to agriculture by providing new biological tools that enable farmers to manage crops in a more sustainable and environmentally sound manner. Examples include biocontrol of pests as a potentially cheaper and more sustainable alternative to chemical pesticides, or the use of biofertilizers as an alternative to petroleum-based synthetic inorganic fertilizers. Other techniques are helping in crop management to improve yield and to avoid losses during the growing season. For example, improved DNA-based surveillance methods are enabling farmers to detect and then act promptly to prevent potential disease outbreaks before they cause significant crop damage.

---

species, such as yeast or *E. coli* that are already well adapted for rapid growth in bioreactors, can be transformed by the addition of new genes that enable them to produce a huge range of novel products. Some of these newer applications using transgenic microbes are now known as red biotechnology.

Medical (red) biotechnology

Red biotechnology can be regarded as an updated form of white biotechnology that is used mainly for medical applications. Transgenic microbes or animal cell cultures are now being used as production platforms for the manufacture of a new generation of recombinant pharmaceuticals. These include high-value proteins or peptides such as human insulin, blood clotting factors, vaccines and antibodies that can be manufactured in bulk at the highest levels of purity. One of the limitations of traditional microbial bioreactor systems is that most prokaryotes and lower eukaryotes are unable to faithfully introduce the full range of post-translational modifications, such as *N*-glycosylation, that are present in many human and animal proteins. This has led to the development of transgenic human and animal cell cultures for the expression of such recombinant proteins.

In many cases, these compounds were previously obtained from human or animal cadavers or from donated blood. Use of such sources carries the risk of the product becoming contaminated, especially by small organisms such as pathogenic viruses that may have been present in the original donor material. For example, during the late 20th century there was widespread contamination of medications derived from donated human blood. Some blood donors carried viruses such as hepatitis B or HIV and their contaminated blood was mixed with other blood into large batches prior to extracting the blood products. Therefore, blood from a single infected donor might contaminate the combined blood samples of hundreds of other non-contaminated donors. Thousands of people, such as haemophiliacs who depend on human blood-clotting factors, received contaminated products and many became ill and died. Thanks to red biotechnology, however, it is now possible to engineer cells that have been transformed with the appropriate human genes so that recombinant blood products can be manufactured in sterile, disease-free systems.

A more recent extension of red biotechnology uses transgenic animals, rather than microbes, as bioreactors to make useful products. As we will see in Chapter 12, this is analogous to the process of biopharming whereby transgenic plants or plant extracts are engineered to manufacture valuable products. As with biopharming, this version of red biotechnology has advantages over bacterial production systems when it comes to synthesizing recombinant proteins that require post-translational modifications that can only be performed by eukaryotic cells. It is mostly targeted at domesticated livestock engineered to produce recombinant proteins in accessible locations such as milk glands (mammals) or eggs (poultry).

For example, transgenic cattle expressing a human lactoferrin gene in their mammary glands have produced milk containing 0.3–2.8 g/l lactoferrin, a protein that improves iron absorption and protects against intestinal infections. Transgenic goats can produce blood-clotting factors in their milk with a herd of 100 animals able to generate an annual $200 million worth of recombinant protein. Transgenic chickens can produce eggs containing commercial quantities of easily purified monoclonal antibodies. Red biotechnology is now a mainstream part of medicine but, as explored in Chapter 11, it was initially regarded in some quarters with as much suspicion as the use of genetically engineered crop plants.

Plant (green) biotechnology

Green biotechnology is sometimes used to describe the employment of advanced biotechnological tools in land plants, algae or cyanobacteria. It is synonymous with plant biotechnology as defined in this book (see above) and is therefore not limited to the development of transgenic organisms but also includes other examples of advanced plant-breeding techniques as discussed in detail in Chapters 8–10.

Other areas of biotechnology

Many transgenic and non-transgenic biotechnologies are also being developed in the livestock and fisheries sectors. These include advanced tissue culture, artificial insemination, reproductive cloning, mutagenesis/TILLING, marker-assisted selection and micropropagation. These animal methods are now using the new genomic technologies as described for plants in Chapter 10. Reproductive cloning is a well-known example of a non-transgenic biotechnology being applied to animals. Whereas most plants can be readily propagated as genetically identical clones from tissue cuttings, it was not until 1996 that the first higher animal was cloned by humans, namely Dolly the sheep. Since then dozens of animal species have been cloned and several clonal lines of superior livestock genotypes have been created for commercial farming. As with plant transgenesis and some methods of plant cloning, animal cloning can result in undesirable pleiotropic effects due to epigenetic changes induced during tissue culture (see Chapter 5). Some of the controversies surrounding biotechnologies such as animal cloning and transgenesis are discussed in Chapter 11.

Transgenic technologies are also being applied to commercially harvested fish, in what is sometimes called blue biotechnology. These methods are mostly being applied in fish farms rather than in open marine or freshwater fisheries. For example, transgenic fish have been produced with elevated levels of growth hormones that enable them to grow faster under culture conditions. In 2010, Atlantic salmon (*Salmo salar*) containing a growth hormone gene from a Chinook salmon (*Oncorhynchus tshawytcha*), became the first transgenic fish to be approved for human consumption in the USA. Other research is focused on reducing or eliminating the growing threat of diseases and parasites in fish farms. Over the past decade, farmed fish have becoming much more important food sources because many populations in wild fisheries have severely declined due to overfishing and other man-made environmental factors. Compared with plants, animal biotechnologies are at a relatively early stage but progress has been rapid and it is likely that such methods will be applied to many more forms of livestock and wild species in the future.

1.6 Summary Notes

- Complex land plants, and plant-like organisms such as algae and cyanobacteria, can be defined as autotrophic organisms using oxygenic photosynthesis.

- Humans have always exploited plant resources, initially by hunter-gathering and management of wild plants, but during the last 10,000 years these strategies have been almost completely replaced by agriculture.
- Agriculture requires various forms of technology for the management and improvement of crop systems. Some technologies are mechanical (ploughs, harvesters), some are chemical (fertilizers, pesticides), while others are biological (breeding).
- There are several forms of modern biotechnology: white biotechnology involves conventional microbial systems; red biotechnology involves recombinant microbes (and more recently animals) for medical applications; green biotechnology is sometimes used to describe plant-related technologies.

Further Reading

Plant–human interactions

Bharucha, Z. and Pretty, J. (2010) The roles and values of wild foods in agricultural systems. *Philisophical Transactions of the Royal Society B* 365, 2913–2926.

Harlan, J.R. (1992) *Crops and Man*, 2nd edn. American Society of Agronomy, Madison, Wisconsin, USA.

Keeley, L.H. (1995) Protoagricultural practices among hunter-gatherers. A cross-cultural survey. In: Price, T.D. and Gebauer, A.B. (eds) *Last Hunters-First Farmers*. School of American Research Press, Santa Fe, New Mexico, USA, pp. 243–272.

Levetin, E. and McMahon, K. (2008) *Plants and Society*. McGraw-Hill, New York.

Murphy, D.J. (2007) *People, Plants, and Genes: The Story of Crops and Humanity*. Oxford University Press, UK.

Weiss, E., Kislev, M.E. and Hartmann, A. (2006) Autonomous cultivation before domestication. *Science* 312, 1608–1610.

Agricultural technologies

Piperno, D.R., Weiss, E., Holst, I. and Nadel, D. (2004) Processing of wild cereal grains in the Upper Palaeolithic revealed by starch grain analysis. *Nature* 430, 670–673.

Revedin, A., Aranguren, B., Becattini, R., Longo, L., Marconi, E., Lippi, M.M., Skakun, N., Sinitsyn, A., Spiridonova, E. and Svoboda, J. (2010) Thirty thousand-year-old evidence of plant food processing. *Proceedings of the National Academy of Sciences USA* (www.pnas.org/cgi/doi/10.1073/pnas.1006993107).

Biotechnologies, ancient and modern

Chrispeels, M.J. and Sadava, D.E. (2003) *Plants, Genes, and Crop Biotechnology*. Jones and Bartlett, Sudbury, Massachusetts, USA.

Nicholl, D.S. (2008) *An Introduction to Genetic Engineering*. Cambridge University Press, UK.

Renneburg, R. (2006) *Biotechnology for Beginners*. Academic Press, New York.

Smith, J.E. (2004) *Biotechnology*. Cambridge University Press, UK.

2 Photosynthesis and the Evolution of Plants

2.1 Chapter Overview

In this chapter, we will survey the evolution of photosynthetic organisms. Oxygenic photosynthesis created the oxygen-rich atmosphere and ozone layer that enabled complex aerobic eukaryotes to evolve in the oceans and eventually colonize the land. Some cyanobacteria became endosymbionts within aerobic eukaryotes and went on to produce the photosynthetic algae and their land plant descendants. Land plants gradually developed a series of adaptations to terrestrial existence, including genomic and developmental flexibility and metabolic complexity, which played a large part in their evolutionary success. As we will see in later chapters, many researchers are currently attempting to manipulate various aspects of plant, algal and cyanobacterial development and metabolism to improve the efficiency of the bio-production of renewable raw materials. In order to achieve these aims, it is important to understand the evolutionary origins of plants and their relatives.

2.2 Origins of Photosynthetic Organisms

The major terrestrial ecosystems are dominated by a diverse flora that consists mainly of multicellular, vascular plants. These plants are the basis of the food chains upon which animals depend for their existence. The origins of land plants lie with some of the earliest organisms that arose on earth. In the highly anaerobic conditions of the early earth several forms of photosynthetic organisms evolved to use reductants (electron donors) such as hydrogen sulfide and hydrogen gas in order to produce complex carbon compounds. More than 3 billion years ago, however, a new group of prokaryotes, the cyanobacteria, adopted a much more efficient form of photosynthesis. Cyanobacteria employ water as a reductant, enabling them to use light energy to synthesize organic carbon at a much faster rate

than earlier photosynthetic organisms. Moreover, this process also produced gaseous oxygen as a by-product, which would eventually open up a vast range of new possibilities for life on earth.

Prior to 3 billion years ago, oxygen was virtually absent from the atmosphere and this highly reactive gas was toxic to almost all organisms of the time, which were obligate anaerobes. The steady build-up of atmospheric oxygen increased the domination of cyanobacteria in global ecosystems as their anaerobic competitors either died out or became restricted to specialized oxygen-free niches such as oceanic and volcanic vents. Eventually, new types of heterotrophic cells arose that contained nuclei and mitochondria. These were the first eukaryotes and many of them ingested cyanobacteria as a food source. Some of the engulfed cyanobacteria were not digested but became symbiotic partners, and then organelles, within these eukaryotic cells. This process gave rise to the various types of algae. About 550 million years ago, variants of some green algae began to colonize the land to produce lower plants such as mosses and liverworts. These soon evolved into the higher plants, including the gymnosperms and angiosperms that now dominate our terrestrial vegetation.

The earth itself was formed about 4.5 billion years ago as a sphere of mostly molten rock with an iron-rich core (see Table 2.1). Within a few hundred million years the planet had cooled down sufficiently to form solid rocks at its surface, and between 4.4 and 3.9 billion years ago the oceans were formed. The first living organisms, probably chemoautotrophs, arose soon afterwards, closely followed by several forms of non-oxygenic photosynthetic bacteria. The latter eventually gave rise to a new and more complex group of oxygenic photosynthetic prokaryote, the cyanobacteria, that are the direct ancestors of modern plants. There are putative fossils of ancient filamentous bacteria from Archaean rocks of Western Australia, from 3.5 billion years ago, some of which may be colonial,

Table 2.1. Timeline of plant life on earth.

Date (years ago)	Event or process	Significance
4.5 billion	Formation of earth	Life impossible on early molten earth
3.9 billion	Formation of oceans	Water is prerequisite for carbon-based life
3.8 billion	Earliest life	Life arises soon after water is available
3.5 billion	Earliest cyanobacteria?	Radical innovation: more efficient photosynthesis, faster growth rates, and oxygen evolution
2.4 billion	Traces of atmospheric oxygen	Aerobes gradually favoured over anaerobes
2.1 billion	Endosymbiosis between cyanobacterium and eukaryote host	Produced the first photosynthetic eukaryotes – the unicellular algae
1.6 billion	Earliest multicellular algae	Recognizable types of modern green and red algae
1.3 billion	Atmospheric oxygen at 3%	Aerobes dominant, multicellular eukaryotes favoured
1.0 billion	Increase in the numbers and diversity of marine algae	Greater potential to raise atmospheric oxygen concentrations
750 million	Complex multicellular green algae, such as *Cladophora*	Beginnings of cellular specialization and tissue formation
550 million	Atmospheric oxygen at 12%	Sufficient oxygen to create ozone layer and shield land from harmful UV radiation
>500 million	Earliest land plants, charophycean green algae	First eukaryotic colonization of land surface
470 million	Earliest spores of small liverwort-like organisms	Liverwort-like plants are the oldest group with members still extant today
428 million	Plants such as *Cooksonia*, 2–3 cm tall, simple body plan, no true roots or vascular system	Bryophytes with dominant haploid generation remain small and restricted to damp habitats
400 million	Seedless vascular plants such as club mosses, horsetails and true ferns	Emergence of larger pteridophytes with dominant diploid generation and wider habitat range
360 million	Earliest fossil gymnosperms	Pteridophytes dominate, gymnosperms remain a minor group for another 130 million years
300 million	Possible earliest angiosperms	Angiosperms remain a minor group for another 200 million years
230 million	Radiation of gymnosperms	Gymnosperms become dominant flora for over 100 million years
140 million	Earliest definite angiosperm fossils	Polyploidy of early angiosperms may have facilitated subsequent diversification
120 million	Major angiosperm groups such as eudicots, monocots and magnoliids	Major angiosperm groups are now present but species numbers and range remain restricted
70–100 million	Beginnings of 'great angiosperm radiation'	Steady increase in species numbers and habitat range
85 million	Earliest ancestral cereals	Basal meristem in many grasses is an effective adaptation to continuous herbivory
68 million	Earliest legumes	Adaptation to nitrogen-deficient soils
65 million	KT (Cretaceous–Tertiary) mass extinction event	Rapid radiation of angiosperms and colonization of wide range of habitats
40 million	Divergence of pooid from panicoid grasses, earliest rice	Global spread of grasses
10–20 million	Divergence of panicoid groups such as sorghum and maize and pooid groups such as barley, rye and the wheats	Establishment of the major families of cereal crops, emergence of C4 photosynthesis in several angiosperm groups
2–3 million	Increasing emergence of grassland habitats as climate cools and becomes more variable	Grassland and savannah habitats favour larger herbivores and several branches of hominin, including *Homo* spp.

Continued

Chapter 2

Table 2.1. Continued.

Date (years ago)	Event or process	Significance
11,000	Earliest domesticated crop plants	Beginnings of significant human impacts on plant evolution to favour 'domestication' traits
500	Globalization of crop plants	Worldwide spread of domesticated crops and increasing restriction of wild plant habitats
300	First man-made hybrid plants	Deliberate creation of new plant species by humans
100	Start of scientific plant breeding	Increasingly effective manipulation of crop genomes and phenotypes by humans
30	First transgenic plants engineered	Ability to deliberately incorporate completely novel genes into plant genomes

non-oxygenic photosynthetic organisms or possibly very early forms of oxygenic cyanobacteria.

The origin of plants therefore dates back to at least 3 billion years ago when the cyanobacteria developed a more effective form of photosynthesis that used water as a reductant and produced oxygen as a by-product. Cyanobacteria are among the easiest microfossils to recognize because morphologies in the group have remained much the same for billions of years. Cyanobacteria also leave behind characteristic 'chemical fossils' in the form of breakdown products of pigments such as phycoerythrocyanin and phycoerythrobilin. Around 2.1 billion years ago, some cyanobacteria were engulfed by larger eukaryotic cells and became transformed into endosymbiotic plastidial organelles within their host cells. These newly photosynthetic eukaryotic cells eventually developed into the various forms of photosynthetic algae.

Some algae, such as the seaweeds, became relatively large and complex multicellular organisms. Between 500 and 600 million years ago, members of one group of seaweeds, the Chlorophyta or green algae, began to colonize the land and created the rich terrestrial biosphere we are so familiar with today. The key attribute of all free-living plants is the ability to photosynthesize but, as we will now see, the process of photosynthesis was only ever developed by bacteria. Eukaryotes such as algae and land plants are only secondarily capable of photosynthesis because they contain ancient bacterial symbionts. These once free-living cells have now been converted into the diverse class of subcellular organelles known as plastids, of which the best known are the chloroplasts, which are the sole sites of photosynthesis in all plants.

2.3   Photosynthesis

Photosynthesis is the use of certain wavelengths of electromagnetic radiation (typically between about 400 and 700 nm) to synthesize complex organic molecules, such as carbohydrates, lipids and amino acids, from simple inorganic compounds such as water and CO_2. The so-called 'light reactions' of photosynthesis use this light energy to drive the conversion of a reductant, such as water or hydrogen sulfide, to the energy-rich compound $NADPH_2$. The same light energy drives the formation of the high-energy intermediate, ATP. The 'dark reactions' of photosynthesis then use ATP and $NADPH_2$ generated by the light reactions to reductively fix gaseous CO_2 into carbohydrates such as starch and sucrose. The latter are either catabolized to generate energy or converted into other metabolic intermediates and end products such as amino acids, proteins, lipids, cellulose and secondary compounds.

Evolution of photosynthesis

Molecular evidence suggests that all forms of photosynthesis are derived from a single lineage of ancient bacteria similar to present-day Heliobacteria. These anoxygenic photosynthetic bacteria were some of the earliest living organisms on earth. They evolved over 3.8 billion years ago in a gradually cooling, highly volcanic and anaerobic world with almost no free oxygen and where seas of liquid water had only recently formed. Early photosynthetic bacteria used reductants such as hydrogen sulfide and organic acids to convert atmospheric CO_2 into complex carbohydrates. These bacteria possessed only one of the two types of the photosynthetic reaction centre (either RCI or RCII) complexes now found in plants. These reaction centre complexes use light energy harvested by separate pigment-protein complexes to split a reductant molecule, hence releasing energized electrons and protons that can be used to generate ATP and $NADPH_2$.

Several families of anoxygenic photosynthetic bacteria with single reaction centres can still be found today. For example, iron-sulfur reaction centres (RCI) are present in the Heliobacteria (Firmicutes) and Chlorobi (green sulfur bacteria), while pheophytin-quinone reaction centres (RCII) are present in the Proteobacteria and Chloroflexi (green non-sulfur bacteria). There are striking similarities between RCI and RCII in their subunit structures and the mechanisms of charge transfer, which indicate that they were originally derived from a common ancestor. Current evidence (see Fig. 2.1) suggests that the original photosynthetic reaction centre complex evolved from a cytochrome b-like protein to create the kind of iron-sulfur reaction centre (RCI) found in green sulfur and Heliobacteria. It is proposed that the RCI gene then underwent a duplication followed by loss of the iron-sulfur centre and formation of the kind of pheophytin-quinone reaction centre (RCII) that is now found in green filamentous and purple bacteria. All of these photosynthetic bacteria contain either RCI or RCII (never both reaction centres) and are relatively slow-growing anoxygenic organisms that are unable to use water as an electron donor.

After several hundred million years, a new group of photosynthetic organisms, the cyanobacteria, emerged.

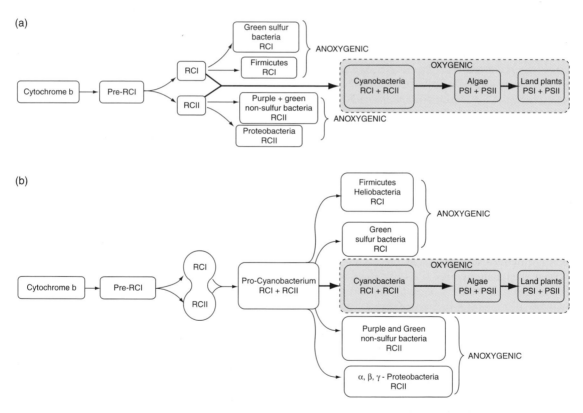

Fig. 2.1. Evolution of photosynthetic organisms. Two alternative evolutionary models of photosynthesis are shown. These models are based on sequence-based phylogenetic information for the reaction centre apoproteins (RCI and RCII) and Mg-tetrapyrrole biosynthesis enzymes. In each model, a cytochrome b-like protein gives rise to the RCI precursor. (a) Pre-RCI then diverged to produce RCI and RCII. Separate lineages of bacteria containing only RCI gave rise to green sulfur bacteria and Firmicutes, while other groups containing only RCII produced purple and green non-sulfur bacteria and Proteobacteria. A separate lineage containing both RCI and RCII gave rise to cyanobacteria. (b) Pre-RCI diverged to produce RCI and RCII, which were combined to produce a pro-cyanobacterium from which the other groups of photosynthetic bacteria are derived.

These blue-green bacteria contained both RCI and RCII complexes, which are now known respectively as photosystem I and photosystem II. Sequence data suggest that cyanobacteria probably acquired their two reaction centre complexes from organisms similar to the Heliobacteria (for RCI) and Proteobacteria (for RCII). The mechanism for this is unknown but is likely to have involved the kind of large-scale horizontal gene transfer between unrelated organisms that is discussed in more detail in Chapter 3. Because they possess two reaction centre complexes rather than one, cyanobacteria are able to generate much greater amounts of useful energy from sunlight than the older forms of photosynthetic bacteria. Indeed, cyanobacteria produce sufficient energy to split water molecules. This reaction releases two protons from each water molecule, plus two high-energy electrons that can pass through a chain of carriers and eventually form ATP and $NADPH_2$.

The development of oxygenic photosynthesis by cyanobacteria freed living organisms from their dependence on the relatively limited types of reductant, such as hydrogen sulfide, that had been used previously. The new reductant, water, was available in almost unlimited quantities in the vast oceans and river systems that had accumulated on earth around 3 billion years ago. Thanks to its improved efficiency, oxygenic photosynthesis enabled global production of organic carbon to increase by between 100- and 1000-fold. It marked a quantum leap forward in the potential biological carrying capacity of the earth. Even more momentously for life on earth, however, this new form of photosynthesis also involved the release of gaseous oxygen as its main by-product. The development of oxygenic photosynthesis went on to fuel a huge increase in the diversity and complexity of living organisms, as well as radically altering the composition of the atmosphere in a way that eventually enabled life to emerge from the seas to colonize the land.

There is good evidence that microfossilized oxygenic cyanobacteria were present before 2.7 billion years ago and they are probably considerably older. By producing an oxygen-rich, ozone-protected atmosphere, cyanobacteria enabled the evolution of the large multicellular plants and animals that now dominate much of the biosphere. Although there were detectable traces of atmospheric oxygen as early as 2.4 billion years ago, free oxygen accumulated very slowly for another billion years (see Fig. 2.2). This was because most of the oxygen released by cyanobacteria was rapidly absorbed by the vast quantities of iron deposits close to the surface of the early earth. This process created the massive banded iron oxide sediments now found around the world. It continued until the supply of free iron was eventually exhausted and oxygen gas could accumulate in useful quantities for terrestrial life.

The concentration of atmospheric oxygen then increased steadily to reach 3–5% by 1.3 billion years ago. Shortly before 600 million years ago there was a dramatic acceleration in the rate of oxygen release, which reached 10–16% by 550 million years ago. By this time there was sufficient atmospheric oxygen to create the stratospheric ozone layer that now shields the land surface from the harmful UV radiation. Prior to this, the extremely high levels of mutagenic UV radiation had prevented the emergence of complex terrestrial organisms. Very shortly after the ozone layer was formed, the first land plants began to colonize the earth's surface.

The light reactions of oxygenic photosynthesis

The basic chemistry of oxygenic photosynthesis in algae and plants is essentially unchanged from that employed by cyanobacteria for billions of years. In both cases, photons of light energy in the region from 400 to 700 nm are intercepted by large arrays of light-harvesting, pigment-protein complexes that contain numerous chlorophyll molecules and accessory pigments such as carotenoids and xanthophylls. The light-harvesting complexes then pass on their absorbed energy via resonance transfer to the two reaction centre complexes, namely photosystem I or II.

As shown in Fig. 2.3, the channelling of light-derived energy to the reaction centre of photosystem II enables molecules of water to be split to create high-energy products (protons and electrons) plus oxygen gas. In algae and plants, the two reaction centres are linked by a third iron-containing protein complex, called cytochrome b/f, to form a chain of electron carriers leading from water to $NADPH_2$. This electron transport pathway, from photosystem II to cytochrome b/f and on to photosystem I, is known as the Z-scheme. The Z-scheme uses light energy and water as its inputs and its products are protons, $NADPH_2$ and oxygen. A further protein complex,

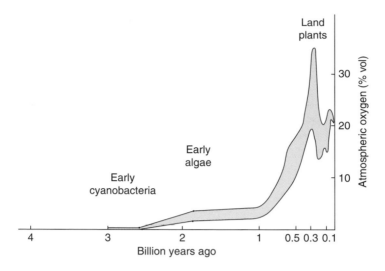

Fig. 2.2. The oxygenation of the atmosphere. Production of oxygen gas by cyanobacteria may have started well over 3 billion years ago, but for almost 2 billion years almost all the released oxygen was absorbed by iron deposits on the earth's surface. After this iron was oxidized, oxygen was able to accumulate in the atmosphere and reached 3–5% (vol/vol) by 1.3 billion years ago. After this there was considerable acceleration of oxygen release as algae joined cyanobacteria as major photosynthetic producers. By 550 million years ago oxygen had reached 10–16%, which was sufficient to generate a stratospheric ozone layer capable of shielding the earth's surface from much of the solar UV radiation. Colonization of the land by plants and climatic warming further accelerated oxygen release, which reached over 30% by the Carboniferous era about 300 million years ago. Subsequent cooling and reduction in plant productivity resulted in a decrease to present-day values of atmospheric oxygen, which are in the region of 20%. The shaded area indicates the range of estimates of oxygen concentrations, which are still somewhat uncertain.

ATP synthetase, uses the proton gradient generated by water splitting to catalyse the formation of ATP. All of the light reactions occur on membrane-bound protein complexes located within the thylakoid membranes of the chloroplast. As discussed in Box 2.1, advances in our knowledge of the mechanism of photosynthesis could eventually lead to the development of new forms of biologically derived solar-powered devices.

The dark reactions of oxygenic photosynthesis: C3 and C4 plants

The thylakoid membrane-based light reactions described above produce two soluble energy-rich compounds, $NADPH_2$ and ATP, which are released into the aqueous stroma of the chloroplast. Here, $NADPH_2$ and ATP provide the chemical energy that enables soluble enzymes of the reductive pentose phosphate pathway (sometimes called the Calvin cycle) to catalyse the fixation of atmospheric CO_2 to produce complex carbohydrates, such as sugars and starches (see Fig. 2.4). The primary enzyme of CO_2 fixation is ribulose bisphosphate carboxylase, or Rubisco, which combines ribulose bisphosphate with CO_2 to form two molecules of 3-phosphoglycerate. Because Rubisco can use either CO_2 or O_2 as substrates, the ratio of these two molecules determines its efficiency as a carboxylase. Some of the problems associated with Rubisco and some potential biotechnological solutions are outlined in Box 2.2.

In all photosynthetic eukaryotes, CO_2 is converted into carbohydrate via the reductive pentose phosphate pathway. So-called C3 plants use this pathway alone, but the process becomes less efficient as the ratio of CO_2 to O_2 decreases. About 16 million years ago several angiosperm groups, including numerous grass species, independently developed the CO_2-concentrating mechanism known as C4 photosynthesis. C4 photosynthesis is particularly useful at higher temperatures where the ratio of CO_2 to O_2 in solution becomes markedly decreased. Because it also results in increased water-use efficiency, C4 photosynthesis is advantageous in arid or semi-arid conditions. By 5–7 million years ago, a combination of lower atmospheric CO_2 levels, higher temperatures and increasing

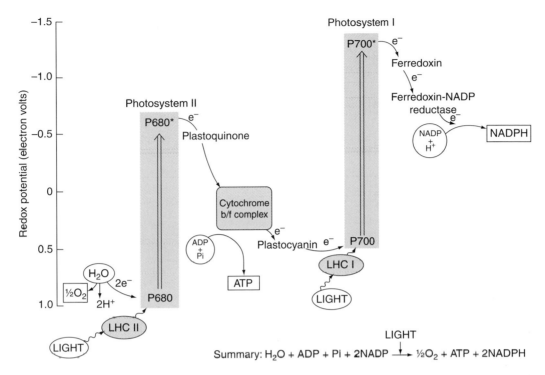

Fig. 2.3. Mechanism of oxygenic photosynthesis. Oxygenic photosynthesis proceeds via the Z-scheme, whereby water molecules are split by photosystem II using energy absorbed by the light-harvesting II (LHCII) pigment-protein complex. Electrons from photosystem II are passed on to a chain of carriers including the cytochrome b/f complex and photosystem I, becoming re-energized via the light-harvesting I (LHCI) pigment-protein complex. Finally, the electrons are used to reduce NADP to $NADPH_2$. A proton gradient generated by water splitting is also used to generate the formation of ATP on an ATP synthetase complex. *, activated forms of P680 and P700.

aridity favoured the rapid global spread of C4 photosynthesis.

The main advantage of C4 photosynthesis is its ability to generate high concentrations of CO_2 in the vicinity of Rubisco, which greatly increases the efficiency of the enzyme. This is achieved by initially fixing CO_2 in a separate leaf compartment, the bundle sheath, via phospho*enol*pyruvate (PEP) carboxylase to form C4 acids such as malate and aspartate. The C4 acids then move to bundle sheath cells where they are decarboxylated to generate local concentrations of CO_2 as high as 1000–2000 ppm. This contrasts with the current average atmospheric concentration of about 390 ppm. The vast majority of our major crops are C3 plants although some high-yielding species, including maize, millets and sugarcane, are C4 plants.

If major C3 crop species, such as rice, wheat, barley and potatoes, could carry out C4 photosynthesis they might grow faster, have higher yields and be more efficient in their use of water and nitrogen. Several breeding approaches have been attempted but one of the main problems is the separate and independent genetic control of C4 plant anatomy and the expression of C4 pathway enzymes. The most common transgenic strategy has been to overexpress introduced PEP carboxylase genes from a C4 plant such as maize in a C3 plant such as rice. In rice plants, PEP carboxylase genes have been successfully overexpressed more than 100-fold compared with control plants. Unfortunately, however, the enzyme was dephosphorylated and hence inactivated in light-exposed leaves of the transgenic rice plants.

In order to circumvent this problem, forms of PEP carboxylase that lack phosphorylation sites were overexpressed in potatoes and *Arabidopsis*. But this led to a reduction in the supply of aromatic amino acids and phenolics (for which PEP is a precursor) and the transgenic plants were severely stunted.

Box 2.1. The quest for artificial photosynthesis

Oxygenic photosynthesis is remarkable in its ability to use solar energy to split water molecules at room temperature. Plants are then able to harness the chemical energy released from water splitting to make complex carbon-based compounds. In contrast, man-made efforts to capture solar energy tend to be relatively inefficient, and the splitting of water can only be achieved at very high temperatures or by using large amounts of electrical energy. This has led to attempts to use hybrid systems such as pigment protein complexes immobilized on artificial supports to assemble structures capable of carrying out photosynthesis *in vitro*. In all cases these technologies are based on our knowledge of the detailed mechanism of plant photosynthesis.

Some examples of artificial photosynthesis currently under development include:

- NADP$^+$/NADPH$_2$ Coenzyme methods employing light-generated hydrides and catalysts such as ruthenium for carbohydrate production, ideally from CO_2.

- Photoelectrochemical cells using catalysts based on rhodium, cobalt or ruthenium have been used to mimic enzymes such as the oxygen-evolving complex of photosystem II and hydrogenases.
- Dye-sensitized solar cells using silicon as a photoelectron source and a dye for charge separation and current generation.

To date, these and other approaches to artificial photosynthesis have had mixed results. Major outstanding problems include durability of catalysts, difficulty in scaling up the processes, and stability of the cells. Two promising new approaches involve titanium dioxide (TiO_2) and cobalt oxide (CoO)-based catalysts. In the future, bio- and nanotechnology may enable us to recreate more stable artificial chloroplasts and biomimetic approaches will allow more faithful copies to be made of existing plant-based systems. The ultimate goal is a new generation of solar-powered devices capable of efficient, cheap synthesis of complex carbohydrates, hydrocarbons and hydrogen for use as fuels, foods and industrial materials.

To date there have been no reproducible reports of increased photosynthesis in transgenic plants overexpressing PEP carboxylase genes and it seems unlikely that single-gene insertions will succeed. New approaches are focusing on the introduction of groups of C4 genes, although the transfer of both the C4 anatomy and C4 biochemistry to C3 crops remains a distant prospect.

2.4 Endosymbiosis and the Origin of Algae

Eukaryotes arose about 2.7 billion years ago and were characterized by more complex forms of intracellular organization than prokaryotes. The earliest eukaryotes were unicellular heterotrophic organisms with endomembrane systems, including a nuclear envelope enclosing their major genome, endoplasmic reticulum, Golgi apparatus, and numerous transport and processing vesicles, all linked by cell-wide cytoskeletal networks. The earliest eukaryotic cells did not contain either mitochondria or plastids. The latter organelles were acquired subsequently by the process of endosymbiosis whereby bacterial cells were

ingested intact and retained as symbionts. A combination of endosymbiosis and horizontal gene transfer (see Chapter 3) was probably responsible for the evolution of the major groups of eukaryotic organisms, including algae, plants, fungi, oomycetes and animals.

Mechanisms of endosymbiosis

Thanks to genomic and biochemical studies, it is now possible to trace the likely course of events whereby ingested bacteria were transformed into mitochondria and plastids (see Fig. 2.5). Mitochondria were captured very early in eukaryote evolution and all subsequent eukaryotic lineages are derived from such cells. It is thought that mitochondria are derived from an aerobic bacterium that was engulfed by a simple unicellular heterotrophic protoeukaryote. The new endosymbiont-containing cell would have been capable of more efficient aerobic respiration in comparison with cells that lacked mitochondria. This was especially advantageous as global oxygen levels increased due to the activity of cyanobacteria, as described above.

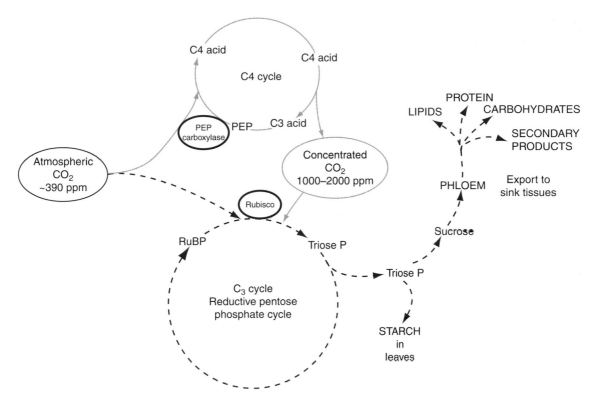

Fig. 2.4. CO_2 fixation mechanisms. Most plants fix CO_2 using Rubisco via the C3 reductive pentose phosphate cycle. The fixed carbon is either stored in leaves as starch or exported via the phloem as sucrose. Some plants use an additional mechanism, known as the C4 cycle, to concentrate CO_2 by as much as three- to fivefold. This involves an initial carboxylation to a C4 acid using PEP carboxylase and subsequent release of CO_2 close to the active site of Rubisco.

Box 2.2. Engineering a more efficient form of Rubisco

Because Rubisco is responsible for the fixation of CO_2 to produce all of the complex carbon compounds in plants, it is arguably one of the most important enzymes on earth. Unfortunately, however, the enzyme suffers from the following two serious drawbacks that limit its efficiency and reduce the potential productivity of plants:

1. In addition to its primary carboxylase activity, Rubisco can act as an oxygenase, using molecular oxygen to convert its substrate, ribulose bisphosphate, into 3-phosphoglycerate and 2-phosphoglycolate in an energy-wasteful side reaction.

2. Its catalytic rate is one of the slowest of any enzyme yet measured and is so sluggish that plants need to produce vast quantities of the enzyme, which constitutes as much as half the soluble protein in a typical mature leaf.

Because of its slow catalytic rate and its poor substrate specificity, this enzyme has both puzzled and fascinated researchers for many decades. Several biotechnological approaches are currently being taken to improve the efficiency of Rubisco in order to improve photosynthetic rates and hence, possibly, to increase crop yields. For example, site-directed mutagenesis is being used to modify the active site of Rubisco in model organisms such as the cyanobacterium *Synechococcus* and the alga *Chlamydomonas reinhardtii*. While these and other approaches have had some limited success, some researchers believe that Rubisco is already highly adapted to its present subcellular environment and that genetic engineering is unlikely to produce more than relatively modest improvements in its catalytic efficiency and plant growth.

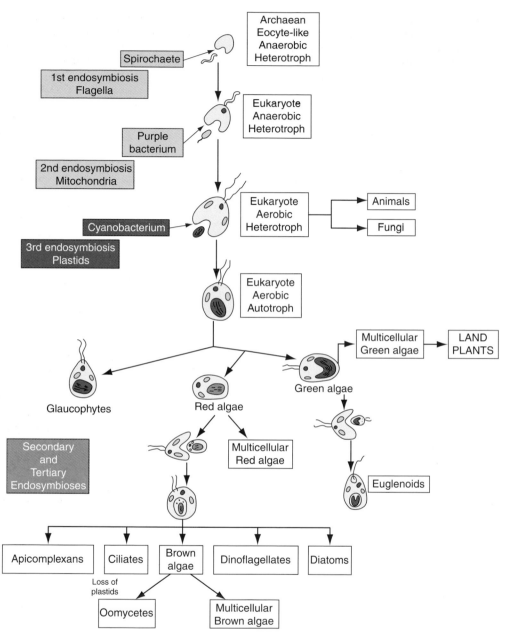

Fig. 2.5. Endosymbiosis and plastid evolution. Repeated endosymbiosis is one of the key processes in the evolution of eukaryotes and especially plants. Algae are probably derived from simple Archaean anaerobic heterotrophic cells that underwent three serial endosymbiotic events involving the capture of a spirochaete, purple bacterium and cyanobacterium that gave rise to flagella, mitochondria and plastids, respectively. Most green algae and higher plants are descendants of a single ingested cyanobacterium. However, some green algae, such as the Euglenoids, and many red and brown algae are derived from additional endosymbiotic events involving the capture and sequestration of entire algal cells by large heterotrophic cells. Finally, in several algal-derived lineages the plastids have become much reduced or lost altogether as the organisms have adopted heterotrophic (often parasitic or pathogenic) lifestyles. Examples include several major oomycete crop pathogens such as *Phytophthora*, and important apicomplexan human pathogens such as *Plasmodium* and *Toxoplasma*.

The original α-proteobacterial-like endosymbiont was probably related to the present-day *Rickettsia*, which are obligate intracellular parasites descended from free-living bacteria. Instead of being digested by the host cell or functioning as an intracellular parasite, the engulfed bacterium was recruited into an endosymbiotic relationship with its host. The host provided nutrients for the bacterium while exploiting its capacity for respiration to produce energy-rich compounds such as ATP and NADPH. Very early in this relationship, the bacterial endosymbiont lost its autonomy, including most of its DNA, the majority of which was transferred to the nuclear genome of the host. The relict bacterium eventually became what we now know as a mitochondrion. Mitochondria still contain a much-reduced amount of bacterial-like DNA. They also possess bacterial-like 70S ribosomes, and have a permeable outer layer complete with porins similar to those of contemporary aerobic bacteria.

The process of organelle formation is still occurring in some organisms. One example where we can observe the process in operation comes from the bacterium, *Candidatus carsonella ruddii*. This organism lives in the gut of the sap-sucking psyllid insect *Pachpsylla venusta*. Insects like psyllids have a restricted diet almost entirely consisting of plant sap (i.e. phloem), which is especially deficient in certain vitamins and original amino acids. The symbiotic bacteria in the psyllid gut make up these deficiencies by synthesizing the missing compounds. In some cases, this has led to genome reduction in the resident bacteria, which has been taken to an extreme in the case of the symbiont of *P. venusta*. This bacterium has a tiny residual genome of only 160 kb (kilobase pairs or 1000 base pairs) that encodes a mere 182 predicted genes. The thousands of missing genes from its genome have either been transferred to the host nuclear genome or lost completely. This remnant of a bacterium has now lost almost all of the functions that would define it as a living organism. Instead it can be regarded as undergoing a transformation from an endosymbiont into a new class of subcellular organelle specializing in the biosynthesis of vitamins and amino acids.

The origin of plastids

The plastid group of organelles is found in all algae and plants. Plastids arose when one of the mitochondria-containing heterotrophic cells, formed as described above, engulfed another bacterial cell, but this time the ingested cell was a photosynthetic cyanobacterium. Like the precursors of mitochondria, the ingested cyanobacterium gradually lost its autonomy and much of its DNA. Whereas the genomes of free-living cyanobacteria contain several thousand genes, their plastid descendants contain only 100–220 protein-encoding genes. Today, all plant-like eukaryotes contain a diverse group of plastidial organelles all derived from a cyanobacterial endosymbiont. One of the earliest known algae is *Grypania*, a giant unicellular organism dating from 2.1 billion years ago. In unicellular algae, all of the plastids in the cell tend to be photosynthetic, but the plastids of higher plants have now developed numerous additional non-photosynthetic functions depending on the tissue in which they reside.

The best-known plastids are the photosynthetic chloroplasts found in green tissues such as leaves and some fruits. Chloroplasts contain chlorophyll, stacked thylakoid membranes, and carry out photosynthesis in a similar manner to their cyanobacterial ancestors. Other types of plastid include: the pigment-rich chromoplasts found in flower petals and some fruits; the starch-rich amyloplasts found in storage organs such as seeds and tubers; and the lipid-rich elaioplasts found in several colourless tissues. These non-green plastids do not photosynthesize but they are bounded by the same kind of double envelope membrane and carry the same cyanobacterial-like genome as chloroplasts. In addition to their photosynthetic and storage roles, plastids perform other functions in plant cells, such as *de novo* fatty acid biosynthesis and various aspects of hormone and nitrogen metabolism (see Chapter 4).

Diversification of the algae

It is commonly agreed that all algae and land plants are descended from a single endosymbiotic event. As shown in Fig. 2.5, some groups of algae then underwent secondary and even tertiary episodes of endosymbiosis by engulfing other algae. By 1.6 billion years ago, some algal groups had developed into multicellular forms, including recognizable types of modern green and red algae. The Bangiophyte red algae date from 1.2 billion years ago. After 1 billion years ago, there was a steady increase in the numbers and diversity of algae. Some groups diverged into diatoms, dinoflagellates, and euglenoids, which could be either unicellular or colonial. Some algae acquired flagella, became motile, and developed into facultative heterotrophs.

As levels of oxygen rose in the atmosphere and in the oceans, the evolution of ever more complex multicellular organisms became possible. By 700 to 800 million years ago, early types of multicellular green algae, such as *Cladophora*, were present in many of the oceans and shallow seas that covered much of the earth. Between 400 and 600 million years ago, there was an explosive growth of new forms of red, brown and green algae. Brown algae are derived from non-algal cells that ingested smaller unicellular red algae and converted them into endosymbionts. Recent DNA analysis suggests that some brown algae lost their plastids and acquired groups of fungal genes via horizontal gene transfer. Some of these chimeric species developed into fungus-like organisms, including the oomycetes or water moulds, which are among the most serious pathogens of plants, including many major crops (see Chapter 6). Another recent report has shown that the horizontal transfer of many specialized genes from bacteria is responsible for the ability of nematode parasites to penetrate plant tissues (see Chapter 3).

Within the green algae, some groups remained either unicellular or colonial while other lineages diversified into true multicellular organisms with some degree of tissue differentiation. Multicellular green algal forms included branched or unbranched filaments and membranous or tubular thallus structures. By 600 million years ago, the green algae were as important as the cyanobacteria in their role as biological primary producers, and as contributors to the still-increasing accumulation of oxygen in the atmosphere, which had reached more than 10%. Marine algae continued to play a dominant ecological role with a continual process of extinction of some forms and the appearance of new forms in response to environmental changes. For example, dinoflagellates underwent a major radiation during the mild climate of the Jurassic Era about 200 million years ago, while vast numbers of algal species were lost during the mass extinction event at the Cretaceous/Tertiary boundary that also killed off the dinosaurs about 65 million years ago.

Today, the oceanic green algae and cyanobacteria together make up the phytoplankton, which is the basis of the entire marine food web. The phytoplankton ultimately supports all aquatic animals

from octopuses to whales, plus the many land-based animals that feed off sea life. This includes millions of people for whom marine organisms are an important but increasingly threatened food resource. Some phytoplankton species, such as the cyanobacterium *Spirulina*, have been harvested as protein- and vitamin-rich foods for many centuries. Nowadays biotechnological methods are being used to mass-produce *Spirulina* for human and livestock consumption. Transgenic methods are also being developed to engineer algae and cyanobacteria as sources of biofuel (see Chapter 12). Meanwhile, other microalgae such as *Haematococcus pluvialis* and *Nannochloropsis oculata* (formerly called *Chlorella*) are being grown in large-scale bioreactors to produce high value products like astaxanthin, a pigment that has beneficial health effects on humans that is also used to supplement rations for farmed salmonid fish.

2.5 The Evolution of Land Plants

We have seen that the earliest recognizable land plants date to over 500 million years ago. Prior to this, the lack of a stratospheric ozone layer meant that large amounts of UV radiation penetrated to the earth's surface. A few UV-resistant cyanobacteria were able to survive in this highly irradiated environment, and some of these organisms eked out a precarious existence in terrestrial habitats as early as 1 billion years ago. But no multicellular eukaryotic organisms, with the possible exception of one or two simple lichens, could live under such harsh conditions. Thanks to oxygenic photosynthesis, however, a substantial stratospheric ozone layer was established by 500 million years ago. This led to a massive reduction in ground-level UV radiation and enabled the first land plants to emerge shortly thereafter. Soils suitable for plant establishment also arose about this time due to a combination of physical weathering of rocks and the presence of microbial mats, mainly consisting of cyanobacteria.

An additional factor that favoured the evolution of land plants during this period was the gradual drying up of the extensive shallow seas that covered large low-lying areas of the principal land masses. The earliest plants were probably derived from branched filamentous algae living in the steadily shrinking lagoons, lakes and ponds. Because these bodies of water were subject to periodic drying, a strong selection pressure would have been imposed on their resident algae that would have favoured variants able to tolerate a lack of water and an increased exposure to the air. The algal group that eventually made a successful transition to dry land was similar to present-day charophycean green algae. It is believed that all of the land plants are a monophyletic group descended from a single charophycean-like Streptophyte ancestor that emerged on the supercontinent of Gondwana (see Fig. 2.6).

In contrast to other green algae, the Charophyte group has a long list of relatively complex cellular and tissue features that are also found in land plants. These include plant-like microtubules, plasmodesmata, apical meristems, intricate branching thalli, tissues made up of three-dimensional arrays of related cells produced by asymmetric cell divisions, sporopollenin-enclosed spores, and a placenta

that nourishes a retained diploid generation. As they began to colonize the land, charophycean green algae with these useful characteristics rapidly developed into small liverwort-like organisms with a dominant haploid generation. Fossilized remains of their spores, resembling those of modern liverworts, have been dated to more than 470 million years ago and the earliest land plant colonists may have emerged as early as the Cambrian Era, 488–542 million years ago.

Bryophytes (non-vascular land plants)

An early innovation of the first land plants was the extension of the diploid phase of their life cycle compared with their algal ancestors. One of the key features of even the simplest land plants is the alternation of generations between a haploid

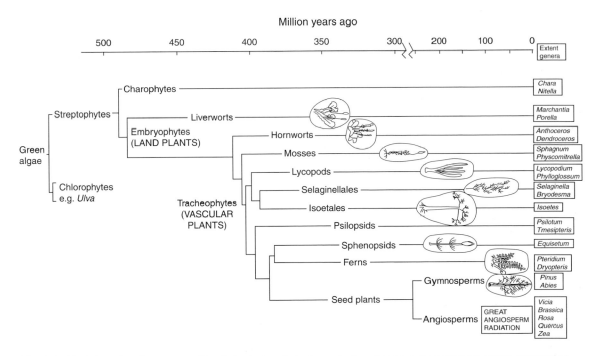

Fig. 2.6. Evolution of major groups of land plants. Land plants are believed to be a monophyletic group descended from a lineage of green algae known as Streptophytes that includes extant Charophyte marine species such as *Chara* and *Nitella*. The earliest land plants, or Embryophytes, probably resembled contemporary liverworts but they rapidly branched to produce another group that gave rise to mosses and hornworts. Soon after this, another branch emerged that gave rise to the vascular plants, or Tracheophytes. Within a relatively short period, the Tracheophyte branch had diversified to produce the ancestors of the major groups of higher plants including the ferns and seed plants. The earliest seed plants date from about 370 million years ago and the gymnosperms may have occurred as early as 360 million years ago. The earliest definite angiosperm fossils date from about 140 million years ago, but DNA sequence data suggest that the group may be as much as 250–300 million years old.

gametophyte and a diploid sporophyte. As plants increased in size and complexity, their diploid phase become more extended and dominant. In contrast, the diploid phase in the charophytic green algae from which plants are descended is very short, with meiosis occurring soon after fertilization and zygote formation. According to the 'antithetic hypothesis' of plant evolution, the diploid sporophyte generation was interpolated into the life cycle of Charophytes by imposing an increasingly lengthy delay before the onset of meiotic cell divisions after fertilization. A further innovation in bryophytes was their increased metabolic versatility. In particular, bryophytes could synthesize a far wider range of secondary compounds, such as phenolics and flavonoids, including those that provide structural support, water resistance, and a degree of protection from the relatively harsh external environment on the land.

The earliest land plants were prostrate organisms that lay directly on their growth substrate and overgrew each other. Plants such as *Cooksonia*, dating from 428 million years ago, were at most a few centimetres tall, with a simple body plan and no true roots or vascular system. Competition for light would have provided a powerful selective pressure for the evolution of upright shoots that were also able to form branches. Land plants soon diversified into the multiple bryophyte lineages that include all the extant mosses, liverworts and hornworts as well as several extinct forms. However, for several reasons, the bryophytes were restricted in their ability to occupy more than a relatively narrow range of ecological niches.

One of their key constraints was an inability to increase their body size beyond a limited extent. There were several reasons for this. A dominant haploid generation means that bryophytes are unable to increase in size or complexity as rapidly as a diploid organism. Their lack of roots or a vascular system hampers efficient nutrient uptake and assimilate partitioning beyond a small maximum size. Their relatively unstrengthened bodies also preclude growth beyond a few centimetres in height or length. Finally, their lack of a continuous impermeable cuticle restricts bryophytes to moist, shady habitats close to water, which they also require for reproduction. In order to achieve a more extensive colonization of the wider land surface apart from wetlands, numerous modifications to the bryophyte body plan were required.

As described in Box 2.3, the vascular plants, or tracheophytes, developed a diverse range of characteristics that enabled them to colonize most terrestrial environments, from hot tropics to cool boreal regions, and from arid deserts to regions of almost constant rainfall.

Vascular plants

Bryophytes diverged very soon after the colonization of the land and the lineage that eventually led to the vascular plants was already present as early as 420 million years ago (Fig. 2.6). The first vascular plants were the pteridophytes, which thanks to a dominant diploid generation and the beginnings of lignification were able to produce much larger forms than the bryophytes. As shown in Fig. 2.7, pteridophytes were the dominant flora for almost 150 million years. Some of the earliest vascular plants were the Psilophytes, which had no roots or leaves and consisted mainly of photosynthetic stems that looked rather similar to the still-extant whisk fern, *Psilotum nudum*. By 400 million years ago, several groups of seedless vascular plants such as the club mosses, horsetails and true ferns had started to occupy new niches away from the shady damp habitats favoured by the bryophytes. Some of these early tracheophytes, such as *Pertica*, may have been over 3 m tall. The period between 360 and 400 million years ago marked the beginning of an extensive colonization of much of the global land surface by plants. Because large herbivorous animals had yet to evolve, vast forests of tree lycopods, tree ferns and horsetails, later joined by seed ferns and pro-gymnosperms, were able to spread virtually unmolested across the landscape.

The earliest fossil gymnosperms date from 350 to 365 million years ago, but the last common ancestor of all extant seed plants probably lived about 300 million years ago. Between 290 and 360 million years ago there was a massive radiation of tracheophytes of all types. These plants benefited from relatively warm, moist climates across most of the world during this period. The consequent rise in plant productivity led to a massive increase in atmospheric oxygen levels to reach about 35% (in contrast to today's 21%). This period is known as the Carboniferous Era and marks the time when most of our current fossil fuel reserves, such as coal, oil and natural gas, were deposited from the

Box 2.3. How vascular plants colonized the world

Some of the most important changes that enabled plants to colonize a greater variety of terrestrial habitats include the following:

- Development of a well-defined sporophytic apical meristem allowed for the production of organs. The capacity for shoot meristem proliferation enabled branching of the sporophyte body as seen in modern vascular plants and some fossil groups known as pre-tracheophytes. This growth habit gave vascular plants far better access both to subterranean resources (nutrients and water) and aerial resources (light and CO_2).
- Sporophyte branching enabled plants to increase their body size, productivity and reproductive potential as well as the capacity to continue growth if some stem cells were damaged or lost (e.g. by herbivory). Multiple sporophytic growth points also permitted the specialization of branch systems to form megaphyllous leaves, cones and flowers. The ability to form branches facilitated the transition towards dominance of the sporophyte generation and led to the substantial reduction in size and complexity of the gametophytic body of higher plants (see Chapter 5).
- A dominant diploid generation was especially important for vascular plant evolution because the dual set of genes enabled a larger number of alleles to persist in a population or species. In any organism the vast majority of phenotypically visible mutations involve partial or complete loss of gene function and therefore tend to be recessive and deleterious. However, mutated alleles will often be masked in a diploid, giving them the chance to mutate further as evolutionarily neutral alleles and perhaps to become favourable as conditions change. In Chapter 3, we will see how higher plants, including many crop species, have taken this process of genetic diversity even further by repeated polyploidization to create tetraploids, hexaploids and even octoploids. Modern breeders routinely create polyploid interspecific hybrids in order to introduce new genetic diversity into crops (see Chapter 8).
- Vascular plants developed complex metabolic networks that enabled them to synthesize a greatly extended range of compounds such as phenolics and flavonoids (see Chapter 4). This allowed them to develop water-impermeable cuticles, tough lignified cells, and more extensive differentiation into a wide variety of tissue types.
- Vascular plants also developed many novel regulatory mechanisms involving processes such as responses to oxidative stress and heat shock (see Chapter 6).

---

decaying remains of the prolific land vegetation. During the period between 400 and 280 million years ago, the vast majority of the early terrestrial animals, most of which were arthropods, were either predators or detritivores. Until the appearance of the large vertebrate herbivores, therefore, most energy flow from plants into the animal components of terrestrial ecosystems came via decomposers rather than direct herbivory.

The spread of larger and better-adapted herbivores, such as dinosaurs, after about 260 million years ago, posed new challenges for terrestrial plants. Vascular plants responded by using their ability to form lignin to increase their height and extent of protective secondary thickening, as well as innovations such as thorns. They also used their formidable metabolic versatility to synthesize an increasingly complex range of antifeedants and toxins in order to make their tissues less palatable to prospective herbivores. The first true gymnosperms, including cycads, Ginkgos, Gnetophytes and conifers, date from this period. Gymnosperms became highly successful and remained the dominant form of land vegetation for about 150 million years until the great angiosperm radiation, which gave rise to the characteristic pattern of global flora that we still see today (see Fig. 2.7).

Radiation of the angiosperms

Angiosperms, the flowering plants, first appear definitively in the fossil record about 140 million years ago, although evidence from recent genome sequencing studies suggests that they might date back as much as 250–300 million years. The genomic data also suggest that a genome duplication event immediately preceded the origin of the angiosperms. This led to the duplication of several

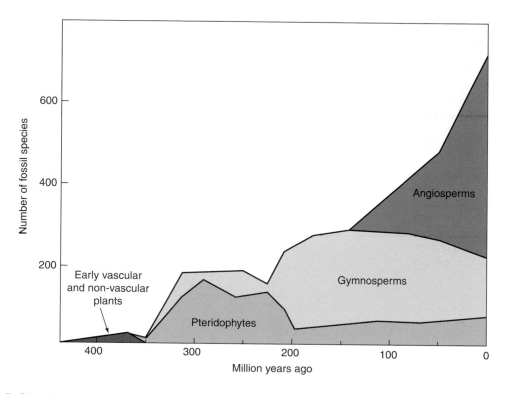

Fig. 2.7. Diversification of the major groups of land plants. The earliest terrestrial flora was dominated by non-vascular plants, such as liverworts and mosses, but simple vascular plants such as lycopods were also increasingly prominent. Although seed plants had already appeared by 360 million years ago, the period from 370 to 230 million years ago was dominated by Pteridophytes, and especially the ferns. Cooler conditions then favoured the spread of a gymnosperm-dominated flora that lasted until the angiosperm radiation, which began about 100 million years ago and accelerated after the KT event some 65 million years ago.

homeotic genes that went on to become the floral organ identity genes responsible for that most characteristic feature of angiosperms, the flower. The divergence between the two major angiosperm lineages, the monocots and dicots, might date from as early as 200 million years ago. By 125 million years ago, major groups such as the magnoliids, eudicots and monocots were well established. Despite the relatively early emergence and diversification of angiosperms, however, the terrestrial flora remained largely dominated by ferns and gymnosperms for many tens of million years after this time.

By 105 million years ago, angiosperms still only accounted for 5–20% of local floras and they appear to have been limited to a relatively narrow range of habitats, such as aquatic, very dry or recently disturbed sites. Then, around 100 million years ago, there was a proliferation of angiosperm taxa, which soon accounted for 80–100% of regional floras. This is the so-called 'great angiosperm radiation', the causes of which were famously described by Charles Darwin in the 19th century as an 'abominable mystery'. The reasons for the rapid spread and global dominance of angiosperms after over 100 million years as a relatively peripheral group of plants have yet to be fully elucidated but there are several possible causes.

One factor is the coevolution of flowering plants and pollinating insects, both of which diversified at about the same time. Insect pollinators provided a more effective reproductive mechanism for flowering plants, and this was further improved by novel methods of seed dispersal that enabled angiosperms to move rapidly into new areas. It is proposed that flowering and insect pollination may have coevolved in an isolated, possibly insular, location that initially involved a single specialized partner, such as a wasp. Subsequently, both angiosperms' floral structures and insect pollinators radiated considerably, leading to the enormous diversity that we see today.

In addition to efficient pollination mechanisms, angiosperms are capable of much higher growth rates than gymnosperms, especially in nutrient-rich soils, which angiosperms are able to create for themselves due to their faster decomposition rates. In contrast, gymnosperms tend to grow relatively slowly and prefer poorer soils than angiosperms. The slowly decomposing, nutrient-poor litter of gymnosperms tends to maintain these low nutrient levels. Because gymnosperms had arrived first, they had created an environment that favoured their slower growth rates and preference for low soil quality. The world of 70–200 million years ago was therefore characterized by a gymnosperm canopy and a fern-dominated understorey with an impoverished nutrient-poor soil. This would have represented an ecologically stable state that was able to persist over such a lengthy period.

Angiosperms in such a world would only thrive in sites where gymnosperms, and their accompanying nutrient-poor soils, were largely absent. This may have precluded the spread of angiosperms beyond a few isolated locations for tens of millions of years. However, once angiosperms in a particular locality reached a critical abundance, there may have been a positive feedback cycle of increasing soil fertility and increasing growth rates. If isolated angiosperm populations were able to link up with neighbouring populations, this might create a 'tipping point' whereby increasing soil fertility tended to exclude gymnosperms and facilitated further extension of habitat colonization by rapidly growing angiosperms.

Faster growth rates in the more fertile soils would have enabled locally dominant angiosperms to capitalize on their existing reproductive advantages such as more efficient mechanisms for pollen and seed dispersal. This led in turn to their rapid expansion into new habitats and subsequent radiation into the tens of thousands of new angiosperm species that appeared around 70–100 million years ago (Fig. 2.8a). Another key factor behind angiosperm radiation may have been their ability to form polyploids with more complex and versatile genomes (see Chapter 3). This may have enabled angiosperms to adapt more readily to the harsh environmental conditions responsible for the KT (Cretaceous-Tertiary) event that was possibly caused by an asteroid impact and/or a surge in volcanic activity. The KT event resulted in the extinction of about 60% of all plant species, and most animals, about 65 million years ago. However, angiosperm diversity was only slightly reduced by this event and they soon resumed their radiation into the 250,000–400,000 species that are present today.

By 65 million years ago, plants belonging to modern families, such as magnolia, beech, oak and maple, are clearly discernible in the fossil record (Fig. 2.8b). The increasing spread of angiosperms into former gymnosperm-dominated habitats is supported by data from coprolites. These fossilized faecal remains were left by herbivorous dinosaurs that had previously grazed mainly on ferns and gymnosperms. The coprolite data suggest that, by 65 million years ago, many herbivorous dinosaurs had already switched to a more mixed diet that included conifers, cycads, dicots and grasses, indicating that angiosperms were becoming much more abundant. As well as colonizing new habitats where gymnosperms were unable to grow, angiosperms steadily displaced existing gymnosperm- and fern-dominated floral assemblages across much of the world. The dominance of the present-day flora by angiosperms is shown by their very high species diversity. Whereas there are only 700–900 gymnosperm species, the angiosperms include well over 175,000 dicot species and 70,000 monocot species, including 10,000 grass species.

Eventually, most gymnosperms became restricted to cooler montane, coastal or boreal locations where the climatic conditions and poorer, often acidic, soils meant that they could still maintain a competitive advantage over angiosperms. This is the world we live in today. It has been formed by the interaction of climatic, geological and other environmental changes with biological processes, principally that of evolution driven by natural selection. As we are confronted by a new series of both human and 'naturally' caused biological and environmental changes, we are now attempting to manipulate many useful plant species by substituting human bioengineering and selection for Darwinian natural selection.

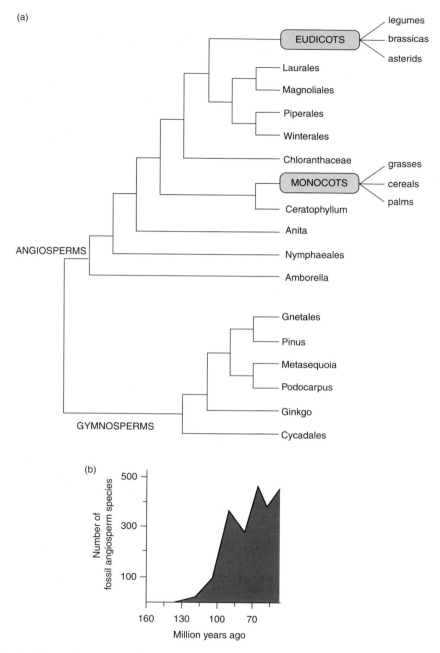

Fig. 2.8. Evolution of the major angiosperm families. Angiosperms are now divided into two major successful groups, the eudicots and the monocots, that include well over 175,000 and 70,000 species, respectively. The other groups of angiosperms, which include relatively primitive groups such as the Magnoliales and Nymphaeales, are much less diverse. Angiosperms are now believed to be a relatively ancient plant lineage, perhaps dating from as much as 300 million years ago, that only achieved global prominence after about 110 million years ago. Today, the eudicots and the monocots dominate most of the terrestrial flora and, while about 250,000 species have already been identified, it is estimated that the true number might exceed 1 million.

2.6 Summary Notes

- Higher plants are the direct descendants of cyanobacteria dating from over 3 billion years ago. Cyanobacteria developed a new form of photosynthesis that used water and CO_2 to produce sugars and the oxygen that created our present atmosphere.

- Some cyanobacteria were captured by larger eukaryotic cells via endosymbiosis and became the plastids that are found in all algae and plants. About 500 million years ago, the Charophytes, a group of multicellular green algae, began to develop into the first land plants.

- Early land plants, such as the bryophytes, were relatively small and moisture loving with a simple haploid adult body form. Subsequent adaptations including adult diploidy, vascular systems, true roots and greater complexity enabled plants such as ferns to increase in size and colonize a wider range of habitats. By 350 million years ago, the first seed-bearing plants, the gymnosperms, had arrived.

- The flowering plants, or angiosperms, may date from as early as 250 million years ago but only achieved their current dominance of terrestrial flora after 100 million years ago. One of the key events in angiosperm radiation may have been their ability to form polyploids, which enabled them to survive the KT mass extinction event 65 million years ago. Today, angiosperms make up well over 90% of land plant species.

Further Reading

Photosynthesis

Collings, A.F. and Critchley, C. (2005) *Artificial Photosynthesis: from Basic Biology to Industrial Application.* Wiley, New York.

Furbank, R.T., von Caemmerers, Sheehy J., Edwards, G. (2009) C$_4$ rice: a challenge for plant phenomics. *Functional Plant Biology* 36, 845–856.

Holland, H.H. (2006) The oxygenation of the atmosphere and oceans. *Philosophical Transactions of the Royal Society B* 361, 903–915.

Olson, J.M. and Blankenship, R.E. (2004) Thinking about the evolution of photosynthesis. *Photosynthesis Research* 80, 373–386.

Xiong, J. and Bauer, C.E. (2002) Complex evolution of photosynthesis. *Annual Review of Plant Biology* 53, 503–521.

Endosymbiosis

Krause, K. (2008) From chloroplasts to 'cryptic' plastids: evolution of plastid genomes in parasitic plants. *Current Genetics* 54, 111–121.

Lynch, M., Koskella, B. and Schaack, S. (2006) Mutation pressure and the evolution of organelle genomic architecture. *Science* 311, 1727–1730.

McFadden, G.I. (2001) Primary and secondary endosymbiosis and the origin of plastids. *Journal of Phycology* 37, 951–959.

Tamames, J., Gil, R., Latorre, A., Peretó, J., Silva, F.J. and Moya, A. (2007) The frontier between cell and organelle: genome analysis of *Candidatus* Carsonella ruddii. *BMC Evolutionary Biology* 7, 181.

Evolution of land plants

Berendse, F. and Scheffer, M. (2009) The angiosperm radiation revisited, an ecological explanation for Darwin's 'abominable mystery'. *Ecology Letters* 12, 865–872.

Graham, L.E., Cook, M.E. and Busse, J.S. (2000) The origin of plants: body plan changes contributing to a major evolutionary radiation. *Proceedings of the National Academy of Sciences USA* 97, 4535–4540.

Hancock, J.F. (2004) *Plant Evolution and the Origin of Crop Species.* CAB International, Wallingford, UK.

Ingrouille, M.J. and Eddie, B. (2006) *Plants: Diversity and Evolution.* Cambridge University Press, UK.

Willis, K.J. and McElwain, J.C. (2002) *The Evolution of Plants.* Oxford University Press, UK.

3 Plant Molecular Genetics and Genomics

3.1 Chapter Overview

This chapter is concerned with the structure, function and organization of plant genomes. Over the past decade or so, the genomes of dozens of organisms have been sequenced and this knowledge is beginning to fundamentally challenge our ideas of what genes are and how their expression is regulated. Examples include the extensive regulation of gene expression by epigenetic modifications and the high incidence of hybridization and polyploidy, which give scope for evolutionary diversification and the enormous metabolic flexibility found in plants. Knowledge of the molecular genetics of plants has underpinned the practical application of modern biotechnology and continues to be vital for efforts to exploit photosynthetic organisms in general.

3.2 Plant Molecular Genetics

In most fundamental respects, the organization and regulation of individual genes in higher plants are similar to other multicellular eukaryotes such as animals and fungi. In particular, most genes are modular structures made up of contiguous stretches of coding and non-coding DNA regions that are packaged within linear chromosomes, which also contain large amounts of non-genic DNA. The processes of plant gene transcription and translation are also very similar to other eukaryotes. One way in which plants differ from animals is the greater contribution of epigenetic mechanisms such as DNA methylation or histone modification to their development and evolution.

According to the 'central dogma' of biology proposed by Crick in the 1950s, one gene encodes one mRNA molecule, which in turn specifies one type of protein. However, we now know that the concept of a gene is much less clear than was initially believed and recent advances in genomics are highlighting how little we really know about the nature and behaviour of genes and genomes. For example, despite their seeming vast differences in complexity, most multicellular organisms, including higher plants, animals (including humans) and fungi, typically contain 20,000–30,000 protein-encoding genes per haploid genome. As we will see, many of these genes can encode more than one protein by means of mechanisms such as alternate splicing or translational braking.

One of the surprising aspects of higher eukaryotes such as plants is that protein-encoding genes typically make up as little as 5–10% of their genomic DNA. Some of the remaining DNA is made up of relatively short segments that are transcribed into RNA but not translated into proteins. There may be as many as 100,000 of these small RNA-encoding DNA segments in a typical eukaryotic genome. The function of these small RNAs has yet to be fully determined but they are known to play roles in processes such as regulating the transcription of other genes and protecting cells from attack by viruses. Finally, a highly variable, but normally large, proportion of most eukaryotic genomes is made up of long stretches of repetitive DNA sequences that are not transcribed but may sometimes perform useful functions in the host organism.

DNA packaging into chromosomes

The well-known double helical arrangement of DNA molecules is shown in Fig. 3.1. In plants and other higher eukaryotes the total length of DNA in each cell nucleus can be several metres. Since plant cells are typically only 10–100 µm in length, their nuclear DNA must be highly folded and condensed. This is achieved by packaging DNA around proteins called *histones* to form a progressive series of higher order structures known as *chromatin*. The basic unit of chromatin is the *nucleosome*, which is an octameric complex made up of eight histone subunits around which the DNA is wrapped. Histones are a family of highly conserved globular

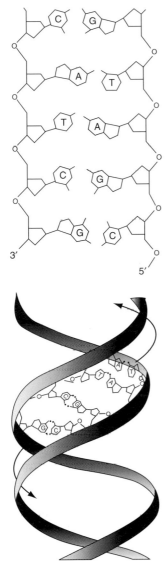

Fig. 3.1. DNA structure. The DNA molecule is made of two antiparallel chains of nucleotides that are arranged on a negatively charged phosphate backbone to form a right-handed double helix. In the centre of the helix, the purine nucleotides (adenine or guanine) form hydrogen bonds with the pyrimidine nucleotides (cytosine or thymine).

wrapped to form a complete nucleosome. As shown in Fig. 3.2, groups of six nucleosomes are coiled to form a chromatin fibre, which is itself coiled further, around scaffold regions. The result is a highly condensed structure where individual regions of DNA can relatively easily become exposed when required for transcription, replication, modification or repair.

Chromatin itself is organized into either loosely packed *euchromatin* or more tightly packed *heterochromatin*. Euchromatin tends to be available for transcription whereas heterochromatin is normally transcriptionally inactive. One factor that determines chromatin packaging is the modification of histone proteins by attachment of small chemical groups to particular amino acid residues. Particular patterns of histone modification are thought to recruit specific chromatin remodelling proteins that direct either heterochromatin or euchromatin formation. A common form of histone modification is N-terminal tail acetylation, where ε-amino groups of phylogenetically conserved lysine residues are acetylated, thereby reducing the positive charge on the histone surface. In mammalian and plant genomes, chromatin packaging is also determined by the attachment of methyl groups to cytosine residues in the DNA by cytosine methyltransferases. Cytosine methylation can be maintained after each round of DNA replication because the template strand of DNA will retain the modification. Patterns of cytosine methylation are often correlated with epigenetic variation (see below), but chromatin changes can also occur independently of methylation.

Regulation of gene expression by DNA methylation arose early in eukaryotic evolution and was already present in the common ancestor of plants, animals and fungi. Land plants and vertebrates retain extensive DNA methylation, especially of transposable DNA elements. As we will see in Chapter 6, the aggressiveness of transposable elements is correlated with sexual outcrossing. This means that land plants and vertebrates are under strong selective pressure to maintain transposable element suppression and hence DNA methylation. In contrast, fungi and some unicellular organisms that primarily reproduce asexually can tolerate much-reduced DNA methylation patterns.

When a region of genomic DNA contains methylated cytosines, it typically becomes assembled into heterochromatin. Methylated DNA recruits methyl-DNA binding proteins, which in turn recruit the histone-modifying enzymes and chromatin-remodelling factors required for heterochromatin formation.

proteins whose N-terminal regions reside on the surface of the nucleosome octomer where they are available for chemical modifications. Each histone octomer comprises two copies each of four different subunits: H2A, H2B, H3 and H4. Around each of these protein complexes about 146 bp of DNA is

---

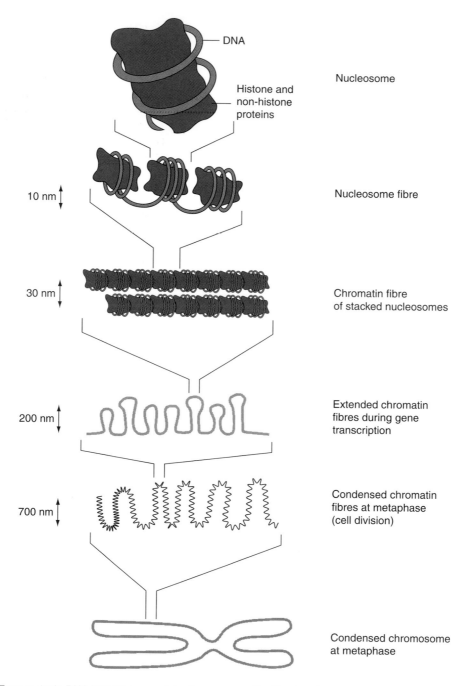

DNA

Nucleosome

Histone and
non-histone
proteins

10 nm — Nucleosome fibre

30 nm — Chromatin fibre
of stacked nucleosomes

200 nm — Extended chromatin
fibres during gene
transcription

700 nm — Condensed chromatin
fibres at metaphase
(cell division)

Condensed chromosome
at metaphase

Fig. 3.2. From a single DNA helix to a complete chromosome. Double-stranded helical DNA is wrapped around an octomeric core complex consisting of two molecules each of the histone subunits H2A, H2B, H3 and H4 to form a nucleosome, which is the basic unit of chromatin assembly. Nucleosomes are separated from one another by short lengths of spacer DNA attached to histone H1. Nucleosomes are coiled in groups of six per turn to produce a 30 nm diameter chromatin fibre, which can be further coiled into loops around scaffold-associated regions enriched in AT bases. During mitosis, chromatin loops are themselves coiled extensively to produce the familiar tightly packed metaphase chromosome.

As a result of such histone or DNA modifications, and the consequent formation of heterochromatin, genes in these regions become silenced. The silencing of significant numbers of genes often accompanies differentiation of tissues or organs so that only a subset of genes related to the particular tissue or organ can then be expressed. One of the challenges for biotechnologists is to reverse this gene silencing in order to propagate new plants from differentiated tissue fragments such as leaf, stem or root explants. Such manipulations are often an integral part of plant tissue culture. For example, they may be necessary as part of the clonal propagation of some types of crop, or the regeneration of explants to create transgenic plants (see Chapter 8–10).

Gene structure and transcription

Eukaryotic genes are modular structures that consist of several interacting elements, each of which is necessary for the correct expression of the gene in question. The generalized structure of a plant gene, which is similar to other eukaryotes, is shown in Fig. 3.3. A typical protein-encoding gene consists of a transcription unit, plus flanking regions involved in regulation of gene function. The most important regulator region, the promoter, is located

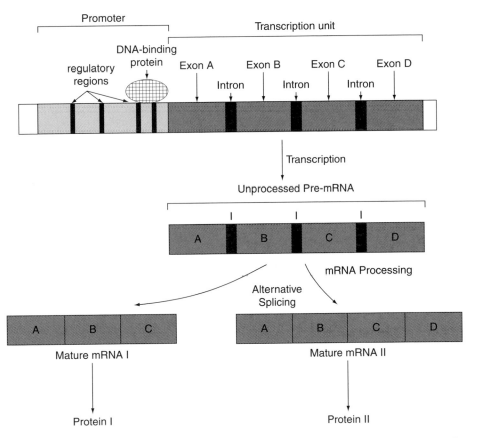

Fig. 3.3. Generalized structure of a plant gene and its transcribed products. A typical plant protein-encoding gene consists of a transcription unit, all of which is transcribed into mRNA, and a 5' regulatory region termed the promoter. The promoter includes specific regulatory elements that determine when and where the gene is transcribed. Often these elements include DNA regions that bind proteins that may either repress or activate DNA transcription. The transcription unit is further divided into 5' and 3' untranslated regions and a central coding region. In most genes the central coding region contains exon and intron regions. Following transcription into mRNA, regions corresponding to introns are spliced out so that only exon regions are ultimately translated into proteins.

upstream (5') of the direction of gene transcription, which is normally depicted diagrammatically as going from left to right (from 5' to 3'). Eukaryotic promoters can be highly variable in length but are typically about 1 kb or more, with several conserved motifs that are involved in the binding of proteins or other regulatory molecules such as hormones. One of the most important binding motifs is the one that enables RNA polymerase II to attach to the promoter and thereby initiate gene transcription.

Transcription is the process of making an RNA 'copy' of the entire core region of the gene, known as the transcription unit. Each complete transcription unit is copied into the corresponding mRNA molecule by RNA polymerase II. As shown in Fig. 3.4, transcription is a complex and highly regulated process that is initiated by the binding of a series of proteins to the TATA box motif immediately upstream from the transcription start site. These DNA binding proteins recruit RNA polymerase II, which unwinds the two strands of the DNA double helix and generates an RNA copy of the template strand of DNA. This produces the pre-mRNA, which is made up of a 5' untranslated region, the central coding region and a 3' untranslated region. The pre-mRNA is processed to remove the 5' and 3' untranslated regions and the new mRNA molecule is stabilized by addition of a cap of 7-methylguanosine at the 5' end and a polyadenylated tail at the 3' end. The stability of mRNAs can also be affected by environmental, metabolite or hormonal signals. For example, the stability of the *Fed1* transcript encoding ferredoxin is greatly increased in the light, while the α-amylase transcript is destabilized in the presence of sucrose. The capping of mRNA also increases the efficiency of its translation following the removal of its introns.

Only part of the transcription unit actually encodes the protein product of a gene. The transcription units of most, but not all, eukaryotic genes are made up of exons and introns. While exons are kept for translation to create the corresponding protein, introns are spliced out of the pre-mRNA and play no part in encoding the protein. Each intron has recognizable 5' and 3' sequences that indicate the exact point at which it is cleaved, allowing the adjacent exons to be joined together to form the mature mRNA. Introns can have useful evolutionary functions, such as separating exons into modular structures that can recombine to form novel proteins. Another example is an intron in the vernalization gene, *FLC* gene: part of this intron encodes a large non-coding RNA called *COLDAIR* that is required for repression of the *FLC* gene and consequent activation of flowering in the spring. In other cases, however, introns can be a liability. For example, heat stress tends to inhibit intron splicing, which means that heat-shock proteins (see Chapter 6) that are specifically expressed at high temperatures often have no introns so that they can be processed properly under such conditions. In other cases, intron-containing heat-shock and other stress-related genes may be processed to form larger protein isoforms that include the translated introns. This is a form of *alternative splicing*, a mechanism that can generate several mature mRNA transcripts and protein products from a single gene (see Fig. 3.5).

Alternative splicing is very common in animals and as many as 80% of human genes may be alternatively spliced. Although it was originally believed to be much less common in plants, alternative splicing is now known to be widespread, with evidence from *Arabidopsis* suggesting that it might occur in >40% of intron-containing genes. In alternative splicing, different combinations of splicing sites can be recognized during pre-mRNA processing. The mature, spliced forms of the mRNA finally leave the nucleus for the cytosol where they bind to ribosomes and serve as templates for translation into proteins. A second way in which two or more proteins can be produced from the same mRNA transcript is via the recently discovered mechanism of translational braking. In this case, if the process of mRNA translation is delayed, the nascent polypeptide can sometimes adopt a different three-dimensional (tertiary) structure with different properties from a polypeptide of identical primary amino acid sequence that was not subjected to translational braking. By regulating the amount of translational braking, a cell can produce several proteins with identical primary structures but different tertiary structures, and therefore with different activities and functions, from a single mature mRNA transcript.

Regulation of gene expression

The expression of eukaryotic genes is primarily regulated via their promoters. The promoter specifies *where*, *when* and *how strongly* the transcription unit to which it is attached is expressed in a plant. For example, genes encoding proteins

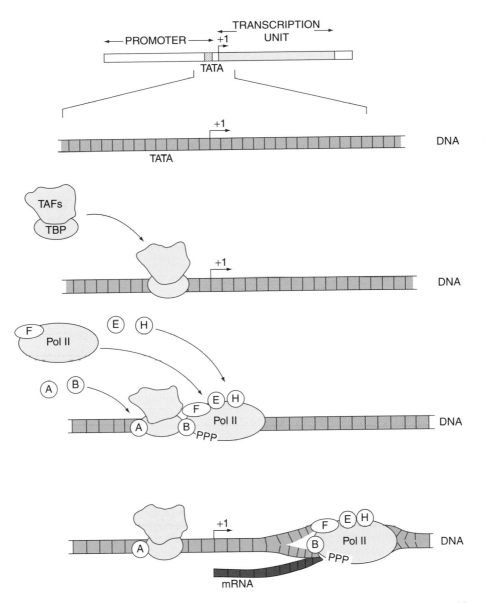

Fig. 3.4. Gene transcription. Most plant promoters possess a TATA box and/or an initiator element, either or both of which may be recognized by the TATA-binding protein (TBP) to form a transcription initiation complex with TBP associated factors (TAFs). Once the TBP/TAFs have bound, other components of the initiation complex assemble as follows: TFIIA (A), TFIIB (B), RNA polymerase II (Pol II) and TFIIF (F), TFIIE (E) and TFIIH (H). The C-terminal region of RNA polymerase II must be phosphorylated (PPP) for transcription to commence. The stability of the transcription initiation complex is enhanced by binding of additional activator proteins to specific DNA motifs called enhancers that are located further upstream in the promoter region. Examples of such enhancers include hormone-regulated motifs such as ABA responsive elements. Other DNA elements in the promoter might bind repressor proteins that suppress transcription. In some cases transcriptional activators and repressors can form direct contacts with the initiation complex, or indirect contacts through a bridging complex called a mediator. In other cases, the structure of the promoter or chromatin may be physically modified to facilitate other interactions. Some regulatory proteins can act as either activators or repressors depending on promoter context and the presence or absence of other proteins.

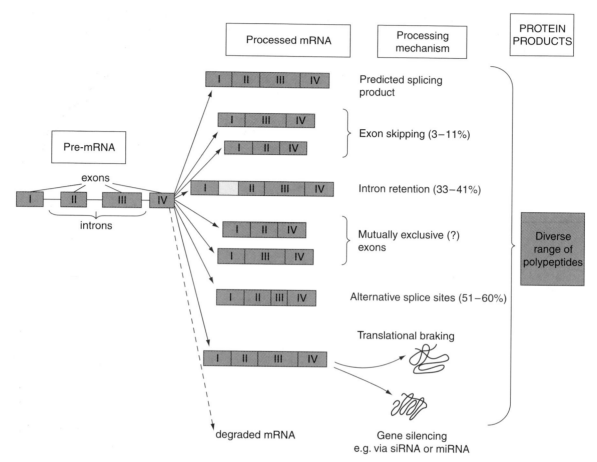

Fig. 3.5. Post-transcriptional processing of mRNA. Many genes can give rise to more than one mature mRNA product following differential processing of a single original pre-mRNA transcript. In this example several alternative mechanisms are shown whereby one pre-mRNA containing four exons can be processed to generate as many as six different proteins. Differential splicing can result in omission of part or all of an exon or the inclusion of part or all of an intron. In plants, the most common forms of differential splicing are the use of alternative sites (51–60% of total) and intron retention (33–41% of total). A less common mechanism is translation braking, which can generate proteins that have different tertiary structures despite having identical primary structures. Finally, pre-mRNA can be degraded by any one of several gene silencing mechanisms, such as those mediated by siRNA or miRNA.

involved in storage product accumulation in plants might be only expressed in the seed or fruit (*where*); they might only be active during the mid-maturation phase of seed development (*when*); and they might be expressed at very high levels compared with most other genes (*how strongly*). Some tissue-specific genes might be expressed at extremely high levels, especially when the gene product is a structural or storage protein required in large amounts, rather than an enzyme that is only needed in relatively small quantities. One of the most highly expressed

groups of genes in plants encodes the seed storage proteins. In some seeds, storage proteins can make up 25% of seed dry weight and they are synthesized during a relatively brief period in the mid-phase of seed maturation. At this stage, storage protein mRNAs can make up 50% of the entire mRNA population of a developing seed, and their gene expression rates might be several orders of magnitude higher than other seed-specific genes.

Hormones such as gibberellins, ABA, auxins and jasmonates often interact with specific *cis*-acting

DNA sequences, or enhancers, in the promoter regions of their target genes. Often these sequences are highly conserved between plant species. Knowledge of the roles of the various *cis*-acting DNA sequences has enabled researchers to redesign gene promoters to achieve precisely regulated gene expression in transgenic plants. As discussed in Box 3.1, the manipulation of gene promoters is one of the core technologies used in plant genetic engineering.

While the majority of genes are expressed at specific times and places during plant development, an important subset of genes is expressed in all tissues throughout development. These are the so-called 'housekeeping' genes that encode key proteins required for the everyday function of all living cells. Such genes are regulated by constitutive rather than tissue-specific promoters, which means

that they are always active in every tissue of the plant. Examples include genes involved in such key processes as respiration, protein synthesis and sub-cellular trafficking. Although constitutive genes are expressed virtually everywhere in a plant throughout its life cycle, their levels of expression are normally relatively low. This is because they tend to encode low-abundance enzymes, or structural proteins that are merely turning over rather than being accumulated at high rates. For this reason, when high level constitutive expression of a transgene is required, non-plant promoters such as those of viruses are commonly used. The most commonly used constitutive promoter and the one that is present in most first-generation transgenic crops is the 35S promoter from the cauliflower mosaic virus, or CaMV (see Chapter 9).

Box 3.1. Biotechnological uses of gene promoters

One of the key discoveries of plant biotechnology was that, in many cases, the promoter of one gene could be attached to a different gene in order to alter the pattern of its expression. This is important because many genes of biotechnological interest originate from other organisms where they might have totally different expression patterns from what is desired in the new host plant. Therefore, if such promoters are transferred to a transgenic plant along with the rest of the gene, they will either not be recognized at all or will direct unsuitable patterns of gene expression. However, if an appropriate promoter from the host plant itself or from a close relative is attached to the new gene, the latter is much more likely to be expressed in the desired manner.

For example, in order to increase ß-carotene formation in rice grains (the so-called 'golden rice project', see Chapter 12), genes encoding the required enzymes were obtained from other plants and bacteria for insertion into rice. These genes were then attached to the existing endosperm-specific glutelin gene promoter of rice so that the newly formed ß-carotene would accumulate at high levels in the grains of the resultant transgenic rice plants.

The most effective tissue-specific promoters for driving exogenous transgenes are normally obtained from the host plant species itself. However, heterologous promoters derived from related plant species can often be effective as well. Heterologous promoters generally function best in moderately closely related species. Hence, *Arabidopsis* promoters generally work

well in rapeseed, which is a fellow member of the Brassicaceae, but legume promoters are much less useful. Looking further afield, promoters from monocots are normally completely ineffective in dicots and vice versa.

Due to their high strength and tissue specificity, some of the most commonly used promoters for driving transgenes intended to be seed-specific are those encoding seed storage proteins. Examples include the rapeseed napin, maize zein and rice glutelin promoters. Useful leaf-specific promoters include those from the Rubisco small subunit and light-harvesting chlorophyll protein genes. There is a wide range of useful gene promoters including those responsive to external induction by chemical reagents or to internal induction by processes such as senescence, wounding, heat-shock and pathogen attack.

An important exception to the rule of using a promoter from the host plant or a close relative is when constitutive rather than tissue-specific gene expression is desired. As noted in the main text, the expression rates of constitutive genes in plants tend to be relatively low. This means that endogenous constitutive promoters are normally not very useful if a high level of gene expression is required, which is usually the case with transgenes. Therefore, exogenous promoters are generally used to drive constitutive expression of transgenes. Examples include the 35S promoter from the cauliflower mosaic virus, as used in dicots, or the *ubiquitin-1* or *actin-1* promoters respectively from maize and rice, as used in monocots.

Translation and post-translational protein processing

Proteins are assembled from amino acids on 80S ribosomes using mRNA as the template and adaptor tRNA molecules similarly to other eukaryotes. Plant cells also synthesize a limited range of proteins in mitochondria and plastids using prokaryotic-like 70S ribosomes. The 5' cap and 3' poly(A) tail of mature mRNA enhances its translation, which is also made more efficient by the presence of specific secondary structures adjacent to the translation initiation codon, AUG. Translation can also be enhanced by the attachment of specific mRNA-binding proteins. Following their *de novo* synthesis on ribosomes, many proteins require various forms of post-translational processing before they become functional and/or reach their ultimate subcellular or extracellular destination(s) (see Table 3.1). Many function-related modifications involve the covalent attachment of small groups such as sugars (glycosylation), fatty acids (acylation) and phosphate residues (phosphorylation). In other cases, formation of a disulfide bond can alter the structure of the protein. Many of these modifications are reversible and can be used to regulate the activity and location of a protein.

Location-related modifications normally involve the cleavage of specific regions called signal peptides that direct the protein to a particular destination. For example, there are several forms of N-terminal transit peptide that direct nascent proteins to specific organelles such as mitochondria or plastids. As the protein is imported into the appropriate compartment, its signal peptide(s) is/are removed to generate a mature protein. Some signal sequences are non-cleavable and are retained in the mature protein. The ability to manipulate signal sequences has been of immense benefit in enabling exogenous proteins to be directed to the organelle of choice in transgenic plants. For example, initial expression of two enzymes involved in biopolymer production in the cytosol resulted in stunting of the transgenic plants. However, by adding a sequence encoding a plastid signal peptide to each transgene, the two enzymes were redirected to plastids where their function was greatly enhanced with no side effects on plant development.

Epigenetics

Epigenetics literally means 'outside genetics' and generally refers to inherited changes that may be acquired during the lifetime of an organism without any change in its DNA sequence. Over the past few decades it has become increasingly clear that not all heritable variation in populations is DNA sequence-based. Instead, novel patterns of spatial and temporal gene expression can result from a range of epigenetic mechanisms. Epigenetic changes are heritable molecular events that alter the phenotype of an organism without involving any change in DNA sequence. A concise definition is: 'the alteration of phenotype, morphological or molecular, without change in either the coding sequence of a gene or the upstream promoter region'. Epigenetics therefore involves the transmission of certain types of information from a cell or multicellular organism to its descendants without that information being encoded in any part of the DNA sequence.

The two principal mechanisms leading to epigenetic changes are DNA methylation and histone modification, both of which can lead to alterations in chromatin architecture. Methyl groups are covalently attached to cytosine residues of DNA by cytosine methyltransferases. Highly methylated regions of genomic DNA are typically assembled into heterochromatin and are therefore less likely to be transcribed. Cytosine methylation is maintained after each round of DNA replication and cell division because the template strand of DNA retains its

Table 3.1. Post-translational modifications of proteins.

Modification	Group added	Common target amino acid residues
Acylation	Acetyl	Lys
	Formyl	N-terminus
	Lipoyl	
	Myristoyl	Gly, other amino acids
	Palmitoyl	Cys
Alkylation	Methyl	Lys, Arg
	Prenyl	Cys
Glycosylation	Glucose	
	Mannose	Asp, Ser
	Fucose	Thr
	Galactose	
Phosphorylation	Phosphate	Ser, Thr
	Pyrophosphate	Tyr, His
Oxidation	None	Cys
Ubiquitination	Ubiquitin	Lys
Proteolytic cleavage	None	Ser, Thr, Cys, Asp, Glu

covalent modification. There are several mechanisms of histone modification, of which the best known is the acetylation or methylation of conserved lysine residues leading to chromatin condensation and gene silencing.

In *Arabidopsis*, it has been found that overexpression of a deacylase can result in the removal of these histone markers, resulting in the ectopic expression of some metabolic pathways and the repression of others in a tissue-specific manner. This shows that histone acylation states can function as regulatory elements that influence gene expression across different tissues and developmental changes. Other forms of histone modification that can lead to epigenetic effects include phosphorylation and ubiquitination (see Table 3.1). In some circumstances, small RNAs such as micro RNA and small interfering RNA can also be involved in epigenetically related histone and DNA modifications. These small RNA molecules can be usefully manipulated as part of plant genetic engineering, as described in Chapter 10 and Box 10.3.

Unexpected epigenetic modifications that sometimes have profound consequences for growth and development are particularly prevalent following the formation of interspecific hybrids and polyploids in plants. Epigenetic modifications can also result from environmental selection or ecological change. The common theme is that the induction of stress within a plant genome can trigger epigenetic change and hence surmount potential evolutionary bottlenecks. In plants, DNA methylation changes can persist throughout development and be inherited between generations. This contrasts with the situation in mammals where DNA methylation sites are reprogrammed in early embryogenesis and altered methylation patterns are more rarely transmitted to progeny. It is now recognized that epigenetic alterations are one of the key driving forces in creating genetic variation, and hence evolutionary change, in plants. Epigenetic alterations occur frequently during the formation of interspecific hybrids and polyploids, both of which have been very common in higher plant evolution.

3.3 Plant Genome Organization

Plants are much more flexible than animals regarding their genome organizations and growth and reproductive mechanisms. Unlike higher animals, many plants will readily hybridize and produce fertile offspring across the species barrier, and often even with plants from separate genera. This characteristic has been of immense use to plant breeders seeking to generate new genetic diversity in crops. The plasticity of the plant genome is also shown by the frequency of polyploidy, namely the existence of more that the normal two (diploid) sets of chromosomes found in most complex multicellular organisms. The basic molecular architecture of higher plant genomes is similar to that of the higher animals, but recent research has revealed that in several key respects plant genomes behave quite differently.

Plant genomes have three major types of DNA

In plants, the nuclear genome is by far the largest of the three cellular genomes and largely controls the function of the plastidial and mitochondrial genomes. As in other eukaryotes, the nuclear genomes of plants contain three principal categories of DNA.

1. The genic DNA makes up classical 'genes', each of which encodes one or more proteins. This DNA also contains non-coding regulatory elements such as promoters and introns. Eukaryotes typically have 20,000–30,000 protein-encoding genes per haploid genome.
2. Relatively short DNA (sDNA) segments that are transcribed into RNA but not translated into proteins. There may be as many as 100,000 sRNA-encoding DNA segments per genome, but their function is still largely unknown.
3. A variable amount of repetitive DNA that is often derived from ancient viral DNA. Some repetitive DNA sequences may be useful to the cell, while others appear to have a neutral effect or may even be harmful.

The total DNA content of a genome and the proportion of each DNA category can vary considerably according to species, and even genotype. As a broad guide, genic (protein-encoding) DNA is typically about 1–10% of the total, while repetitive DNA normally makes up the bulk of most eukaryotic genomes – typically between 40 and 95% of total DNA, and sRNA-encoding DNA is generally less than 1% (see Table 3.2).

A typical flowering plant has about 28,000 protein-encoding genes per haploid genome. However, the total number of genes can often be much higher as a result of the successive duplications and hybridizations, combined with varying extents of subsequent gene loss, that have occurred at various

Table 3.2. Variation in genome size in plants and other organisms.

Systematic name	Common name	Estimated genome size, Mb	Estimated % repetitive DNA	Genome sequenced
	Animal/fungal mitochondria[‡]	0.015–0.080	0	√
	Plant mitochondria	0.18–3.0	0	√
	Plant plastids	0.12–0.16	0	√
φX174[‡]	Virus of *E. coli*	0.005	0	
E. coli[‡]	E. coli	4.6		√
Saccharomyces cerevisiae[‡]	Yeast	12		√
Genlisea margaretae	Genlisea	63		
Arabidopsis thaliana	Thale cress	125	>25	√
Brachypodium distachyon	Brachypodium	270	11	√
Aesculus hippocastanum	Horse chestnut	200		
Carica papaya	Papaya	235		√
Cucumis sativus	Cucumber	250	24	√
Populus trichocarpa	Black cottonwood	300		√
Ricinus communis	Castor bean	320	50	√
Oryza sativa	Rice	390	50	√
Theobroma cacao	Cocoa	430		√
Vitis vinifera	Grapevine	480	40	√
Setaria italica	Foxtail millet	490		
Physcomitrella patens	Physcomitrella	500		√
Musa accuminata	Banana	610	>50	
Sorghum bicolor	Sorghum	660	60	√
Beta vulgaris	Sugarbeet	760	63	
Solanum tuberosum	Potato	844	>70	√
Amborella trichopoda	Amborella	870		
Glycine max	Soybean	1,000	57	√
Elaeis oleifera	Oil palm (American)	1,710		√
Pennisetum purpureum	Napier grass	2,250		
Pennisetum glaucum	Pearl millet	2,350		
Zea mays	Maize	2,500	85	√
Homo sapiens[‡]	Human	3,200		√
Nicotiana tabacum	Tobacco	4,500		
Pisum sativum	Pea	4,800		
Triticum monococcum	Einkorn wheat	>5,000	>80	
Hordeum vulgare	Barley	5,500	80	
Secale cereale	Rye	8,800	92	
Vicia faba	Broad bean	12,000	85	
Allium cepa	Onion	15,100	95	
Triticum aestivum	Breadwheat	17,000		
Pinus elliotii	Slash pine	24,000		
Fritillaria assyriaca	Fritillary	130,000		
Paris japonica		150,000		
Psilotum nudum	Whisk fern	250,000	90	

[‡]non-plant genome

times during their evolution. This is particularly true in relatively recently formed polyploids. For example, the genome of hexaploid breadwheat was formed about 10,000 years ago and contains about 56,000 genes arranged in three recognizably distinct sets of chromosomes originating from the three parental species (see Fig. 3.10). In contrast, maize, which became a tetraploid many millions of years ago, 'only' has about 32,500 genes while the more recent autotetraploid potato has about 39,000 genes.

Over time, there is a tendency for polyploid species to lose many of their redundant genes. However, some redundant genes may acquire important new functions that ensure their retention in the genome. For example, in the polyploid brassica species, such as cabbage and mustard, some of the duplicated genes involved in glucosinolate biosynthesis have now diverged to produce a wider range of these important antifungal and anti-insect compounds. Such biochemical versatility can also be useful to humans, since glucosinolates have beneficial anti-cancer properties as well as being responsible for the different tastes of the various brassica vegetables.

As with other complex eukaryotes, the genomes of higher plants are made up of gene-rich clusters of DNA interspersed with relatively large stretches of highly repetitive DNA. The two major classes of repetitive DNA in plants are DNA transposons and retrotransposons. DNA transposons typically make up between 1 and 14% of genomic DNA and this content is not correlated with genome size. They tend to be associated with gene-rich regions and therefore have a greater potential to affect gene function if they alter their location. In contrast, retrotransposons are the main components of gene-poor areas in the genome and they are both abundant and highly variable in their content. For example, the proportion of retrotransposons in genomic DNA is comparatively low in the small genomes of castor bean (18%) and rice (25%), but is much higher in the larger genomes of sorghum (55%) and maize (76%). One reason for their low levels in rice is that a great deal of repetitive DNA has been lost from the genome of this species over the past few million years. However, for reasons that are not yet understood, in other species such as maize, the retrotransposon content is still increasing. This means that while only about 40% of the rice genome is made up of non-genic repetitive elements, this proportion rises to 85% for maize.

Chromosome architecture

During the period when cells are not dividing, their chromosomes are arranged into highly condensed structures that are visible in the light microscope. The DNA in condensed chromosomes is highly folded in a systematic manner so genes that may be at opposite ends of a linear chromosome may actually be in close physical proximity in the folded state. This means that genes that are apparently not closely linked to each other along a linear chromosome may become clustered together and potentially subject to coordinated regulation. This may be one way in which eukaryotes achieve coordinated regulation of functionally related genes in the absence of the operons (physically linked groups of genes that are transcribed together) found in prokaryotes.

In most plant genomes analysed so far there is a strong separation between gene-rich regions and repeats. Gene-rich regions tend to be associated with euchromatin while most retrotransposon elements are located in and around centromeres. However, recent analysis of the relatively large maize, barley and wheat genomes has revealed some novel features for these important crop species. The genome of maize is atypical in having some retrotransposon elements of the *copia* family occurring in euchromatin while elements of the *gypsy* family are found in the more usual centromeric location. Another unusual feature of these large cereal genomes is that their euchromatin is mostly made up of very small gene islands, 50–60% of which contain just one gene. These compact gene islands have densities of one gene per 5–15 kb and are separated from each other by large stretches of tens or hundreds of kb of repetitive DNA. The physical organization of several chromosomes from the soybean genome is depicted in Fig. 3.6.

Plastid and mitochondrial genomes

The organellar genomes of plants are very different from those of animals. Animal cells only contain mitochondria whereas plant cells contain both mitochondria and plastids. In both plants and animals, the organellar genomes are contained on small chromosomes with many similarities to those of their prokaryotic ancestors. Hence, each plastid or mitochondrion in a plant cell contains many identical copies of its genome with few introns or repetitive sequences. Until recently, organellar genomes were believed to consist of many identical copies of a single circular chromosome. However, new separation techniques have shown that organellar chromosomes can be arranged in other ways, including as branched linear molecules each containing several genome copies in various concatemeric head-to-tail orientations (Fig. 3.7).

Plant mitochondrial genomes are highly variable in size, ranging from 180 to 600 kb, which is much larger than those of animal and fungal mitochondria, most of which are only 14–20 kb. It appears that plant mitochondrial genomes are so much

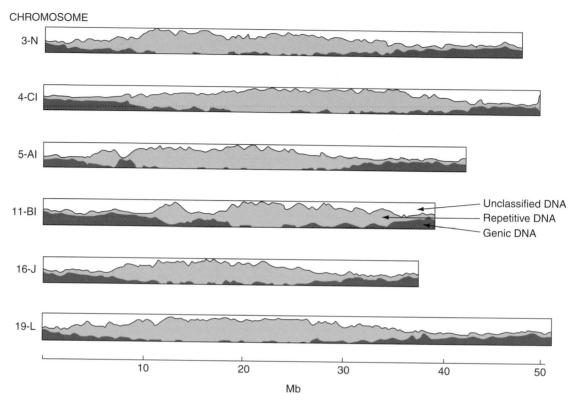

CHROMOSOME

3-N

4-Cl

5-Al

11-Bl
— Unclassified DNA
— Repetitive DNA
— Genic DNA

16-J

19-L

10   20   30   40   50
Mb

Fig. 3.6. Physical organization of part of the soybean genome. Genomic landscape of six of the 20 soybean chromosomes. The three major DNA components of the genome are shown as follows: dark shade (genic regions), light shade (repetitive regions including DNA transposons, *copia*-like retrotransposons, *gypsy*-like retrotransposons) and unshaded regions (unclassified sequences). The respective proportions of the total soybean genome sequence made up by these regions are 18% genic, 60% repetitive and 1% unclassified DNA. Data from Schmutz *et al.* (2010).

larger because of differing evolutionary pressures at the time of the rapid size reduction that followed soon after the original endosymbiotic events. Plant mitochondrial genes encode several electron transport components and proteins involved in transcription and translation. Like those of plastids, the mitochondrial genomes are much-reduced versions of the genome from the original prokaryotic endosymbiont with >90% of this DNA now transferred to the nuclear genome.

The plastid genomes of land plants typically range in size from 120 to 160 kb and contain between 100 and 220 protein-encoding genes. The vast majority of these plastid genes are involved in either photosynthesis or protein synthesis. In contrast, the cyanobacteria from which plastids are derived typically contain between about 1700 and 7300 genes that encode all the proteins necessary

for independent life. Since the capture of the original cyanobacterial endosymbiont, therefore, many genes have been lost from the genome. While some genes were lost altogether others were transferred to the nuclear genome. Examples of genes now in the nucleus include those encoding ferredoxin, plastocyanin and the light-harvesting proteins. Photosynthetic genes remaining in the plastid genome include those encoding photosystems I and II proteins, and cytochromes b6 and f. In the case of Rubisco and the thylakoid ATPase, genes encoding some subunits of each protein complex remain in the plastid genome while others have been transferred to the nucleus. The fact that the composition of the plastid genome is similar in all higher plants suggests that much of the gene loss or transfer to the nucleus occurred comparatively early in their evolution.

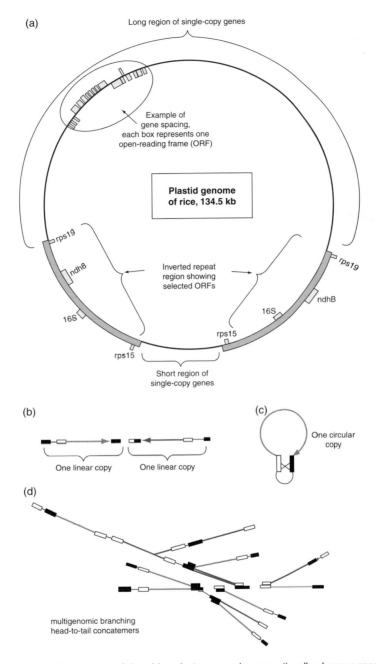

(a) Long region of single-copy genes

Example of
gene spacing,
each box represents one
open-reading frame (ORF)

Plastid genome
of rice, 134.5 kb

rps19

ndh8

rps19

Inverted repeat
region showing
selected ORFs

ndhB

16S

16S

rps15

rps15

Short region of
single-copy genes

(b)

One linear copy   One linear copy

(c)

One circular
copy

(d)

multigenomic branching
head-to-tail concatemers

Fig. 3.7. Plastid genomes. (a) Genetic map of rice chloroplast genome is conventionally shown arranged as a circle (but see also b–d). Genes (grey boxes) shown outside the circle are encoded on the A strand and transcribed counter-clockwise. Genes shown inside are encoded on the B strand and transcribed clockwise. About two-thirds of the genome is made up of single-copy genes arranged in one long region and one short region. The remaining third of the genome is made up of an inverted repeat consisting of two duplicated regions, each about 22 kb in length. (b) It is proposed that plastid genomes can assume different configurations, including a linear arrangement where the end of one monomeric genome recombines with another molecule and initiates replication. (c) In some cases circular forms of the genome can be produced by intramolecular recombination. (d) In other cases a more complex multigenomic structure is formed by recombination-dependent replication. Adapted from Bendich (2004).

Plant nuclear genomes

Plant nuclear genomes are complex, dynamic entities that constantly change their size, organization and DNA composition. Even within a single species, genome size can vary considerably in a way that is not generally seen in animals. Also, the genome size of closely related plant species can be very different, even though they may have similar numbers of chromosomes. For example, rice and maize are both functionally diploid members of the grass family, or Gramineae, with 12 and 10 chromosomes respectively. They also have similar numbers of genes in their genomes, currently estimated at around 28,000 for rice and 33,000 for maize, and both are derived from ancient tetraploid ancestors. However, whereas the rice genome contains only 400 Mb of DNA, the maize genome contains almost sixfold more DNA, totalling 2300 Mb. The two other major cereal crops, barley and wheat, are even more closely related than rice and maize but have even more divergent genomes. Barley is a diploid with only 14 chromosomes and a relatively large genome of about 5000 Mb. In contrast, hexaploid breadwheat has 42 chromosomes and a genome of 16,000 Mb, which is one of the largest crop genomes and more than fivefold larger than the human genome.

The genome size of a single species can vary greatly over time. For example, in the past 8 million years, the rice genome size has more than doubled and then contracted again to lose two-thirds of this additional DNA. The reason for these huge variations in genome size, both within and between species, is that the genomes of many, but not all, plants harbour large quantities of repetitive DNA. It is estimated that as much as 90–99% of the genomes of maize, the wheats and barley consist of such repetitive DNA. Even more remarkably, the genome of a single plant species, such as maize, can vary greatly in size between different individuals. Depending on their climatic adaptations, some varieties of maize have twice the genome size of other seemingly indistinguishable varieties. This very high variability also applies to the sequence of the maize genome, which according to DNA marker studies is several hundredfold more varied than that of any mammal. It is not yet known whether such high levels of structural variation apply to other plants with large genomes. As discussed below, however, these high levels of genomic variability within a single species are just one example of the extreme plasticity of many aspects of plant organization and development compared with animals.

Although plant genomes are highly variable and can appear relatively unstable over evolutionary timescales measured in millions of years, it should be emphasized that this mainly applies to the repetitive regions that make up the bulk of their DNA content. For example, grass genomes have evolved independently for a few tens of millions of years and now show almost no conservation of DNA in non-coding regions that make up most of their genomes. In contrast, protein-encoding DNA regions are highly conserved, not only among grass species but also among less closely related vascular plants that diverged from each other hundreds of millions of years ago. This suggests that protein-encoding DNA sequences are under stronger evolutionary selection to remain relatively constant whereas repetitive sequences have few such constraints.

The range of genome sizes found in higher plants is far greater than in animals (see Table 3.2). Within the vascular plants, genome sizes extend over 2–4 thousandfold from a minute 63 Mb in the carnivorous plant *Genlisea margaretae*, to over 150,000 Mb in the small subalpine species *Paris japonica*, and possibly as much as 250,000 Mb in the whisk fern, *Psilotum nudum*, which is 100-fold larger than the human genome. It has been proposed that all flowering plants originally had small genomes and, although many species still have fairly modest-sized genomes, a sizeable number of species (most notably in the monocots) have acquired many hundredfold more DNA than most plants. Recent evidence suggests that very large genome sizes in some plants may be selectively disadvantageous in the long term, although this has not been a sufficiently powerful constraint to prevent the extreme diversity and constant flux in genome size found in many species today.

Functional organization of plant genomes

The first plant genome to be fully sequenced was that of *Arabidopsis*, published in 2000. This genome contains about 27,400 predicted functional protein-encoding open reading frames (ORFs), about 5000 non-functional ORFs, some of which may be parts of transposons, and more than 1300 sRNA-encoding sequences. Since then, several additional plant genomes have been sequenced including those of the major crops rice, maize potato and soybean.

Analysis of these genomes has revealed many surprises, such as the large number of genes of unknown function, and evidence of extensive genome duplication in most species (even those that behave like diploids), as well as the vast amount of non-coding DNA as described above.

What do genes encode?

One of the striking things about all eukaryotic genomes sequenced to date, including those of plants like *Arabidopsis*, is that 15–35% of their predicted ORFs encode proteins of no known function (see Fig. 3.8). It is possible that some of these 'unknown' genes are incorrectly annotated DNA sequences, but most of them probably encode proteins whose function has yet to be discovered. These proteins may be present at very low levels in cells and/or may only be produced for brief periods under specific conditions. In such cases the genes concerned would rarely be active and therefore are easily

overlooked during screening programmes. Of the remaining putative genes in *Arabidopsis*, the major categories are protein synthesis, protein trafficking and general metabolism. Other sequenced plant genomes show broadly similar patterns of gene function.

Origin and role of repetitive DNA

The vast majority of repetitive DNA in plants and other eukaryotes is of external origin. It was initially thought that such DNA is useless or parasitic, and it is still sometimes erroneously called 'junk' DNA. While repetitive DNA may sometimes be evolutionarily deleterious or neutral in plants, in other cases it can play useful roles. The majority of repetitive DNA consists of regions called LTR (long terminal repeat) retrotransposons or retroelements. The small sections of DNA that make up an LTR can duplicate themselves without excision from the chromosome. Retrotransposons can transcribe

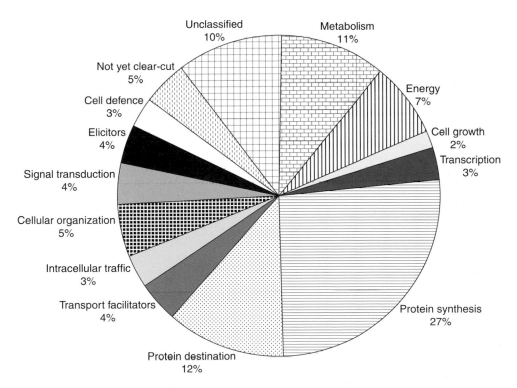

Fig. 3.8. Predicted functional organization of the genome of *Arabidopsis thaliana*. When it was first sequenced in 2000, almost 40% of the expressed protein-encoding genes in the *Arabidopsis* genome were of unknown function. More recently, advances in expression profiling and other technologies have reduced the proportion of unknown or unclear genes to about 15%. Data from Xu (2010).

their DNA into RNA and then use reverse transcriptase to make a new DNA copy that can be inserted into a different genomic location. This often results in the steady accumulation of more and more LTR DNA sequences within the genome. Other transposons cannot replicate themselves but are able to move from one place to another in the genome. Sometimes, a transposon will be inserted into an existing gene resulting in an alteration in its function. This process of 'transposon mutagenesis' can generate useful genetic variation in some plants. For example, transposon movement is responsible for the diversity of pigment patterns found in some traditional maize cobs, and in the variegated flowers of plants such as morning glory (*Ipomoea* spp.) and the ornamental petunias (*Petunia hybrida*).

Some retrotransposons closely resemble retroviruses, and may indeed be derived from such viruses. Instead of replicating within host cells and causing disease, some viruses integrate their DNA into the genome of their host and are then duplicated, along with the host DNA, during cell division and inherited by their progeny. Many retrotransposons are made up of so-called *gypsy*-like or *copia*-like elements that are now known to be ubiquitous in plants. Many of these elements encode very similar gene products and their organization, and mechanisms of mobility are remarkably conserved in plants and animals. This implies that they were already in existence in the last common ancestor of the fungi, plants and animals.

In plants, some repetitive DNA has acquired new functions. For example, it can participate in the DNA repair process after double-strand breakages caused by oxidative or other environmental stresses. Retrotransposons are much more active in monocots (especially cereals) compared with dicots. Transposons and other forms of repetitive DNA are implicated in some of the events that follow hybridization between plant species and the formation of a new allotetraploid species. Such hybridization events have been of crucial importance in the evolution of many crops, most notably the wheat family, but also in many others including brassicas, oats and cotton. Repetitive DNA elements are often exchanged between the two genomes in a new hybrid plant, sometimes silencing genes in one genome and sometimes overwriting them to generate identical sequences in otherwise different regions of the two genomes.

To summarize, the massive amount of (originally) exogenous DNA that accumulates in many genomes is sometimes selected against during evolution and may be gradually removed from the host (*e.g.* in rice). In general, there seems to be an attempt by most organisms to rid their genomes of at least some non-coding DNA. However, some other species, including several major crops, either do not or cannot remove most of their repetitive DNA. Indeed, some exogenous DNA seems to play a key role in certain aspects of plant development. The repetitive DNA may be involved in enhancing genetic variation, hence contributing to the evolution of new varieties and even new species (e.g. the various wheat species). This may be especially true for plants like maize, which is one of the most genomically diverse of all crop species, largely due to the many active transposons that are able to move within and between its chromosomes. Movement of transposons within genomes often causes changes in the expression of genes in the vicinity of their new site of integration.

3.4 The Flexible Genomes of Plants

Plant genomes are highly flexible in terms of their size, organization, function, and ability to combine with other genomes via hybridization to form new polyploid individuals. The processes of intra- and interspecific hybridization in plants have been major drivers of genetic diversity and speciation. A further form of genomic flexibility is the ability to exchange genes with unrelated species via horizontal gene transfer. Recent evidence shows that this process is much more common than previously believed, with many plants harbouring genes originally derived from bacteria and animals.

The species concept in plants

Biological organisms are conventionally divided into species within which members can normally interbreed while breeding between species is generally regarded as unusual. A standard definition of a biological species is: 'A group of organisms sharing a considerable measure of genetic and phenotypic similarity with the potential for interbreeding and producing fertile progeny.' However, the concept of what constitutes a biological species appears much less straightforward for plants than it is for animals. In 2006, an article in the journal *Nature* began as follows: 'Many botanists doubt the existence of plant species, viewing them as arbitrary constructs of the human mind, as opposed to discrete objective entities that represent reproductively independent lineages or "units of evolution".'

For example, many plants can successfully hybridize with other species to produce fertile offspring. Hence Asian rice, *Oryza sativa*, can interbreed with African rice, *Oryza rufipogon*, and also with various wild rice species. However, although *Oryza sativa* is defined as a single species, it contains several races that cannot interbreed with each other. Also, the rice genome has expanded considerably in size and then contracted again, but we still classify all these very different historical forms of Asian rice as a single species. In other cases, species that appear morphologically distinct, such as various oaks and sycamores, are readily able to interbreed. In contrast, a single species might have seemingly identical diploid and polyploid members that are reproductively isolated from one another. These examples challenge conventional notions of what constitutes a biological species. In particular, the ease with which plants can form fertile interspecific hybrids makes it difficult to regard plant species as relatively fixed breeding units as tends to be the case for animals.

Polyploidy: the multiple genomes of plants

Polyploidy (meaning 'many-fold') denotes the presence of more than two sets of chromosomes in the genome of an organism. It is one of the key phenomena in plant genetics that is responsible for the creation of new genetic variation and the formation of new species. In contrast to animals, the vast majority and quite possibly all plant species are believed to have undergone multiple rounds of polyploidy during their evolution. Molecular evidence suggests that there was an ancient genome duplication about 320 million years ago in the ancestor of all seed-bearing plants, followed by a second genome duplication about 190 million years ago in the common ancestor of all extant angiosperms. As shown in Fig. 3.9, there is also evidence of subsequent polyploidy events during monocot and dicot evolution. For example, it appears that all dicots may be descended from tetraploid or hexaploid ancestors.

Some present-day angiosperms have as many as 400 chromosomes and some ferns have more than 1000 chromosomes due to repeated polyploidy. Although plant polyploidy may sometimes be evolutionarily neutral in the sense of having little selective advantage, there is little doubt that polyploidy has been important in the evolution and domestication of many major crop plants. For example, emmer or durum wheats are tetraploid (four sets of chromosomes) while spelt, breadwheat and oats are all hexaploid (six sets of chromosomes) (see Fig. 3.10). Most apparently diploid species with relatively small genomes, such as rice and *Arabidopsis*, are actually ancient polyploids that have undergone genome reduction. Even-numbered plant polyploids can still be fertile as long as each chromosome can line up with an appropriate partner during cell division.

Autopolyploidy and allopolyploidy

An autopolyploid contains almost exactly the same genome as its parental species, except that it has become duplicated. Therefore autopolyploids tend to remain similar to their diploid ancestors over many generations. Compared with its diploid parent(s), however, an autotetraploid has an additional copy of every gene in its genome. As new autotetraploid individuals reproduce and are subject to mutation and evolutionary selection, many duplicated genes will eventually be lost or will diverge in structure and/or function from the original versions. As a result, the descendants of an autotetraploid organism will increasingly tend to differ from its diploid cousins, but this process might take many generations. Of course, the relatively long timescale of phenotypic divergence in autopolyploids is only true in relative terms. In terms of evolutionary changes that span millions of years, both autopolyploid and allopolyploid divergences are actually quite rapid processes and are major drivers of plant speciation.

In contrast to autopolyploids, allopolyploids are more likely to give rise to radically different species from their parents over a short timescale, possibly as brief as a few generations, or even immediately after they are formed. This is because allotetraploids are hybrids with a complete set of genomes from two dissimilar parents of different species. The dramatic reshuffling of its genome means that an allotetraploid organism will automatically constitute a new species, carrying a mixture of characteristics from each parental species. Sometimes, equivalent genes derived from the respective parental species that encode the same protein (such genes are called homeologues) will 'specialize' almost immediately after polyploidization. For example, one of the two homeologous genes might be silenced in one set of tissues and organs, while the other is silenced in a different set of organs during plant development. In newly formed allopolyploids, there

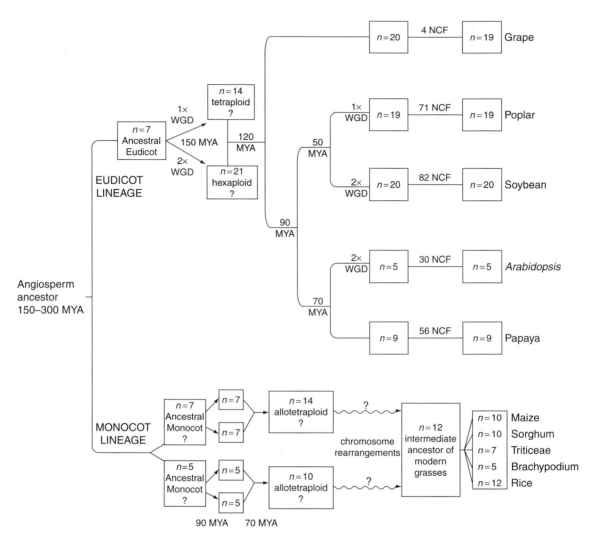

Fig. 3.9. Model for the evolution of angiosperm genomes. In this model, the eudicots are derived from an ultimate ancestor with seven chromosomes that underwent one ($n = 14$) or two ($n = 21$) whole genome duplication (WGD) events about 150 million years ago (MYA). Evidence of triplications in all eudicot genomes sequenced to date favours the latter option and means that all eudicots may be derived from a single $n = 21$ hexaploid intermediate ancestor. This genome was subsequently modified by extensive nested chromosome fusion (NCF) and further WDG events to generate the diversity of present-day eudicot genomes. For example, the *Arabidopsis* genome evolved from an $n = 21$ intermediate followed by two specific WGD and 30 NCF events to produce its present $n = 5$ genome. Monocot genomes are derived from either an $n = 5$ or an $n = 7$ ancestor, which would have generated allotetraploid intermediates of $n = 10$ or $n = 14$, respectively. In the case of the modern grasses, further chromosome rearrangements led to a common intermediate ancestor of $n = 12$. This is the genome from which major cereal crops such as rice, maize and wheat are derived. Adapted from Abrouk *et al.* (2010).

is also an element of 'cross-talk' between the two parental genomes that now coexist in the new hybrid plant.

There are several important genetic processes that occur in both auto- and allopolyploids above the organizational level of their duplicated (or homeologous) genes. These include the sort of intergenomic 'cross-talk' mentioned above, but also such phenomena as saltational variation, intergenomic invasion and cytonuclear stabilization. Saltational variation (from the Latin *saltus*, 'leap') is the occurrence of a sudden inherited

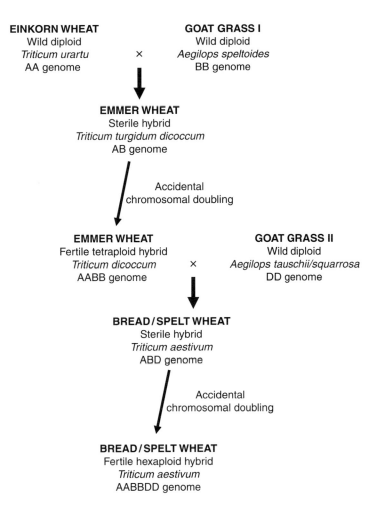

Fig. 3.10. Recent evolution of the domesticated wheats. The various domesticated wheats are originally derived from wild diploid grasses of the *Triticum* (AA genome) and *Aegilops* (BB or DD genomes) genera. One of the earliest cultivated wheats was the diploid species einkorn (AA), but this soon gave way to the more versatile and productive allotetraploid hybrid species emmer (AABB). A further round of hybridization and chromosome doubling gave rise to the hexaploid bread and spelt wheats (AABBDD), which remain our major temperate cereal crops to this day.

change from one generation to the next. Such non-Darwinian changes can arise as a consequence of various forms of polyploidization or from the mutation of a key regulatory gene. Intergenomic invasion involves the colonization of one genome in a polyploid organism by DNA sequences from the other genome. This has been studied in detail in cotton, where the D genome has been extensively colonized by repetitive elements from the A genome.

One of the most important problems facing a newly formed polyploid is the inappropriate pairing of homeologous, rather than homologous, chromosomes at meiosis. Such illegitimate chromosomal pairings can lead to sterility and successful allopolyploids have developed at least two mechanisms to avoid them. The first mechanism is the selective elimination of large amounts of non-coding DNA from different parts of the genome so that homeologous chromosomes no longer resemble one another enough to form such illegitimate pairings. Remarkably, tens of millions of DNA base pairs and as much as 15% of the entire genome can be rapidly lost in this manner. A second mechanism involves direct suppression of homeologous pairing, such as the system regulated by the *Ph1* locus in polyploid species of wheat.

Exploiting polyploidy in crops

In some crops, like breadwheat and oats, an allotetraploid species has hybridized with a diploid species to create a new hexaploid species that now contains six sets of chromosomes. There are more extreme types of polyploidy involving eight, ten or even 12 sets of chromosomes. For example, members of the wheat group include the octoploid *Elytricum fertile* and decaploid *Agropyron elongatum*, while members of the birch genus, *Betula*, can be dodecaploid. In the case of breadwheat and oats, the new hexaploid species that arose after domestication were just as fertile and vigorous as their diploid parents. As a bonus, these new hexaploid cereals also tended to be higher yielding; they produced better quality grain; and they had a greater tolerance to cold and drought than their parental species. For these reasons, such cereal polyploids that arose via spontaneous hybridization events were selected by early farmers and became favoured crop varieties. However, polyploids do not always make better crops. Barley is a diploid plant and polyploid versions, whether spontaneous or man-made, do not have any increased performance as crops and all modern varieties of barley remain diploid. Therefore, although polyploidy can be advantageous for a crop, this is not always the case and there do not seem to be any universal rules that apply to all species.

Hybrids with odd numbers of chromosomes are normally unable to form complete pairs and therefore cannot reproduce sexually. Despite their sterility, plants or tissues with triploid genomes can sometimes be useful. For example, there are two interesting examples of triploidy, both relating to well-known edible plants. The first example is somatic triploidy, which is when a single tissue is triploid, rather than the entire plant. All flowering plants produce a nutritive tissue called the endosperm (see Chapter 5), which helps to sustain the growth of the young embryo during seed development in an analogous manner to the mammalian placenta. The endosperm is triploid due to the fusion of two haploid maternal nuclei with a haploid pollen nucleus during fertilization. In some plants the endosperm persists to form a starch-rich store that can occupy most of the volume of the mature seed. Most of the grain in cereal crops is made up of a starchy endosperm, and it is this triploid tissue that is by far the major source of calories for human societies around the world. The second example is of whole-plant triploidy in the commercial banana, *Musa acuminata*. Virtually all internationally traded bananas in the world consist of a single triploid clonal variety, called 'Cavendish'. These bananas, i.e. the large-fruit varieties that can be bought in a typical shop, are genetically identical to each other and hence exceptionally uniform. Because they are triploid these bananas are sterile and must be propagated vegetatively, and are highly susceptible to new diseases.

Horizontal gene transfer

As we saw above, many plants can readily form hybrids outside their species or genus, which gives considerable scope for the exchange of genetic information, normally in the form of entire chromosomes or segments thereof. Until recently, the transfer of individual genes from one species to another was believed to be limited to prokaryotes. Thanks to genome sequencing programmes, however, it is now evident that DNA can move between the genomes of plants, animals and prokaryotes in a process called horizontal (or lateral) gene transfer. In the case of plants, gene transfer to and from the plant can occur via several mechanisms, including the following:

- *Host–parasite exchange* – examples include gene transfer to an endophytic parasite in the Rafflesiaceae from its hosts in the genus *Tetrastigma*, as well as gene transfer in the opposite direction, from members of the parasitic angiosperm order, the Santalales (including the sandalwoods and mistletoes), to a non-angiosperm host plant, the rattlesnake fern (*Botrychium virginianum*). See also Box 3.2 for a discussion of the case of the parasite striga and its apparent 'theft' of genes from its monocot hosts.
- *Transfer via or from a plant virus* – this is relatively common and in one case, the genome of the non-parasitic, tropical shrub *Amborella trichopoda*, was found to contain no less than 26 foreign genes.
- *Transfer via pathogenic or mycorrhizal fungi* – fungal–plant gene transfer in the context of mycorrhizal associations (which may affect >90% of soil-growing plants) is a powerful and pervasive mechanism for horizontal gene flow between almost all plant species.

- *Transfer from a plant to a herbivore* – one of the few examples is a group of small freshwater invertebrates, the bdelloid rotifers, that harbour numerous 'foreign' genes in their genomes, including a UDP glycosyltransferase and ß-D-galactosidase of apparent plant origin.
- *Transfer from a plant to a symbiont* – the simple aquatic animal *Hydra viridis* contains a functional ascorbate peroxidase gene transferred to it from a former symbiotic partner, the alga *Chlorella vulgaris*.

In conclusion, it is becoming ever more apparent that inter-organism gene transfer between unrelated species is a lot more common than previously suspected. This also implies that various forms of transgenesis have been occurring in nature for billions of years rather than being a new invention of plant biotechnologists.

As well as moving between species, genes and other DNA elements can also move to different locations within a cell. For example, entire genes or clusters of genes can be transferred between mitochondrial or plastid genomes and the main nuclear genome. Individual genes and larger DNA segments from the plastid genome are continually being transferred to the larger nuclear genome. When this process began, many of the transferred plastid genes were successfully integrated into the nuclear genome, with the result that much of the original plastid genome (which is of bacterial origin) now resides in the nucleus. However, it seems that the process has now reached some sort of limit and further gene transfer from plastid to nucleus is no longer favoured. Gene transfer to the nucleus still occurs at a surprisingly rapid rate, but newly integrated plastid genes are normally broken up and eventually eliminated from the nuclear genome. In other words, the balance of DNA between plastid and nuclear genomes has now reached a state of dynamic equilibrium whereby further transfer appears to be maladaptive.

Box 3.2. Can plants act as genetic engineers?

There is increasing evidence that higher plants are capable of exchanging genes with a variety of organisms, including fungi, insects and other plants. This raises the question of whether plants might be able to acquire functional genes from other organisms, including other plants, in a way that enables them to build up a new 'transgenic' genotype that might have a selective advantage in terms of evolution. If this speculation is true, then such plants could be regarded as (unconscious) genetic engineers that can sometimes acquire novel and useful genes from unrelated organisms.

In 2010, a group of Japanese researchers reported that the genome of the parasitic plant *Striga hermonthica* contained at least one protein-encoding gene that had originated from one of its host plants, sorghum. Like other members of the *Striga* genus, *S. hermonthica* is a virulent parasite that specializes in infecting members of the grass family of monocots, including crops such as maize, rice and sorghum (see Chapter 6). It had already been reported that mitochondrial genes could be exchanged between plant hosts and parasites. In some cases, the genes may have been exchanged via fungal symbionts acting as intermediary vectors, but there is also evidence that angiosperms and ferns can directly transfer genes to one another.

Because *S. hermonthica* is a dicot, most of its genes are different from the monocot genes of its hosts. Thanks to recent genome sequencing programmes, the signature sequences of many monocot and dicot genes are known in detail and can therefore be distinguished from one another. This led the Japanese group to search for monocot-like sequences in the *S. hermonthica* genome. They soon found a nuclear gene that encoded a 448-amino acid protein of unknown function that closely resembled a gene from the monocot sorghum, but was completely different from other dicot-like genes in the *Striga* genome. The authors speculate that *Striga* might have originally acquired the transgene from sorghum as an mRNA. This could have occurred when a *Striga* plant formed an invasive haustorium that connected it to the vascular system of the sorghum host. As discussed in Chapter 5, the phloem of higher plants contains many nutrients, such as sugars, but also larger molecules including proteins and mRNAs. Indeed, there were remains of what might have been a poly-A tail from an original mRNA in the DNA sequence of the putative transgene in *Striga*. Having 'captured' this putative foreign gene, *Striga* may have retained it providing it conferred a selective advantage, thereby acting as its own 'genetic engineer'.

3.5 Genomics of the Major Crop Plants

Recent advances in technology, computing power and bioinformatics have led to the sequencing of several important crop genomes. These sequencing successes have been complemented by studies of gene expression patterns in crops that are beginning to provide useful information about the regulation of such key processes as pathogen resistance, nitrogen use and plant architecture.

Cereals

The cereals of the grass family are the most important plant group in agriculture. The cereals are particularly interesting from an evolutionary perspective because of the high degree of variation found in overall body size, ploidy and chromosome number as discussed above. Genome sizes range from 17,000 Mb in hexaploid breadwheat, about 5000 Mb in diploid einkorn wheat and barley,

2500 Mb in maize and a mere 400 Mb in rice. In comparison, the human and mouse genomes each contain about 3200 Mb. Major genome sequencing and characterization projects are currently under way for the monocots rice, maize, sorghum, foxtail millet (*Setaria italica*) and brachypodium (*Brachypodium distachyon*).

Recent research suggests that the ancestral grass plant from which all cereals are derived had either five or seven haploid chromosomes. Around 85 million years ago, this plant underwent a whole-genome duplication and by 50–70 million years ago a series of breakage and fusion events had led the haploid chromosome number to stabilize at $n = 12$, as in modern rice. This plant was the common ancestor of all extant grasses. While rice has retained its ancestral chromosome number, it has been reduced in the wheat, maize and sorghum genomes, following a pattern that may be typical of plant chromosome evolution in general. As shown in Fig. 3.11, the earliest rice species date

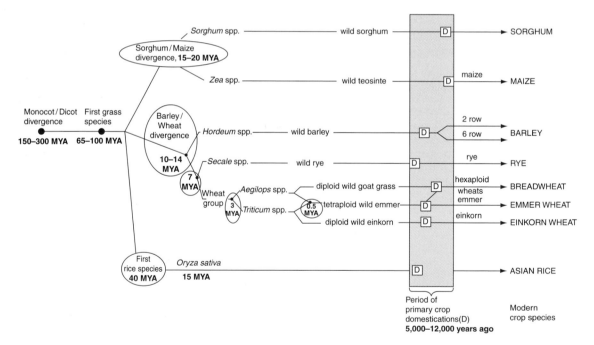

Fig. 3.11. Evolution of cereal crops. The first grasses date from almost 100 million years ago and the most ancient group of cereals are the rices, which appeared about 40 million years ago. By 15 million years ago several groups of rice in the *Oryza* genus had developed in South and East Asia. In Western Asia, the major group of cereals were wheat, barley and rye, which diverged from a common ancestor about 10–14 million years ago. The ancestors of maize and sorghum split about 15 million years ago and wild sorghum species became important edible plants in Africa, while teosinte served the same purpose in Mesoamerica. The evolution of domesticated (D) forms of these cereals did not occur until about 12,000 years ago, and was determined by a new suite of human-created environmental conditions such as cultivation and harvesting rather than Darwinian natural selection. MYA, million years ago. Adapted from Murphy (2007).

back 40 million years, which is also about the time when the pooid grasses diverged from the panicoid species. Major panicoid groups include sorghum and maize, whose ancestors diverged about 15–20 million years ago. Major pooid groups include barley, rye and the wheats, whose ancestors diverged about 10–14 million years ago. Modern cereal crops are therefore the products of a highly complex and diverse genetic legacy. The understanding of this genomic legacy is helping to explain many aspects of the performance of cereal crops as well as giving us powerful new tools for their manipulation.

Legumes

Legumes are members of the Rosidae, which is the largest group of eudicots. Their characteristic feature is the formation of specialized nodules that support nitrogen-fixing bacteria, enabling legumes to grow better than other plants on relatively nitrogen-limited soils (see Chapter 4). Several legume species have been domesticated, including major crops such as pea, lentil, soybean, lupin, clover and lucerne (alfalfa). Annual grain legume crops are often known as *pulses*. After cereals, legumes are the most important group of food crops. In most human cultures, starch-rich cereals and protein-rich legumes are grown alongside one another to provide a complementary mix of nutrients that can, if necessary, sustain human life without the need for animal protein (see Chapter 1).

Several important legume crops have 14 haploid chromosomes (peas and lentils) or are recently derived from such species (broad beans). In contrast, the relatively large soybean genome has 20 chromosomes ($2n = 40$) containing about 1000 Mb of DNA. In 2010, publication of the draft DNA sequence of soybean indicated that it has undergone two rounds of genome duplication, at about 59 and 13 million years ago, respectively, followed by a more recent hybridization with a related species to create an allopolyploid. One of the unusual aspects of soybean evolution is that it has retained many of the extra genes created during these duplication and hybridization events so that it now contains more than 46,400 predicted genes. Almost 75% of the genes in soybean are present in multiple copies, allowing the plant to become more diverse and resilient in its morphology, development and metabolism. Preliminary annotation shows that protein-encoding genes make up 18% of the genome, which also contains over 60% repetitive DNA with the remainder as yet unclassified.

Much of our knowledge about legume genomics has come from the analysis of the two model species, lotus (*Lotus japonicus*) and medicago (*Medicago truncatula*), both of which behave as diploids with relatively small genomes of about 500 Mb. These studies have demonstrated large-scale synteny between the genomes of the two species, which will be a considerable help in locating genes of interest for crop breeding. Wider comparisons with the *Arabidopsis*, poplar and rice genomes have also revealed some limited but potentially useful genomic synteny across angiosperms in general. Other important legumes for which large-scale genomic projects are under way include groundnut (*Arachis hypogaea*), chickpea (*Cicer arietinum*), common bean (*Phaseolus vulgaris*) and cowpea (*Vigna unguiculata*).

Comparative analysis of the different legume genomes is now enabling researchers to uncover the phylogenetic relationships between species that will enable knowledge about traits and genes gained in model plants to be applied to the more complex genomes of the major crop species. For example, the disease resistance gene, *RCT1*, has been isolated from the crop legume, lucerne, using genomic information from medicago. Other important genomic regions where genomic synteny is facilitating gene identification and isolation include those involved in the key symbioses with rhizobia (for nitrogen fixation) and mycorrhizae (for enhanced nutrient uptake).

3.6 Summary Notes

- The nuclear genomes of higher plants consist of linear chromosomes containing DNA and proteins packaged into condensed structures known as chromatin. Gene expression normally occurs in the more loosely packed euchromatin, while tightly packed and methylated heterochromatin is less transcriptionally active. Chromatin modification, e.g. by methylation, can induce inherited epigenetic changes in gene expression without altering the DNA sequence. Epigenetic modifications are common in plants and can have had profound evolutionary consequences.
- Plant genes are modular structures containing a promoter region that regulates the expression of a transcription unit made up of introns and exons. Gene expression involves formation of a

pre-mRNA molecule corresponding to the entire transcription unit. Pre-mRNA is spliced to remove intron-derived regions and the mature mRNA is translated into protein. Several proteins can be produced from a single gene via mechanisms such as alternative splicing and translational braking.

- While the basic organization of higher plant genomes is broadly similar to that of animals, there are many detailed differences that reflect their differing lifestyles. Plants have a third, cyanobacterial-derived, genome in their plastids although most genes from the original cyanobacterial endosymbiont are now part of the nuclear genome.

- All higher plants are probably descended from polyploid ancestors, and formation of new polyploids via interspecific hybridization or genome duplication is still relatively frequent. Polyploidy has been a remarkably successful strategy in angiosperm evolution and diversification and has been a major factor in the selection of some major crop species.

- Plants have highly flexible genomes that can differ greatly even within a single species. Individual genes can move between plants and other organisms via various horizontal transfer mechanisms.

- The genomes of several major crop plants have recently been sequenced. This is giving valuable information about crop evolution and will have great utility in facilitating crop improvement.

Further Reading

Molecular genetics

Bender, J. (2002) Plant epigenetics. *Current Biology* 12, R412–R414.

Grant-Downton, R.T. and Dickinson, H.G. (2005) Epigenetics and its implications for plant biology. 1. The epigenetic network in plants. *Annals of Botany* 96, 1143–1164.

Heo, J.B. and Sung, S. (2011) Vernalization-mediated epigenetic silencing by a long intronic noncoding RNA. *Science* 331, 76–79.

Meyer, P. (2005) *Plant Epigenetics*. Blackwell, Oxford, UK.

Rapp, R.A. and Wendel, J.F. (2005) Epigenetics and plant evolution. *New Phytologist* 168, 81–91.

Zemach, A., McDaniel, I.E., Silva, P. and Zilberman, D. (2010) Genome-wide evolutionary analysis of eukaryotic DNA methylation. *Science* 328, 916–919.

Genome organization

Abrouk, M., Murat, F., Pont, C., Messing, J., Jackson, S., Faraut, T., Tannier, E., Plomion, C., Cooke, R., Feuillet, C. and Salse, J. (2010) Palaeogenomics of plants: synteny-based modelling of extinct ancestors. *Trends in Plant Science* 15, 479–487.

Bendich, A.J. (2004) Circular chloroplast chromosomes: the grand illusion. *Plant Cell* 16, 1661–1666.

Sterck, L., Rombauts, S., Vandepoele, K., Rouzé, P. and Van de Peer, Y. (2007) How many genes are there in plants (… and why are they there)? *Current Opinion in Plant Biology* 10, 199–203.

Varshney, R.K., Nayak, S.N., May, G.D. and Jackson, S.A. (2009) Next-generation sequencing technologies and their implications for crop genetics and breeding. *Trends in Biotechnology* 27, 522–530.

Vazquez, F., Legrand, S. and Windels, D. (2010) The biosynthetic pathways and biological scopes of plant small RNAs. *Trends in Plant Science* 15, 337–345.

Xu, Y. (2010) *Molecular Plant Breeding*. CAB International, Oxford, UK.

Polyploidy

Comai, L. (2005) The advantages and disadvantages of being polyploid. *Nature Reviews Genetics* 6, 836–846.

Dubcovsky, J. and Dvorak, J. (2007) Genome plasticity a key factor in the success of polyploid wheat under domestication. *Science* 316, 1862–1866.

Jiao, Y., Wickett, N.J., Ayyampalayam, S., Chanderbali, A.S., Landherr, L., Ralph, P.E., Tomsho, L.P., Hu, Y., Liang, H., Soltis, P.S., Soltis, D.E., Clifton, S.W., Schlarbaum, S.E., Schuster, S.C., Ma, H., Leebens-Mack, J. and dePamphilis, C.W. (2011) Ancestral polyploidy in seed plants and angiosperms. *Nature* 473, 97–100.

Levy, A.A. and Feldman, M. (2005) Genetic and epigenetic reprogramming of the wheat genome upon allopolyploidisation. *Biological Journal of the Linnean Society* 82, 607–615.

Rieseberg, L.H., Wood, T.E. and Baack, E.J. (2006) The nature of plant species. *Nature* 440, 524–527.

Wood, T.E., Takebayashi, N., Barker, A.S., Mayrose, I., Greenspoon, P.E. and Rieseberg, L.H. (2009) The frequency of polyploid speciation in vascular plants. *Proceedings of the National Academy of Sciences USA* 106, 13875–13879.

Flexible genomes

Andersson, J.O. (2005) Lateral gene transfer in eukaryotes. *Cellular and Molecular Life Science* 62, 1182–1197.

Broothaerts, W., Mitchell, H.J., Weir, B., Kaines, S., Smith, L.M.A., Yang, W., Mayer, J.E., Roa-Rodríguez, C.

and Jefferson, R.A. (2005) Gene transfer to plants by diverse species of bacteria. *Nature* 433, 629–633.

Syvanen, M. and Kado, C.I. (2002) *Horizontal Gene Transfer*. Academic Press, London.

Yoshida, S., Maruyama, S., Nozaki, H. and Shirasu, K. (2010) Horizontal gene transfer by the parasitic plant *Striga hermonthica*. *Science* 328, 1128.

Crop genomics

Devos, K.M. (2010) Grass genome organization and evolution. *Current Opinion in Plant Biology* 13, 139–145.

Murphy, D.J. (2007) *People, Plants, and Genes: The Story of Crops and Humanity*. Oxford University Press, UK.

Paterson, A.H. *et al.* (2009) The *Sorghum bicolor* genome and the diversification of grasses. *Nature* 457, 551–556.

Sato, S., Isobe, S. and Tabata, S. (2010) Structural analyses of the genomes in legumes. *Current Opinion in Plant Biology* 13, 146–152.

Schmutz, J. *et al.* (2010) Genome sequence of the palaeopolyploid soybean. *Nature* 463, 178–183.

The Potato Genome Sequencing Consortium (2011) Genome sequence and analysis of the tuber crop potato. *Nature* 475, 189–195.

4 Plant Metabolism

4.1 Chapter Overview

This chapter provides an overview of the major metabolic processes that occur during plant growth and development. The principal focus is on classes of compounds that are important to plants themselves and/or have practical uses. The most important edible plant compounds are the carbohydrate, protein and oil storage products of seeds, fruits and roots. These are also the three macronutrients that form the basis of our diet, as well as being used for livestock feed, industrial raw materials and biofuels. Other important plant products include structural compounds such as cellulosic and lignified fibres, which provide clothing, building materials and fuels. Finally, there is the vast range of secondary compounds such as phenolics, alkaloids and terpenoids that provide products such as spices, beverages, medicines and biocides. All of these plant-based resources are actual or potential targets for biotechnological manipulation.

4.2 Carbohydrates

All our major food sources are derived from plant storage tissues, of which the most important are the carbohydrate-accumulating cereal grains and tubers. As shown in Table 4.1, protein and oil macronutrients are also available in plant sources such as protein-rich legume seeds and oil-rich seeds and fruits.

The major plant carbohydrates, in terms of their importance both for plant structure and function and for human utilization, are sucrose, starch and cellulose. Plants also synthesize many other types of carbohydrate, including hemicelluloses and pectins that help stabilize the cell wall. In addition, plants make a diverse range of sugar residues that are found in thousands of compounds such as glycoproteins, glycolipids, and glycosylated secondary compounds such as glucosylated alkaloids and terpenoid cardiac glycosides. The disaccharide trehalose and glycoalcohols such as mannitol and pinitol also function as osmolytes, cryoprotectants and energy sources in plants.

While sucrose is a chemically simple disaccharide, starch and cellulose are complex polymers of the monosaccharide glucose. The composition of starches and celluloses can vary greatly, even within the same plant, and they are often associated with other compounds *in vivo*, which can further alter their physiochemical properties. The major uses of sucrose and starch are as foodstuffs, although both are used increasingly for non-food applications including as biodegradable structural materials and for conversion to ethanolic biofuels. Cellulose is the most abundant organic molecule on earth and has numerous uses from clothing to paper. Various forms of cellulosic plant biomass are being developed as feedstocks for bioconversion to ethanol or hydrogen in the manufacture of so-called 'second generation' biofuels (see Chapter 12).

Sucrose

Sucrose is a disaccharide made up of one residue each of glucose and fructose. It is the major form of carbohydrate that is transported from source (photosynthetic) to sink (non-photosynthetic) tissues. The majority of photosynthetically fixed CO_2 is exported from chloroplasts as triose phosphates and is converted into sucrose in the cytosol. Some of this newly formed sucrose is stored in the vacuole but most is exported via the apoplast and phloem to other parts of the plant. Key enzymes that regulate the formation and breakdown of sucrose are sucrose phosphate synthase, sucrose phosphatase, sucrose synthase and invertase (Fig. 4.1). There are several forms of these enzymes in different tissues and subcellular compartments and their activity is closely regulated in response to a variety of external and

Table 4.1. Storage product compositions of selected crops.

Species		Dry weight* (%)		
		Carbohydrate	Lipid	Protein
Seed crops				
Rice	*Oryza sativa*	86	3	9
Wheat	*Triticum aestivum*	84	2	12
Maize	*Zea mays*	83	5	11
Barley	*Hordeum vulgare*	81	3	14
Pea	*Pisum sativum*	69	2	26
Lentil	*Lens culinaris*	60	1	25
Linseed	*Linum usitatissimum*	38	37	21
Soybean	*Glycine max*	34	21	40
Almond	*Prunus amygdalus*	21	53	22
Pumpkinseed		19	49	26
Rapeseed	*Brassica napus*	11	49	32
Safflower	*Carthamus tinctorius*	20	43	19
Tef	*Eragrostis tef*	81	2	10
Edible nuts	Various species	1–4	43–67	8–22
Sunflower	*Helianthus annuus*	18	50	21
Root and tuber crops				
Potato	*Solanum tuberosum*	82	<1	10
Cassava	*Manihot esculenta*	90	1	3
Yams	*Dioscorea* spp.	70	<1	9
Sweet potato	*Ipomoea batatas*	81	<1	4–7
Taro	*Colocasia esculenta*	>80	<1	7

*These values are typical for most varieties of each crop but in some crops, such as soybean and maize, additional varieties, e.g. with high oil contents, have also been bred.

internal signals. For example, sucrose phosphate synthase and sucrose phosphatase activities are modulated by protein kinases that themselves respond to light/dark cycles, osmotic stress and metabolite concentrations. As discussed in Box 4.1, several efforts to alter carbohydrate composition in plants involve the manipulation of these enzymes.

In the majority of plants, sucrose is the most abundant component of the phloem, or sap. The presence of this sweet substance attracts a wide range of sap-sucking pest species such as aphids and psyllids. In some plant families, additional carbohydrates are transported in the phloem. Examples include sorbitol in the Rosaceae and raffinose or stachyose in the Cucurbitaceae. Sucrose transported to vegetative tissues, such as growing shoots or roots, tends to be unloaded from phloem sieve elements directly into cells via plasmodesmata, i.e. using a symplastic route. In contrast, sucrose destined for storage tissues, such as developing seeds, generally uses an apoplastic route, i.e. it is unloaded from phloem sieve elements into an extracellular compartment through which it travels until it is taken up by cells in the storage tissue.

Once it reaches its destination, sucrose can be used for several purposes. It might undergo catabolism via glycolysis and the tricarboxylic acid cycle and thereby serve as an energy source. Alternatively it might undergo anabolism and be used for the further biosynthesis of complex carbohydrates or other metabolic products. Sucrose has a key role in osmoregulation and the maintenance of turgor in plants. It can also be used as part of the response to external stresses such as wounding, pathogen infection and cold. The sugar composition of many fruits and storage organs is also regulated by sucrose. Finally, sucrose can modulate the expression of several genes involved in plant growth and development. This is superficially similar to the action of many plant hormones, but sucrose is not normally classified as a hormone because it only has an effect at relatively high concentrations (>100 mM) compared with true hormones (<1 mM).

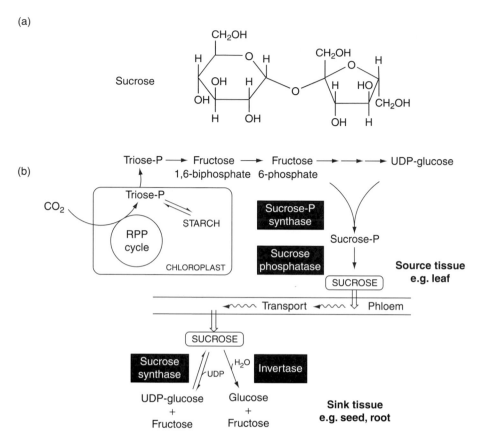

Fig. 4.1. Structure and metabolism of sucrose. (a) Sucrose is a disaccharide consisting of a glucose residue (left) esterified to a fructose residue (right). (b) In leaf cells, sucrose is synthesized in the cytosol from fructose 6-phosphate and UDP-glucose derived from the primary photosynthetic assimilate triose phosphate. Sucrose is then exported via the phloem network to sink tissues where it can be metabolized directly or converted into starch for storage.

The ability to synthesize sucrose is only found in plants and cyanobacteria, but very few plants accumulate large amounts of sucrose in their tissues. Instead, starch is by far the most common storage carbohydrate due to its higher energy content on a fresh weight basis. Among the small number of crops that are grown for their sucrose content are sugarcane and sugarbeet. However, sucrose may also be a flavour component of some sweet-tasting fruits and vegetables.

Starch

Starch is the principal storage carbohydrate in plants and is a major product of photosynthesis with up to 30% of fixed CO_2 being made directly into starch. Starch is also the single most important source of dietary calories for humans, providing about two-thirds of average worldwide energy intake. Starch has additional uses as a highly versatile material with a host of industrial and medical applications (Table 4.2). Starch is the major long-term storage product accumulated by plants and the existence of calorie-rich starchy tissues is the primary reason for the domestication of most of our staple crops. Starch is especially abundant in the grains of cereals and legumes, and in the underground storage organs of tuber and root crops such as potatoes, cassava and yams.

Each plant species synthesizes a different type of starch and starches can vary greatly in their physical properties and end uses. For example, some types of wheat make a floury starch that can be used for bread-making; other wheats produce a starch more useful for making biscuits; while in others the starch is much more glutinous and is

Box 4.1. Biotechnology targets for carbohydrates

Sucrose: Major targets include increasing sucrose levels in crops such as sugarcane and sugarbeet, plus the manipulation of sucrose metabolism in other crops where this may have indirect benefits. Enzymes such as sucrose phosphate synthase, sucrose synthase, sucrose phosphatase and invertase are key targets for manipulation not only because they are involved in sucrose metabolism itself but also because they regulate aspects of starch and cellulose composition in some tissues. For example, transgenic cotton plants overexpressing a spinach sucrose phosphate synthase had improved cellulose fibre quality under certain growth conditions. In another study, transgenic potatoes overexpressing an endogenous sucrose synthase gene had increased tuber size and starch content.

Stachyose: In some legume crops it is desirable to reduce stachyose levels in grains. For example, soybeans contain relatively high levels of stachyose and raffinose, which reduces the digestibility of soybean meal and leads to unpleasant side effects such as bloating and flatulence. The sequencing of the soybean genome (see Chapter 10) has helped to identify useful mutations in genes such as *rsm1* (raffinose synthase) that can be used to select for low-stachyose-containing soybean lines. Some low-stachyose/low-raffinose varieties have the added benefit of elevated sucrose levels, which can greatly improve the taste and digestibility of soybean products.

Amylose-free starches: Highly branched, amylose-free starches are uses as stabilizers, thickeners and emulsifiers. One of the few examples of a commercial transgenic amylose-free crop variety is the Amflora™ potato developed by BASF and approved for industrial applications in the EU in 2010. These potatoes have a *waxy* phenotype due to the anti-sense mediated down-regulation of the granule-bound starch synthase responsible for amylose formation. Amylose-free starch from these potatoes has widespread uses for making paper products, packaging and adhesives. In the future, similar starches might have various applications in the food industry. For example, they readily form clear pastes that are useful as thickeners and stabilizers in a wide range of products from ketchup to yoghurt.

High amylose starches: These less branched starches are often relatively sweet and have high gelling strengths, which makes them useful ingredients in confectionary products. Their film-forming ability also allows them to be used for food coatings. Transgenic high-amylose potatoes have been created by anti-sense mediated down-regulation of genes encoding the two isoforms of starch branching enzyme. The resulting potatoes had improved texture due to their higher water content but have yet to be commercialized. One reason for this is that, so far, all crops bred to produce high-amylose starches have greatly reduced yields compared with similar non-transgenic varieties.

---

used to make foods such as pasta or couscous. Some starches are more suitable for non-food applications such as in packaging, adhesives, cosmetics, detergents, or as components of composite structural materials. As with sucrose from sugarcane, crop-derived starch, e.g. from maize, is increasingly being used for conversion to ethanol for the manufacture of biofuels (see Chapter 12).

Starch biosynthesis occurs throughout development in all tissues of the plant body. In the chloroplasts of photosynthetically active leaves, much of the fixed CO_2 is temporarily stored during the day as starch grains. This transient store of starch is then broken down at night to provide the plant with substrates for respiration. Any excess photosynthate not required by the leaves is exported as sucrose, which is transported via the phloem network to the rest of the plant. Once the sucrose reaches its destination, it either is immediately

catabolized or is stored locally as starch grains within specialized plastids termed amyloplasts. With the exception of some specialized roots and tubers, starch deposits in most vegetative organs are relatively transient and are not very significant biotechnologically. However, many plants accumulate large amounts of more useful long-term storage starch in fruits and seeds.

The basic biochemistry of starch biosynthesis is relatively straightforward. A series of glucose molecules are linked together via (α-1,4) and (α-1,6) glycosidic bonds to form a polymer with varying degrees of branching (Fig. 4.2). Starch synthase transfers a glucosyl unit from ADP-glucose to the growing starch chain to form a 1,4-α-glucosidic bond with the release of ADP. This results in a linear extension to the growing starch chain. Different isoforms of starch synthase have specialized roles in the formation of amylose (unbranched) and

Table 4.2. Industrial uses of starch.

Type of starch	Industrial sector	Use of starch or modified starch
High amylose	Adhesives	Glues, adhesives
	Agrochemicals	Seed coatings, fertilizers, pesticide delivery, mulches
	Animal feed	Dry feed pellets
	Beverages	Alcohol, beer, soft drinks, coffee
	Building	Mineral fibre tiles, gypsum board, plaster, concrete
	Confectionary	Ice cream, boiled sweets, jam, jelly gums, marshmallows
	Cosmetics	Face and talcum powders
	Detergents	Surfactants, bleaching agents, bleaching activators
	Energy	Starch-based bioethanol fuels
Low amylose	Food	Bread, baby food, salad dressings, mayonnaise, soups
	Medical	Plasma extender/replacers, absorbent sanitary products, transplant organ preservation
	Oil drilling	Viscosity modifier
	Paper and board	Binding, sizing, coating, cardboard, paper, printing paper, packaging, corrugated board
	Pharmaceuticals	Diluent, binder, drug delivery
	Plastics	Biodegradable fillers
	Purification	Flocculant
	Textiles	Sizing, finishing, fire-resistant textiles, fabrics, nappies

Data adapted from Bowsher *et al.*, 2008.

amylopectin (branched) forms of starch. Therefore starch synthase has been the target of many attempts to manipulate starch composition in crops. Starch branching enzyme cleaves an existing α-1,4 glycosidic bond and creates an α-1,6 bond instead, with consequent branching of the chain.

Amylose is a mostly unbranched polymer of about 600–3000 glucose units. Amylopectin is much larger and more highly branched with 6,000–60,000 glucose units and branch points every 20–25 units. The type of starch synthesized in a particular plant, and therefore its physiochemical properties and usefulness to people, depends critically on the nature and extent of its branching and the manner of its assembly into the complex semicrystalline structure that is a starch grain. Starch grains typically contain about 20–30% amylose and 70–80% amylopectin, plus small amounts of lipid, protein and phosphate (Table 4.3). Amylopectin formation occurs on the surface of the starch grain by means of enzymes located on or within the grain. The growing branched oligosaccharide chains pack together as parallel layers of left-handed double helices to give the starch grain its semicrystalline structure. The amylose content largely determines the shape and size of the grain with lower amounts producing more spherical grains and higher amounts producing oval and other shapes.

After grinding and processing the starches of cereal grains become associated with other components, such as storage proteins and lipids, to produce the unique flours and pastes associated with each type of starchy crop. Non-seed starches, such as potato or cassava, generally have much larger grains and lower protein and lipid contents than cereals. This enables them to be processed to bland-tasting, clear pastes with many applications in the food industry, such as in thickeners or gelling agents.

Fig. 4.2. Structure and metabolism of starch. (a) Starch molecules consist of long chains of linear α-1,4 linked glucose units with α-1,6 glucose units forming branch points. (b) Amylose is a mostly unbranched chain of 600–3000 α-1,4 linked glucose units. (c) Amylopectin is highly branched with α-1,6 branch points every 20–25 glucose units. (d) Physical structure of starch grain showing the different growth rings made up of crystalline and amorphous lamellae. (e) Biosynthesis of starch. This highly simplified pathway shows the roles of starch synthase and starch-branching enzyme in the formation of branched and unbranched regions of starch molecules.

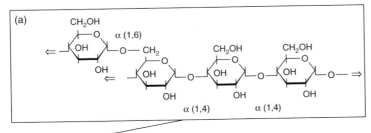

(a)

(b) O-O-O
O-O-O-O-O-O-O-O-O-O-O .... O-O-O-O-O

(c) O-O-O-O-O-O-O-O-O-O-O-O-O

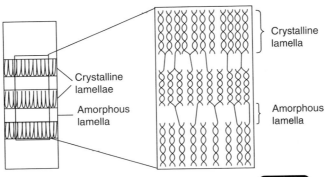

(d)

Semicrystalline growth rings

Amorphous growth rings

Crystalline lamella

Crystalline lamellae

Amorphous lamella

Amorphous lamella

(e)

Soluble starch synthase

ADP-glucose pyrophosphorylase

Starch branching enzyme

Granule-bound starch synthase

Glucose 6-P → Glucose 1-P → ADP-Glucose

AMYLOPECTIN

STARCH GRAINS

AMYLOSE

Table 4.3. Starch grains of major crop species.

Property	Wheat	Maize	Potato	Cassava
Plant location	Grain	Grain	Tuber	Root
Shape	Lenticular or round	Round, polygonal	Oval, spherical	Oval, truncated
Diameter (μm)	<10, 10–20	5–30	5–100	4–35
Lipid (%w/w)	0.8	0.7	0.05	0.1
Protein (%w/w)	0.4	0.35	0.06	0.1
Phosphate (%w/w)	0.06	0.02	0.08	0.01
Nature of starch	Very fine	Relatively fine	Coarse	Coarse
Uses of starch	High quality bread, biscuits, cakes	Tortillas, cornbread, industrial starches, ethanolic biofuel	Food, industrial starches, ethanolic biofuel	Food

Data adapted from Bowsher *et al.*, 2008.

Cellulose

Cellulose is a glucose-derived homopolymer that forms the basic matrix of plant cell walls and is therefore a key structural component of all plant tissues. Cellulose-based cell walls are one of the defining features of land plants and algae. In contrast, the cell walls of the true fungi contain chitin as the major structural element. As discussed in Chapters 3 and 6, the fungus-like oomycetes have cellulose-based cell walls and these organisms are now thought to be descended from algae that secondarily acquired several fungal genes by horizontal gene transfer. The fact that oomycetes have plant-like cell walls means that some plants find it difficult to recognize and/or respond to invading oomycete pathogens. The oomycetes include some of the most virulent plant pathogens that are still responsible for substantial crop losses. Their similarity to plant cells also makes it more difficult to devise chemical or biotechnological control methods that do not also affect the host plant.

Cellulose is chemically similar to starch in that both are polymers made up of thousands of glucose units. However, due to the strictly linear arrangement of its glucose residues, cellulose is very different from starch in its physical properties. Unlike starch, cellulose polymer chains are unbranched and are composed of long stretches of glucose residues linked by β-1,4 glycosidic bonds (Fig. 4.3a). Cellulose synthesis occurs at the outer face of the plasma membrane using UDP-glucose supplied by sucrose synthase from the inner face of the membrane. Cellulose synthase is a large multi-subunit enzyme complex that is difficult to study *in vitro*.

Cellulose polymer chains can range in length from about 6,000 to 16,000 glucose units. In higher plants, 30–36 chains self-associate via hydrogen bonding to form a series of highly stable extracellular microfibrils, each up to 5 μm in length. The microfibrils are laid down in parallel to create a series of sheets that are cross-linked by other polysaccharides, most notably hemicelluloses and pectins.

Primary cell walls consist of a strong but flexible series of cellulose microfibrils embedded in a matrix of hemicelluloses and pectins. Secondary cell walls are more densely packed with cellulose and their matrix can be modified, e.g. by deposition of suberin, to make them impermeable to water. In some cells, such as xylem tracheids or vessels, lignin is deposited in the matrix where it becomes highly cross-linked with the microfibrils. This prevents access to these cells and results in their death. The dead lignified cells then create a formidably tough and resilient structure that is the basis of the woody tissues that are one of the keys to the evolutionary success of vascular plants. The strength and resilience of cellulose are matched by its chemical stability and no animals possess enzymes capable of breaking it down to glucose. However, although cellulose cannot be digested directly by animals, many herbivores harbour symbiotic cellulase-producing gut flora, including bacteria and fungi, that can break down cellulose to glucose and other more readily digestible compounds.

Like other animals, humans are unable to digest cellulose directly and, since we do not harbour cellulolytic gut flora, dietary cellulose is ultimately excreted. Despite this, cellulosic foods eaten in moderation are useful dietary components because

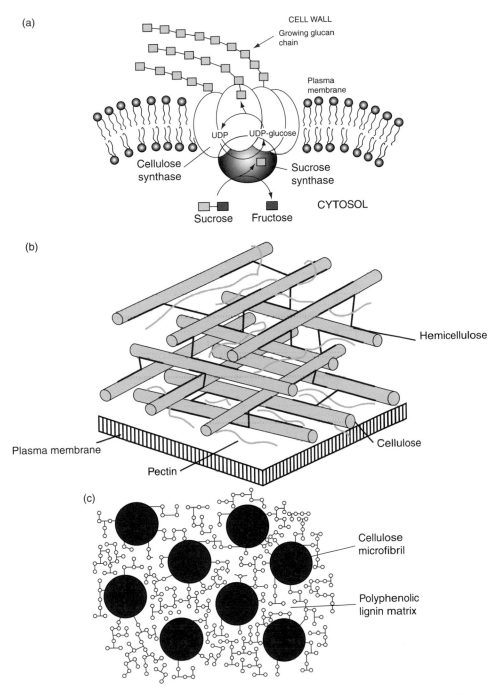

Fig. 4.3. Structure and metabolism of cellulose. (a) Cellulose synthase is a multi-subunit plasma membrane enzyme that works with sucrose synthase to convert glucose monomers into a long linear ß-1,4 linked glycan polymer that is extruded into the extracellular space where it makes up the primary component of the cellulose cell wall. (b) Parallel arrays of cellulose microfibrils form a series of sheets that are stacked on top of each other and are cross-linked by shorter polysaccharides such as hemicelluloses and pectin. (c) Secondarily thickened cell walls contain additional deposits of lignins that form a dense matrix between the cellulose microfibrils.

they form the roughage that provides bulk to foods and helps to maintain regular bowel activity. The domestication of herbivorous livestock, such as sheep and cattle, has enabled people to develop a highly effective method of exploiting the vast amount of potential food resources locked up in cellulosic biomass. Our domesticated animals convert otherwise inaccessible cellulose in plant matter such as grassland pastures into the readily digestible meat and dairy products that are such an important and growing part of the diet of cultures around the world.

Cellulose also has a host of non-food uses, such as in making textile fabrics and paper. Cotton is made from the elongated seed hairs of *Gossypium* spp. and in its dewaxed, non-dyed form it is almost pure cellulose. Linen consists of phloem fibres from the flax or linseed plant, *Linum usitatissimum*, from which the pectins have been removed by microbial action as part of the retting process. Other commercial cellulose-rich fibres include hemp from *Cannabis sativa*, jute from *Corchorus* spp. and coir from *Cocos nucifera*. Hundreds of additional plants are used on a smaller, local scale to produce cellulosic textiles, paper and rope fibres. Some of these fibres have exceptional properties that have the potential to be more widely exploited in the future providing the plants can be domesticated (see Chapter 7).

In addition to cellulose itself, plants synthesize several branched heteropolymeric polysaccharides as accessory cell wall matrix components (Fig. 4.3b). The two main groups are hemicelluloses and pectins. Hemicelluloses such as xlyans, xyloglucans and mannans form relatively short-chain polymers that attach to adjacent cellulose microfibrils via hydrogen bonding, thereby cross-linking the microfibrils and greatly strengthening the structure while maintaining its flexibility. Pectins are acidic polysaccharides containing a high proportion of galactouronic acid residues. Their negative charge, and the resulting presence of cross-linking calcium ions, may be important for the strength and function of the cell wall and its interactions with the adjacent apoplastic intercellular network. Pectin breakdown is one of the most commercially important processes in fruit ripening, which is why pectinases were among the earliest targets for crop genetic engineering, as discussed in Chapter 9. In secondarily thickened cell walls, the cellulose matrix is supplemented by deposits of lignins, which provide additional strengthening and support (Fig. 4.3c).

Other carbohydrates

Fructans are the third most abundant type of storage carbohydrate in plants, after starch and sucrose. These water-soluble polymers are formed by the progressive addition of additional fructose residues to the fructose moiety of a sucrose molecule. Accumulating in the central vacuole, fructans are the major storage carbohydrate in about 15% of angiosperm species, including economically important plants such as cereal crops, forage grasses, vegetables and ornamentals. Fructan accumulation appears to be less sensitive to low temperatures than that of starch, which is useful for early and late season plants. Short-chain fructans have a sweet taste while long-chain versions have a fat-like texture, which has led to their assessment as low-calorie food ingredients and fat substitutes. Although humans cannot digest fructans directly (hence their low-calorie properties), some gut flora can convert fructans into short-chain fatty acids which can be absorbed and may be beneficial.

Trehalose is a disaccharide of glucose that was originally thought to be mainly confined to desiccation-tolerant plants. However, recent genome sequencing studies suggest that most, if not all, higher plants contain a full set of trehalose biosynthesis genes. Trehalose and its metabolites appear to have a variety of specialized functions ranging from osmoprotection during desiccation to signalling in response to biotic and abiotic stresses, as well as in normal plant development. For example, data from transgenic tobacco have shown that trehalose 6-phosphate is part of a signalling pathway regulating photosynthesis and leaf development. Other studies have shown that increased levels of trehalose can protect relatively labile proteins, such as some of the subunits of photosystem II, from photo-oxidation.

4.3 Lipids

In its broadest sense, the term 'lipid' encompasses a wide range of water-insoluble, hydrophobic or amphipathic compounds. The most abundant plant lipids are acyl (fatty acid containing) molecules such as phospholipids, glycolipids, triacylglycerols and oxylipins. Other important lipidic compounds include sterols, carotenoids, tocopherols, phytols and waxes. This section will focus primarily on the major acyl lipids shown in Fig. 4.4. Acyl lipids include the major glycolipids and phospholipids

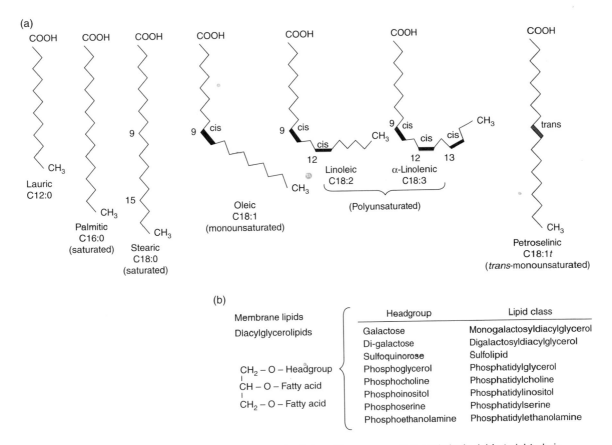

Fig. 4.4. Major classes of plant acyl lipids. (a) The major fatty acid groups are (i) the relatively rigid straight-chain saturates such as lauric (C12:0), palmitic (C16:0) and stearic (C18:0); (ii) the more flexible kinked-chain monounsaturates such as oleic (C18:1c); and (iii) the highly flexible kinked-chain polyunsaturates such as linoleic (C18:2c) and α-linolenic (C18:3c). Most naturally occurring unsaturated fatty acids are kinked and flexible due to their *cis* double bond conformation. In contrast, *trans* fatty acids, such as petroselenic acid (C18:1t), have straight chains and resemble saturated fatty acids in their physical properties. (b) Membrane lipids consist mostly of diacylglycerolipids made up of two fatty acids esterified to a glycerol backbone and a distinctive phospho- or glyco-ester headgroup.

that make up the bilayer matrix of all biological membranes. But the most important plant lipids from the human standpoint are the storage tri-acylglycerols of seeds and fruits. Other plant acyl lipids include oxylipins such as jasmonic acid and the many lipid hydroperoxides that participate in a variety of signalling and defence responses.

Acyl lipids are the second most important source of dietary calories for humans (after starch) and for most of the world's population plant storage triacylglycerols are their principal sources of such lipids. Acyl lipids are also sources of the essential fatty acids, linoleic and α-linolenic. These vitamin-like compounds cannot be synthesized by

humans and are mostly derived from plant oils. In plants, lipids have quite distinct membrane and storage functions. Due to the physical constraints of inserting into a lipid bilayer, membrane lipids have relatively restricted fatty acid compositions and are overwhelmingly made up of unsaturated C16 and C18 species. In contrast, storage lipids are contained in highly flexible droplets with few con-straints on their size and composition. Therefore storage lipids have a vast diversity of acyl chain lengths and acyl modifications such as hydroxyla-tion, epoxidation, conjugation and triple-bond formation. The useful properties of many of these fatty acids means that storage lipids have been a primary target for the transgenic manipulation of plants.

In addition to their triacylglycerol content, many unrefined plant oils contain other lipidic nutrients such as the antioxidant vitamins, A and E. Many unrefined, or 'virgin', plant oils also contain volatile fatty acid peroxides and hydroperoxides formed by partial oxidation of unsaturated fatty acids. These lipidic volatiles contribute significantly to the organoleptic qualities of products such as unrefined olive and cotton oils. Because many traditional food staples are deficient in these vitamins, several biotechnological approaches are being taken to increase their levels in major crops such as rice, maize and sweet potato (see Chapter 12).

Fatty acid biosynthesis and modification

De novo fatty acid biosynthesis in plant cells occurs almost exclusively in plastids. Indeed, in non-photosynthetic cells, fatty acid biosynthesis may be one of the primary functions of plastids. The first committed step is the carboxylation of acetyl-CoA to malonyl-CoA (see Fig. 4.5). The pathway for acetyl-CoA formation is different in photosynthetic and non-photosynthetic tissues. In green photosynthesizing tissues, such as leaves, CO_2 is fixed into triosephosphates by Rubisco in the chloroplast stroma. Triosephosphates are then converted to pyruvate, and hence to acetyl-CoA, by glycolytic enzymes. In largely non-photosynthetic tissues, such as developing fruits and seeds, the carbon source for fatty acid biosynthesis is sucrose imported from photosynthetic organs. Sucrose cannot enter plastids and is therefore converted to hexosephosphates, such as glucose 6-phosphate, in the cytosol. In most developing seeds, cytosolic hexosephosphates are converted into intermediates, such as phosphoenolpyruvate or malate, which are transported into the plastids via specific carriers to serve as fatty acid precursors via acetyl-CoA.

Acetyl-CoA carboxylase, which converts acetyl-CoA to malonyl-CoA, is a highly regulated, multi-subunit enzyme that is one of the major rate-limiting steps in fatty acid biosynthesis. Evidence from flux control analysis suggests that increasing the activity of this enzyme in developing oilseeds should increase the flow of carbon towards fatty acid formation, possibly resulting in higher oil yields. This has stimulated interest in the regulation of acetyl-CoA carboxylase as well as efforts to increase its activity by overexpressing genes encoding its various subunits. The C3 compound malonyl-CoA is the building block for the assembly of fatty acids on the multi-enzyme fatty acid synthetase complex. This prokaryotic type II fatty acid synthetase is derived from the original cyanobacterial endosymbiont, and consists of dissociable subunits, each catalysing a separate reaction. In contrast, animals have a type I fatty acid synthetase made up of a single large multifunctional polypeptide.

Fatty acid biosynthesis is initiated by the addition of a C3 malonyl-CoA unit to the synthetase complex, followed by a decarboxylation to generate a C2 acyl group. Addition of a further malonyl-CoA unit followed by its decarboxylation generates a C4 acyl group. The cycle of progressive acyl chain extension by C2 units continues until its length reaches C16 or C18 at which point the acyl group dissociates from the complex. In most of the major oil crops, both membrane and storage lipids are overwhelmingly made up of C16 and C18 acyl groups. One of the key discoveries that made possible the transgenic manipulation of fatty acid profiles in plant oils is that chain elongation can be interrupted to release intermediate length acyl groups, i.e. C4 to C14. This requires the action of one of several specific thioesterases found in the small number of plants that accumulate shorter-chain fatty acids in their seed oils. Expression of these thioesterase genes in transgenic oil crops leads to premature acyl chain termination and the accumulation of short-chain fatty acids in their seeds. In contrast, some plants are able to elongate storage lipid acyl groups beyond C18 up to C24 by means of additional fatty acid elongases. The expression of exogenous elongase genes in crop plants may increase the accumulation of very long-chain fatty acids in seeds.

While chain length is one of the key attributes that determine the properties of a fatty acid, the other attribute is the way in which the saturated hydrocarbon chain can be modified after *de novo* synthesis. This is often referred to as the *functionality* of the acyl chain. The most common functionality introduced into an acyl chain is unsaturation, namely presence of one or more double bonds. As listed in Table 4.4, other fatty acid functionalities include triple bonds, and hydroxy, epoxy and oxo groups. Most of the enzymes responsible for introducing such functionalities into fatty acid are

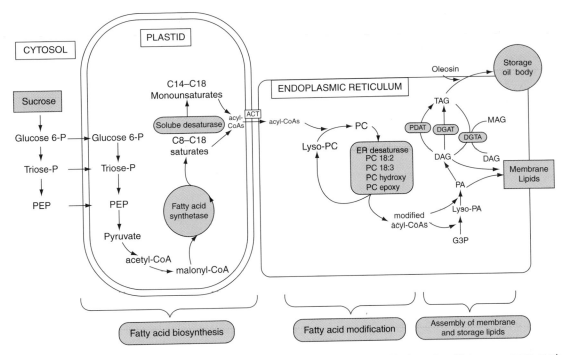

Fig. 4.5. *Acyl lipid biosynthesis and fatty acid biosynthesis*: sucrose is converted in the cytosol into precursors, such as glucose 6-phosphate and phosphoenolpyruvate, for onward transport into plastids for production of fatty acids. Acetyl-CoA and malonyl-CoA are the precursors for assembly of C8 to C18 saturated fatty acyl-ACPs on a plastidial fatty acid synthetase complex. Plastids are also the site of the insertion of the first double bond by a soluble desaturase to form monounsaturates. Fatty acids are exported via an acyl-CoA transporter (*ACT*) from plastids to the endoplasmic reticulum for further processing. *Fatty acid modification*: plastid-derived acyl-CoAs can be modified in the endoplasmic reticulum by a variety of enzymes to produce some of the hundreds of different fatty acids found in naturally occurring seed oils. Most fatty acid modification reactions occur via membrane-bound phosphatidylcholine (PC)-specific ER desaturases, or desaturase-like enzymes such as hydroxylases or epoxidases. Acyl-CoAs are then assembled into complex lipids on the endoplasmic reticulum. Whereas storage oil bodies can accumulate virtually any type of fatty acid, the biological functions of membrane and signalling lipids require that they only contain a small range of C16 and C18 fatty acids. One of the challenges to producing oilseeds with novel acyl compositions is therefore to maintain the segregation of exotic fatty acids away from pools of membrane or signalling lipids. *Triacylglycerol assembly*: triacylglycerols are assembled via sequential acylation of glycerol plus extensive 'acyl editing' via PC-dependent desaturases or desaturase-like enzymes. The final conversion of diacylglycerol (DAG) into triacylglycerol (TAG) can occur via at least three enzymes: *DGAT*, acyl-CoA-dependent diacylglycerol acyltransferase; *PDAT*, phosphatidylcholine-dependent acyltransferase; or *DGTA*, diacylglyceroltransacylase. Nascent TAG droplets are coated with a phospholipid monolayer into which is embedded an annulus of specific proteins, such as oleosins and caleosins, to form mature storage oil bodies that are released into the cytosol. G3P, glycerol 3-phosphate; MAG, monoacylglycerol; PA, phosphatidic acid.

derived from mutated versions of fatty acid desaturases. The resolution of the crystal structure of several fatty acid desaturases and the sequencing of genes encoding many modified desaturases has led to the engineering of novel versions of these enzymes that may be able to carry out industrially useful reactions.

Plastids are the site of the initial desaturation reactions that convert saturated fatty acids to monounsaturates. For all subsequent long-chain fatty acid modification reactions, saturated and monounsaturated fatty acyl-CoAs synthesized *de novo* in plastids are exported to the endoplasmic reticulum via specific carriers on the plastid envelope. Such modification reactions include acyl-chain elongation up to C24, as well as numerous additional desaturase and desaturase-like reactions to form epoxy and hydroxy fatty acids, plus conjugated

Table 4.4. Diversity of fatty acyl composition of selected plant storage oils.

Chain length/functionality[a]	Common name	FA in oil (%)	Plant species	Uses
8:0	Caprylic	94	*Cuphea avigera*	Fuel, foods
10:0	Capric	95	*Cuphea koehneana*	Detergents, foods
12:0	Lauric	94	Betel nut laurel[b]	Detergents, foods
14:0	Myristic	92	*Knema globularia*	Soaps, cosmetics
16:0	Palmitic	75	Chinese tallow	Foods, soaps
18:0	Stearic	65	Kokum	Foods, confectionary
16:1 9c	Palmitoleic	40	Sea buckthorn	Cosmetics
18:1 9c	Oleic	78	Olive[b]	Foods, lubricants, inks
18:1 6c	Petroselinic	76	Coriander[b]	Nylons, detergents
18:1 9,10me, 9c Cyclopropene	Sterculic	50	*Sterculia foetida*	Insecticides, herbicides
18:2 9c, 12c	Linoleic	75	Sunflower[b]	Foods, coatings
18:3 9c, 12c, 15c	α-Linolenic	60	Linseed[b]	Paints, varnishes
18:3 6c, 9c, 12c	γ-Linolenic	25	Borage[b]	Therapeutic products
18:1 9c, 12OH	Ricinoleic	90	Castor[b]	Plasticizers, lubricants
18:1 9c, 12epx	Vernolic	70	Ironweed	Resins, coatings
18:2 9c, 12trp	Crepenynic	70	*Crepis alpina*[b]	Coatings, lubricants
18:3 4-oxo, 9c, 11t, 13t	Licanic	78	Oiticica	Paints, inks
18:3 8t, 10t, 12c	Calendic	60	Calendula	Paints, coatings
18:3 9c, 11t, 13t	α-Eleostearic	70	Tung	Enamels, varnishes, resins, coatings
18:3 9c, 11t, 13c	Punicic	70	Pomegranate	Varnishes, resins, coatings
20:0	Arachidic	35	Rambutan[b]	Lubricants
20:1 11c	Eicosenoic	67	Meadowfoam[b]	Polymers, cosmetics
20:1 11c, 14OH	Lesquerolic	70	Lesquerella	Lubricating greases, resins, waxes, nylon, coatings, cosmetics
22:0	Behenic	48	Asian mustard	Lubricants
22:1 13c	Erucic	56	Crambe[b]	Polymers, inks
20:1/22:1	Jojoba wax	95	Jojoba[b]	Cosmetics, lubricants, wax
24:1 15c	Nervonic	24	Honesty[b]	Pharmaceuticals

[a]c, *cis* double bond; t, *trans* double bond; epx, epoxy group; trp, triple bond; me, methylene.
[b]Indicates that genes have been isolated for synthesis of these fatty acids.

and non-conjugated polyunsaturated fatty acids, and very long-chain wax esters. In the case of membrane lipids, where the major fatty acids are polyunsaturated C18 species, these fatty acids are exported from the endoplasmic reticulum to the other membrane systems of the plant cell, such as the organelles and plasmalemma.

In photosynthetic tissues, by far the most abundant membrane system is the chloroplast thylakoids, which can make up over 80% of the acyl pool of a leaf cell. The most abundant membrane lipids are mono- or di-galactolipids. These are synthesized from polyunsaturated C18 fatty acids originating from the endoplasmic reticulum, which are transferred to a glycerol backbone to form

diacylglycerol. On the chloroplast envelope, one or two galactose groups are added to diacylglycerol to form the galactolipid molecule that is inserted into the thylakoid membrane. Galactolipids contain almost entirely C18:2 and C18:3 acyl groups and this unusual composition is required for the efficient functioning of the thylakoid membranes that are the sites of the photosynthetic light reactions and ATP formation.

Storage lipids

Plant storage lipids are important energy reserves in many seeds and pollen grains and are a major component of the human diet. They are also sources of thousands of valuable industrial products. At present, only about 20% of globally traded plant oils have non-food uses. In the future, as non-renewable petroleum feedstocks become scarcer and more expensive, renewable, carbon-neutral oleochemicals derived from plant lipids will become increasing important. Eventually, once the non-renewable fossil-based hydrocarbons have run out, many familiar petroleum-based products such as plastics, nylons, paints, lubricants and fine chemicals will have to be manufactured from plant oils. Recently, plant oils have been increasingly exploited as sources of renewable fuels. While some plant oils can be used directly to power engines, it is usually more efficient to convert their constituent fatty acids into methyl ester derivatives, which can be used either on their own as a high quality diesel fuel or as a blend with regular diesel (see Chapter 12). Growing demands for different oils for a range of food, oleochemical and biofuel applications have created huge interest in the biotechnological manipulation of acyl compositions of plant oils.

Storage lipids in plants typically accumulate in the cytosol as fluid triacylglycerol or wax ester droplets termed oil bodies. Each oil body is surrounded by a phospholipid monolayer, and often by an additional annulus of specific lipid-binding proteins such as oleosins and caleosins. These complex macromolecular structures are assembled on the endoplasmic reticulum. Within the endoplasmic reticulum bilayer, acyl-CoAs produced by fatty acid modification enzymes as described above are transferred to a glycerol backbone, resulting in the formation of mono-, di- and triacylglycerols. These acylation reactions are catalysed by distinct acyltransferases (see Fig. 4.5). The differing specificities of the acyltransferases for different acyl-CoAs mean that these enzymes play an important role in regulating the final acyl composition of storage lipids. The mechanism of storage lipid assembly in plants is similar in many respects to that in animals and microbes. It has recently been shown that storage lipids in all organisms, including plants, may not always be the inert end products of metabolism as was once thought. For example, triacylglycerol-rich oil bodies in leaves and meristematic tissues have additional roles in short-term metabolic processes such as lipid trafficking and inter- and intracellular signalling.

Most cytosolic oil bodies in plant storage organs are about 1–2 μm in diameter, although they can vary from less than 0.5 μm to well over 30 μm.

Small oil bodies of 0.5–2.0 μm tend to be found in seeds that undergo desiccation as a normal part of their development. This includes the vast majority of temperate species. Such oil bodies are invariably surrounded by a continuous protein layer, principally oleosins. These relatively small amphipathic proteins (15–24 kDa) are uniquely associated with plant oil bodies. The ratio of oleosin to triacylglycerol determines the final size of the storage oil body. Oleosins appear to play a critical role in maintaining oil-body stability during the extreme hydration stresses associated with seed imbibition. There are also indications that seed viability during storage may be related to the stability of oil bodies. This may have implications for the establishment of seed banks as germplasm repositories, especially in the case of the many tropical species that contain more labile oleosin-poor oil bodies.

Oil-rich fruits, such as olive, palm or avocado, accumulate relatively large, triacylglycerol-enriched but protein-depleted oil bodies in their fleshy mesocarp. These are not true storage lipids but as the fruits ripen some of the triacylglycerols break down to release attractive odours that serve as a bait to attract fruit-eating animals. The seeds within such fruits normally pass intact through the digestive system of the animal, which thereby acts as a vector for seed dispersal. Oil-rich fruits are now some of the major sources of plant lipids for food, oleochemicals and biofuels. The many different uses of plant lipids coupled with their great diversity have made them targets for biotechnological manipulation, discussed in Box 4.2.

4.4 Nitrogen Metabolism

Unlike animals, plants must synthesize all their nitrogen-containing organic compounds, such as proteins and nucleic acids, *de novo* from inorganic nitrogen. The availability of inorganic nitrogen is frequently rate-limiting for plant growth, which is why fertilizers are so important in optimizing crop yields in agriculture. Plants are unable to assimilate nitrogen directly from organic fertilizers so the latter must first be broken down in the soil to inorganic nitrogen compounds by bacteria and fungi before they can be assimilated. The main groups of nitrogen compounds taken up by plants are nitrates, nitrites and ammonium. In addition, nitrogen gas from the atmosphere can be fixed by leguminous plants thanks to the presence of symbiotic nitrogen-fixing bacteria such as *Rhizobia*.

Box 4.2. Biotechnology targets for acyl lipids

Modified fatty acid composition: Storage lipids were one of the first targets of plant biotechnologists seeking to modify crop composition. The diversity of acyl chain composition in seed oils led to the idea of 'designer oil crops' whereby a few genes could be transferred from a donor plant to a major commercial oilseed crop in order to create valuable new varieties. Major targets were acyl chain length and functionality, especially unsaturation. In the early 1990s the first commercially approved transgenic crop with a modified seed composition was a rapeseed variety called Laurical™. This crop variety contained a thioesterase gene derived from the California Bay plant that resulted in the accumulation of the medium-chain lauric acid in its seed oil. As discussed in Chapter 9, early achievements during the 1990s in producing high-lauric and high-stearic rapeseed were not followed by commercial success. During the 2010s, a new generation of nutritionally enhanced transgenic varieties, such as high-oleic and very long-chain ω-3 polyunsaturates, looks more promising.

Increased oil content: Due to the high value of many plant oils, even a modest increase in crop oil yield is a very desirable target. The best commercial oilseed crops yield about 40–45% of their seed weight as oil, with most of the remainder consisting of lower value protein and fibre. In contrast, some nuts can accumulate >60% of their weight as oil. This has led to efforts to divert more carbon towards oil accumulation during seed development. Conventional breeding approaches have had limited success but there are indications that transgenic strategies might be more fruitful. For example, in transgenic soybean, the upregulation of the final, and apparently rate-limiting, enzyme of triacylglycerol biosynthesis, diacylglycerol acyltransferase, via insertion of extra gene copies from the fungus *Umbelopsis ramanniana* (formerly *Morteriella*), resulted in significant increases in oil content without apparent adverse effects on plant performance. Other studies have also shown that storage oils can be accumulated ectopically in tissues such as leaves. For example, ectopic expression of the transcription factor gene WRINKLED1 resulted in the accumulation of almost sixfold more oil in leaves of *Arabidopsis* compared with wild-type plants. Providing they still photosynthesize normally, oil-rich leaves in crops could become a new source of edible and industrial products or even biofuels.

The majority of plants also have symbiotic associations with extensive networks of root-associated actinomycete mycorrhizal fungi. These symbioses can greatly increase the efficiency of inorganic nitrogen assimilation, even in relatively poor soils.

The key role of nitrogen compounds in plant growth, and especially in determining crop yield, makes the manipulation of nitrogen availability and uptake/metabolism among the most important targets for plant biotechnology. Indeed, the initial success of the Green Revolution in the 1960s (see Chapter 9) was largely due to the selection of a semi-dwarf wheat mutant that was able to increase its grain yield in response to nitrogen fertilizer without excessive height gain.

Nitrate assimilation

The most abundant form of inorganic nitrogen in well-aerated neutral or mildly alkaline soil is nitrate. This is because soil nitrates are produced by aerobic bacteria that require a neutral or alkaline pH. Poorer soils that are anaerobic and/or acidic, such as peat bogs, tend to be high in ammonium, which is a much less accessible source of nitrogen for most plants. Soil nitrate is taken up via specific transporters on the plasma membrane of root epidermal and cortical cells. Higher plants have numerous forms of nitrate carrier, which enables them to adapt to a wide range of soil nitrate concentrations without suffering deficiency or toxicity effects. Absorbed nitrate can be reduced to ammonium in the roots or transported elsewhere in the plant for subsequent storage or reduction.

As shown in Fig. 4.6, nitrate reduction is a two-stage process that yields ammonium as its end product. The first step is catalysed by nitrate reductase, a large, complex homodimer that is highly regulated at both the enzyme and gene levels. The enzyme activity is determined by its phosphorylation status, which is itself affected by factors such as calcium, O_2 and CO_2 concentration, photosynthetic rate and okadaic acid. Transcription of the nitrate reductase gene is regulated by factors such as light, circadian rhythms, sucrose, cytokinin and glutamine. The high degree of control exerted over nitrate reductase is necessary because its product, nitrite, is potentially toxic and is normally kept

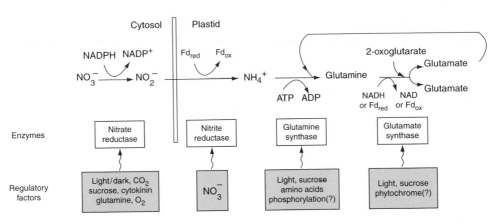

Fig. 4.6. Nitrate assimilation. Nitrate from the soil is efficiently taken up by plasma membrane carriers in root cells and reduced to nitrite by a cytosolic nitrate reductase. Further reduction of nitrite to ammonium occurs via a plastidial nitrite reductase. Ammonium is converted to organic nitrogen via glutamine synthase and glutamate synthase, which yield the amino acid glutamate from which other amino acids can then be derived via transamination.

below 15 nmol/g fresh weight in leaves. The second reduction step is catalysed by nitrite reductase, a ferredoxin-dependent enzyme that reduces nitrite ions to ammonium (NH_4^+) ions.

Inorganic ammonium is converted into amino acids by the enzymes glutamine synthase and glutamate synthase, which effectively add an amino group to 2-oxoglutarate to produce glutamate. Glutamine synthase is an octomeric homodimer with cytosolic and plastidial isoforms that is regulated by a complex set of internal and external factors including light, nutrient availability, metabolites and possibly by phosphorylation. There is evidence of substantial post-transcriptional regulation, as demonstrated by a failure to detect increased enzyme amounts or activity even after large increases in mRNA levels in transgenic plants. Glutamate synthase is generally found as two plastidial monomeric isoforms that respectively use ferredoxin and NADH as reductants. While not as highly regulated as glutamine synthase, its gene expression and enzyme activity are still subject to a wide range of effectors including light and sucrose. Other amino acids, such as aspartate and asparagine, can be formed from glutamate via aminotransferase reactions.

Nitrogen fixation

Some plants, most notably the legumes, can fix atmospheric nitrogen into nitrogenous compounds thanks to symbiotic associations with various bacteria.

Over 10,000 species of legume are able to obtain ammonium from gram-negative bacteria of the Rhizobiaceae. In exchange, the plants house the bacteria and provide them with carbon compounds. Another important symbiosis is that between the water fern *Azolla* and the nitrogen-fixing cyanobacterium *Anabaena*, which supplies much-needed nitrogen to rice in paddy fields. *Rhizobia* bacteria can live independently in the soil and only fix nitrogen when they become symbionts within a legume. Nitrogen-fixing plants actively recruit bacterial symbionts when they grow in soils that are deficient in nitrogen. A potential host plant excretes specific flavonoids that induce formation of lipo-oligosaccharide nodulation factors by soil-resident *Rhizobia*. These nodulation factors cause the plant to take up *Rhizobia* via an infection thread and eventually to form the specialized root nodules that house the bacteria.

Recognizable legumes first appeared about 60 million years ago and had acquired the capacity for nodulation by 58 million years ago. The first type of infection by bacteria may have been at junctions where lateral roots emerged. Later, the plants developed infection threads to confine bacteria and transport them to cells in specialized nodules. Infection threads were a prerequisite for plants to direct bacteria to infect them via their root hairs. In this way, the host plant could better regulate the infection and restrict it to more benign bacterial species. In an alternative mechanism, dating from the same time and persisting in about 25% of legumes, there are no infection threads but bacteria enter a few host cells where they are surrounded by an undefined matrix.

As shown in Fig. 4.7, nitrogen fixation is catalysed by the nitrogenase complex, which is unique to prokaryotes. Nitrogenases from Archaea and Eubacteria are highly conserved, indicating a single ancient origin. The complex consists of two enzymes: an iron-binding dinitrogenase reductase and a molybdenum-iron-binding dinitrogenase. Despite decades of study, many aspects of the chemistry of nitrogenase reactions remain unclear. This applies particularly to the mechanism of dinitrogen (N_2) binding to the intricate iron-molybdenum-sulfur cofactor and its subsequent reduction to ammonia. As with Rubisco (see Chapter 2),

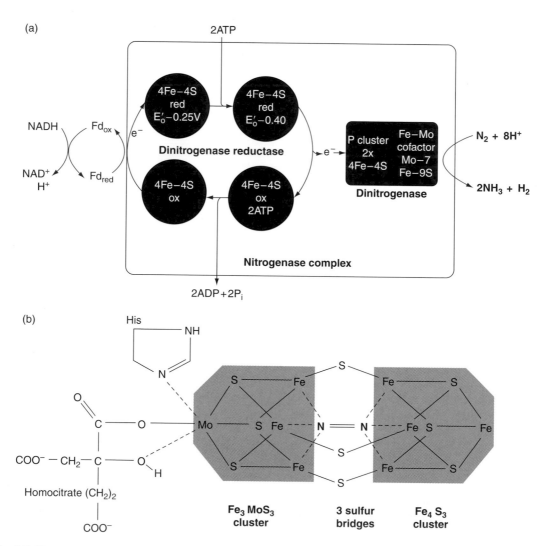

Fig. 4.7. Nitrogen fixation. (a) The nitrogen fixation complex is a highly conserved prokaryotic protein consisting of an iron–sulfur-binding dinitrogenase reductase and a molybdenum–iron-binding dinitrogenase that together convert nitrogen gas to ammonia in a strictly anaerobic reaction that is relatively inefficient and highly energy demanding. (b) The active site of the dinitrogenase responsible for nitrogen fixation is made up of three sulfur bridges formed by a molybdenum–iron–sulfur cluster and an iron–sulfur cluster. The molybdenum atom is covalently linked to homocitrate and also stabilized by a histidine residue. Uptake of electrons from reduced ferredoxin facilitates binding of a nitrogen molecule and its reduction to yield two molecules of ammonia.

nitrogenase appears to be an ancient but relatively inefficient and non-specific enzyme that catalyses a key biological process. In the case of nitrogenase, 30–60% of the energy supplied can be lost in a wasteful side reaction involving the conversion of protons to H_2. The likely anaerobic origins of nitrogenase are demonstrated by its extreme sensitivity to molecular oxygen, resulting in rapid and irreversible inactivation.

For this reason, the nodules of legumes are maintained in a virtually anaerobic state. The oxygen concentration of nodule bacterioids is about 10,000-fold lower than adjacent plant tissues. In addition to drastically limiting oxygen diffusion into nodules (by an as-yet unknown process), legumes can export oxygen via the protein leghaemoglobin. Oxygen concentration within a nodule is one of the key factors regulating nitrogenase gene expression. In some rhizobial species, there is a cascade involving multiple protein phosphorylations and the activation of transcription factors leading to expression of the entire set of nitrogen-fixation genes. Host plants supply assimilated carbon to bacteria in the form of dicarboxylic acids, most commonly as malate formed from sucrose and delivered via the phloem. In return, ammonium produced by bacterial nitrogenase is imported into adjacent plant cells via a specific plasma membrane carrier. The imported ammonium is converted into amino acids by glutamine synthase and glutamate synthase as described above.

Nitrogen fixation is very energy expensive for a plant, consuming up to one-fifth of its total photosynthetic production. The legume–Rhizobium symbiosis is also a highly developed association that involves large numbers of genes in each partner. To date more than 30 bacterial and 60 plant genes have been identified, most of them subject to multiple levels of regulation. Each symbiotic partner can control aspects of genetic activity in the other, for example by producing chemical signals that induce expression of particular genes in its plant or bacterial partner when required. For these reasons the highly desirable goal of engineering nitrogen fixation into cereal crops will probably remain a long-term aspiration for the foreseeable future.

Mycorrhizal nitrogen assimilation

It is estimated that the majority of plants can form symbiotic associations with soil fungi. In most perennial species in 'natural' ecosystems, symbiotic mycorrhizae provide a significant proportion of the nitrogen assimilated by the plants. Some plant–mycorrhizal associations are species-specific while others tend to be more general. In some cases, removal of a plant to a location where its favoured mycorrhizal partner is absent can severely limit its growth. This is especially common with coniferous trees, where presence of suitable mycorrhizae can double the uptake of nitrogen compared with non-mycorrhizal soil. In commercial plantations of pine or spruce, fungal inoculants or 'nurse' trees such as larch are often used to help establish an active plant–mycorrhizal system and thereby increase plant productivity.

In return for nutrients such as nitrogen and phosphorus, mycorrhizal fungi receive carbon assimilates from their host plant. The fungus greatly increases the volume of soil that can be exploited for nutrient extraction and can utilize forms of organic nitrogen that are unavailable to plants. A common source of otherwise inaccessible soil nitrogen is protein from dead invertebrates that fungi can digest to amino acids via secreted proteases. Fungi can also efficiently assimilate inorganic nitrogen as nitrate or ammonium, although not all species respond to fertilizer supplementation. Assimilated nitrogen is transferred from fungus to plant via various complex systems that can differ substantially between species, which makes it difficult to envisage how one might engineer this useful trait into a crop plant in order to increase output.

Storage proteins

Plant storage proteins are major components of the human diet and of livestock feed. In grain legumes such as peas, beans and lentils, storage proteins can make up over 25% of the dry weight of a mature seed. Significant amounts of protein are also present in many other types of seed, including nuts and oilseeds, while much smaller amounts are found in root or tuber storage organs. Storage proteins accumulate either in the lumen of the endoplasmic reticulum or in the central vacuole of endosperm or cotyledon cells during seed development. Storage proteins do not normally have any enzymatic activity although they are probably originally derived from enzymes such as proteases and other hydrolases (see Fig. 4.8). These former enzymes have now become adapted to serve as nitrogen storage compounds that can be packaged

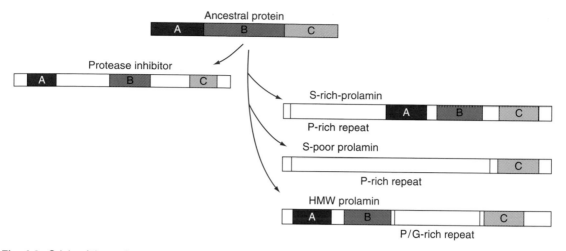

Fig. 4.8. Origin of the major storage proteins in cereals. The major group of cereal storage proteins, the prolamins, shares a common ancestor with protease and α-amylase inhibitors that contained three domains: A, B and C. It is likely that these three domains originated from a single gene that was triplicated followed by divergence to yield the three distinct protein-encoding regions found in modern proteases. In the various classes of cereal prolamins, the original domains A, B and C have become truncated or partially lost and are now fused to proline-rich repeats to form a functional storage protein rather than an active protease enzyme.

efficiently into dense paracrystalline arrays. They tend to be enriched in the di-amino acids, glutamine and asparagine, as a way of maximizing their nitrogen content. Some storage proteins are deficient in sulfur-containing amino acids, such as lysine and methionine, that are essential components of the human diet. Increasing the sulfur amino acid content is one of several biotechnology targets for storage proteins.

The major classes of plant storage proteins are shown in Table 4.5. Vascular plants have a rather restricted range of storage proteins, indicating that they may have originated early in plant evolution. It is likely that gene duplication rendered some hydrolase genes redundant, enabling them to evolve into nitrogen stores rather than active enzymes. As with storage lipids, there are fewer constraints on the detailed structure of storage proteins, as long as they act as efficient nitrogen stores that can be readily mobilized after germination. The two main storage protein groups are the globulins and prolamins, which are present in different amounts in the major crops. Hence, barley, wheat and maize grains contain mainly prolamins, while the majority of legumes, oats and rice contain mainly globulins. Non-seed storage organs such as roots and tubers accumulate a diverse range of less specialized storage proteins, some of which still have enzymatic activity.

One of the most widespread storage protein families is the globulins. The biosynthesis and processing of the major 11S globulin of legume seeds is shown in Fig. 4.9. In most legumes and cereals, storage proteins are initially synthesized as full-length precursors on the endoplasmic reticulum. They are then processed through the normal eukaryotic secretory pathway to the Golgi and transported to a specialized organelle called a protein storage vacuole. These storage vacuoles are distinct from the acidic lytic vacuoles that often take up much of the cell volume in plants. In addition to storage proteins, many seeds also accumulate smaller quantities of defensive proteins, such as chitinases, hydrolases and various toxic proteins within their storage vacuoles in order to deter seed-eating animals. One of the most toxic substances known to humans is the lectin, ricin, which accumulates in the endosperm of castor bean seeds.

4.5 Secondary Compounds

Whereas storage proteins, lipids and carbohydrates in plants are used mainly (but not exclusively) for edible purposes, secondary compounds also have many non-edible uses. Perhaps the most striking feature of plant secondary metabolites is their huge diversity. There are three broad categories of secondary compounds in plants, namely terpenes

Table 4.5. Classes of major plant storage proteins in important crop groups.

Protein class	2S Albumins	7–8S Globulins	11–12S Globulins	Prolamins
Crop groups	Legumes Crucifers Compositae Castor bean Cottonseed Brazil nut	Legumes Palms Cocoa Cottonseed Cereals (minor)	Legumes Compositae Rice Oats Crucifers Cucurbitaceae Brazil nut	Cereals

and terpenoids (>25,000 types), alkaloids (>12,000 types) and phenolics (>10,000 types). These compounds have a host of vital functions in plant growth, development and environmental responses as well as supplying useful products ranging from pharmaceuticals to industrial materials.

One of the major functions of plant secondary compounds is in defence. Plants constantly produce new combinations of defence-related compounds as part of their coevolution with the huge diversity of heterotrophic threats that they face, ranging from large herbivores to pests and diseases. Unlike motile animals, sessile plants are unable to move away from such threats and are forced to 'stand and fight'. The extreme genetic diversity provided by interspecific hybridization and the high incidence of polyploidy in higher plants underpins their huge capacity for biochemical experimentation, which enables plants to synthesize many thousands of secondary compounds. The vast majority of these compounds might have little or no biological function at a given time, but chemical diversity is essential in the face of existing and future threats.

The presence of particular secondary compounds in a plant often determines whether or not it is edible to humans. Throughout our history, we have sought out palatable and nutritious food plants and it is likely that hunter-gatherers routinely ate hundreds of different plants. However, with the development of farming, the potential list of food crops became much smaller, and some potentially nutritious crops still contained dangerous levels of toxic secondary compounds. For example, red kidney beans are highly nutritious legumes but the dry beans contain poisonous lectins and cyanogenic glycosides. These beans must be carefully prepared by prolonged soaking to leach out the toxins, followed by cooking to inactivate residual toxins. Dozens of poisoning episodes still occur each year due to inadequate removal of toxins from raw plant foods. An obvious aim of plant biotechnologists is to reduce or remove such toxins from food crops, although this has to be done without harming the ability of the crop to protect itself from pests and diseases while it is growing in the field.

Phenolics and lignins

Although they all contain one or more aromatic phenyl rings, phenolic compounds are otherwise extremely diverse in their structures and properties. The main classes of phenolics include phenylpropanoids and flavonoids, plus more complex groups such as lignins, tannins and stilbenes (Fig. 4.10). Phenolics have numerous functions including structural, defence, signalling, flavour and UV protection. Their roles in lignin and flower colour are described below but phenolics have many additional functions related to plant interactions with herbivores, pests and diseases. For example, vanillin is an attractive scent that aids pollen and fruit dispersal by animal vectors. Other scent or flavour phenolics, such as gingerols, capsaicin and tannins, are more astringent and can also act as feeding deterrents or toxins.

Despite their roles as dietary deterrents to many herbivores, some phenolics have become important components of the human diet. A typical daily dietary intake of phenolic compounds by humans ranges from 20 mg to 1 g and is derived from sources such as fruits, vegetables, cereals, herbs, spices and beverages. The rather astringent taste of phenolics such as capsaicin and tannins is part of their appeal as flavouring agents in foods and beverages. Such phenolics can also act as antioxidants, which are sometimes used in the promotion of phenolics-enriched foods as part of a 'healthy' diet. The antioxidant function of phenolics is of interest because most vitamin antioxidants are lipidic and

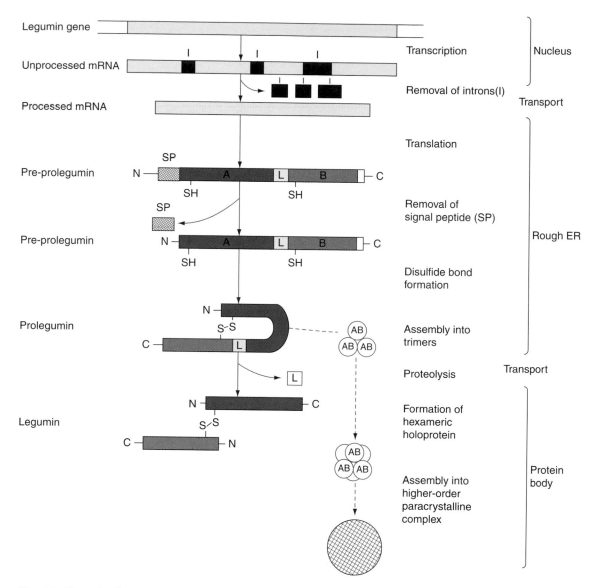

Fig. 4.9. Biosynthesis, processing and targeting of the 11S storage protein of pea seeds. Globulins are major storage proteins in several food crops, including legumes. Following intron removal, the 11S globulin from pea seeds (legumin) is translated as pre-prolegumin, which is targeted to the rough ER where a signal peptide is removed. Formation of a disulfide bond between two conserved cysteine residues results in the hairpin structure of prolegumin, which assembles into trimers. After transport from the ER to the protein storage body, prolegumin is cleaved at an asparagine residue by a protease to generate two polypeptide subunits linked by a disulfide bond. This new conformation enables the protein to form hexamers that can be efficiently packed into the dense paracrystalline complexes found in mature protein storage bodies. Adapted from Jones *et al.* (2000).

are therefore mainly found in specialized lipid-rich structures such as oilseeds, which reduces their dietary availability. Because they are water soluble, phenolic antioxidants are present in a wider range of plant tissues and are more easily incorporated into the diet. For example, palm oil is enriched in the lipidic antioxidants, tocopherols, tocotrienols and carotenoids but palm fruits are also enriched in

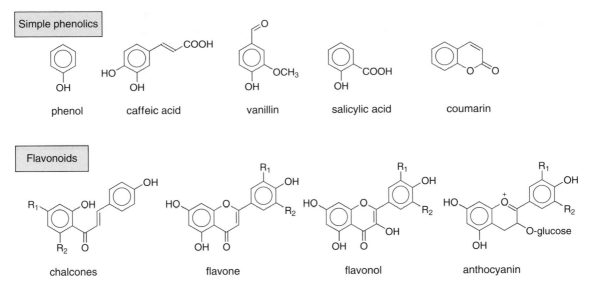

Fig. 4.10. Examples of plant phenolics. The simple phenolic compounds such as vanillin are based on a single phenol group while flavonoids such as anthocyanin consist of more complex multi-ring structures.

water-soluble phenolic antioxidants, such as caffeoylshikimic and hydroxybenzoic acids.

Phenolic antioxidants have been shown to have cardioprotective, antidiabetic and anticancer effects in animal studies, although these results have yet to be confirmed by human clinical trials. The three main groups of simple phenolics are the simple phenylpropanoids, coumarins and benzoic acid derivatives. Simple phenylpropanoids include caffeic, ferulic, p-coumaric and cinnamic acids. Coumarins include coumarin, umbelliferone and scopoletin. Benzoic acid derivatives include vanillin and salicylic acid as well as the tannins, gallic and erusic acids. An outline of the biosynthesis of some important phenolics is shown in Fig. 4.11 and some biotechnology targets are discussed in Box 4.3.

Lignins are the most important group of complex phenylpropanoids and are largely responsible for the structural strength of wood. Lignins are formed from a highly branched and variable group of hetero-polymers based on several types of phenylpropane unit. The three most important phenylpropane precursors of lignin are coumaryl alcohol, coniferyl alcohol and sinapyl alcohol. The type of lignin found in a particular plant will differ in the composition and cross-linking of its constituent phenylpropanoids. Lignin composition is highly diverse across different plant groups and can even vary between and within the individual cells of a given plant. All of the chemical bonds that form the three-dimensional molecular network of lignin are covalent (Fig. 4.12). This makes lignin a much more rigid and less permeable polymer than cellulose, which is largely stabilized by weaker hydrogen bonds (Fig. 4.3).

The type of phenylpropane units and degree of lignin branching and cross-linking largely determine the strength of a particular type of wood and hence its economic uses. Ferns have a rather basic form of lignin consisting mainly of coniferyl alcohol units and do not make true wood. Therefore, although tree ferns including popular ornamentals such as *Dicksonia* spp. can grow up to several metres in height, their lignified tissues are far weaker than those of seed-bearing plants and they remain relatively susceptible to lack of moisture. In contrast, the lignin of gymnosperms is more heterogeneous and stronger than ferns, enabling groups such as the conifers to grow into very tall trees capable of colonizing a wide range of habitats. However, the lignin of gymnosperm wood tends to be less highly branched and more resinous than that of angiosperms. This means that gymnosperms tend to produce relatively weak softwoods compared with many angiosperm trees, which form highly branched and cross-linked lignin that produces hardwoods suitable for heavy structural uses such as in ship manufacture or building construction.

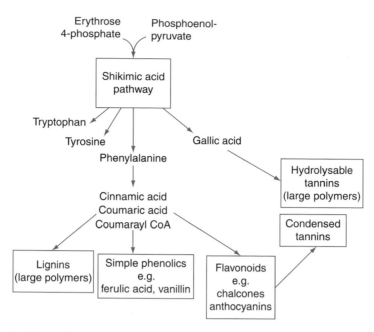

Fig. 4.11. Biosynthesis of the major plant phenolic compounds. Most plant phenolics are derived from phenylalanine formed via the shikimic acid pathway. Phenylalanine provides the phenol ring and three-carbon side chain found in all phenylpropanoids. Subsequent modifications lead to formation of simple phenolics, flavonoinds and highly complex polymers and condensed structures such as lignins and tannins.

Box 4.3. Biotechnology targets for phenolics and lignin

Lignin up/down-regulation: The annual global market for paper and pulp from wood is worth over $800 billion. Reduction of the lignin content of trees grown for paper production would greatly simplify wood processing at paper mills, saving energy and reducing pollution from the many processing chemicals used, as well as reducing costs. Lignin formation involves complex biosynthetic pathways with many enzymes, often encoded by large gene families. This metabolic complexity, plus the redundancy and overlapping function of its many enzymatic components, makes it a formidable challenge even to understand the basic processes of lignin formation, let alone to manipulate them via transgenesis.

Despite these challenges, some success in lignin manipulation has been achieved with poplar, although it has proved more difficult in economically important gymnosperms such as pine, larch and spruce. Up-regulation of cellulose and reduction of lignin content in biomass crops such as switchgrass, miscanthus and sorghum can reduce processing costs and increase bioethanol yields, and several preliminary transgenic breeding lines are now being assessed.

For example, when expression of the cinnamyl alcohol dehydrogenase gene is down-regulated in hybrid poplar trees, coniferyl alcohol formation is reduced (see Fig. 4.12). As a result, the lignin becomes more chemically reactive and transgenic wood can be pulped with fewer chemicals. In contrast, suppression of caffeic acid *O*-methyl transferase gene expression renders lignin more chemically resistant, requiring more pulping chemicals. Despite these changes in lignin structure, the field performances of both transgenic types appear to be similar to those of wild-type poplars.

Flavonoids: Anthocyanin composition is being altered to manipulate colour in some ornamental flowers, including long-standing attempts to produce blue roses. As early as 1996, a mauve-coloured transgenic carnation (*Dianthus* spp.) containing an introduced flavonoid-3'5'-hydroxylase was being marketed by Florigene. The same gene was used to produce blue flowers in transgenic carnations. More recently, manipulation of other anthocyanin/flavonoid pathway genes has led to a range of different flower colours in ornamentals, including petunias, carnations and roses.

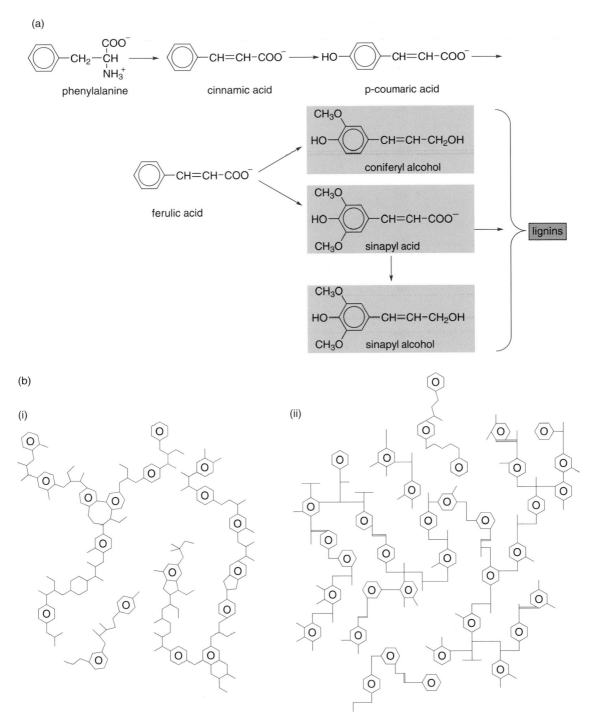

Fig. 4.12. Structure and biosynthesis of lignin. (a) Simplified pathway for the biosynthesis of the three most common lignin precursors from phenylalanine. (b) Structures of the relatively unbranched lignin found in softwoods such as conifers (i) compared with the denser and more highly branched lignin found in hardwoods such as oak (ii). Adapted from Glazer and Nikaido (1995).

Flavonoids are a diverse group of compounds made up of three-ringed, water-soluble phenolic molecules that accumulate in the vacuoles of plant cells. They have many important functions ranging from pathogen responses (e.g. chalcones, isoflavonoids) to flower and fruit coloration (e.g. anthocyanins, flavones). Anthocyanins in particular are responsible for a vast range of flower colours such as red, orange and pink colours from pelargonidins, magentas and reds from cyanidins, and purple and blue hues from delphinidins.

Alkaloids

Alkaloids are an extremely diverse group of compounds (see Fig. 4.13), most of which are synthesized from amino acid precursors via highly complex pathways that are often localized in specific tissues and subcellular compartments. Alkaloids consist of multiple ring structures containing nitrogen and carbon atoms and nearly all are alkaline in solution (hence their name). As with many compounds in higher plants, the huge range of alkaloids has evolved due to processes such as gene duplication (via hybridization and polyploidy) to create redundant enzymes that subsequently diversify through mutation.

Almost all alkaloids are injurious or toxic to specific target animals, especially pests or herbivores. Some animals have adapted by learning to avoid certain plants in which they detect alkaloids thanks to their characteristic bitter taste. Other animals have developed an ability to detoxify or sequester potentially toxic alkaloids. In a few cases, such as poisonous frogs, animals sequester toxic plant-derived alkaloids in their skin or flesh as a defence against predators. The vast majority of alkaloids are synthesized from one of a small number of amino acids, with arginine, lysine, phenylalanine, tryptophan and tyrosine acting as the most common precursors. The five most abundant and important groups are the purine, pyrrolizidine, tropane, isoquinoline and terpenoid indole alkaloids, as outlined below.

Tropane alkaloids are esters of tropane and carboxylic acids that include scopolamine, hyoscyamine and atropine. Nicotine is a related compound formed from tropane alkaloid precursors. Many tropane alkaloids are toxic due to their anticholinergic properties, namely their ability to inhibit transmission of nerve impulses by blocking the binding of acetylcholine to its receptor. In low doses, they can have medicinal uses. Hence extracts of deadly nightshade, *Atropa belladonna*, were formally used cosmetically as eye drops to dilate the pupils (hence, belladonna – beautiful lady). In higher doses the same extracts can be lethal. Similarly, cocaine from *Erythroxylon coca* can be used as a topical anaesthetic and stimulant at low doses but can be addictive and potentially toxic at higher doses.

Many tropane alkaloids cause hallucinations and such extracts have been used, and sometimes abused, as mind-altering agents by human societies for millennia. Nicotine has a different mode of action from other alkaloids in this group as it affects the vegetative ganglia of the brain, initially as a stimulant but then as a blocker that can cause convulsions and respiratory paralysis. Nicotine is also a formidable insecticide (probably its natural function) and nicotine 'bombs' are an effective method of pest control in greenhouses.

Pyrrolizidine alkaloids probably originally arose in plants due to the duplication of a gene involved in an entirely unrelated process. One of the fundamental mechanisms of eukaryotic cell growth is the addition of an amino butyl group to the translation initiation factor, eIF5A, by the ubiquitous NAD-dependent enzyme deoxyhypusine synthase. This gene appears to have been duplicated in plants and in a few species the redundant copy underwent a series of mutations that slightly altered the function of its encoded enzyme to add an amino butyl group to spermidine instead of eIF5A. This new enzyme, which is only found in a few angiosperm families, is homospermidine synthase, the first committed step in the biosynthesis of pyrrolizidine alkaloids.

These alkaloids are only toxic to humans when broken down to pyrroles in the liver. Pyrroles form cross-links with DNA with the resulting fibrous lesions leading to liver failure and death. Toxicity can be sudden if ingested in large amounts or may be cumulative if smaller amounts are gradually ingested. Pyrrolizidine alkaloid poisoning from plants such as groundsel, *Senecio vulgaris*, or ragwort, *Senecio jacobaea*, is still a common cause of death or birth defects in livestock such as cattle and horses. Some phytophagous insects are able to sequester and modify

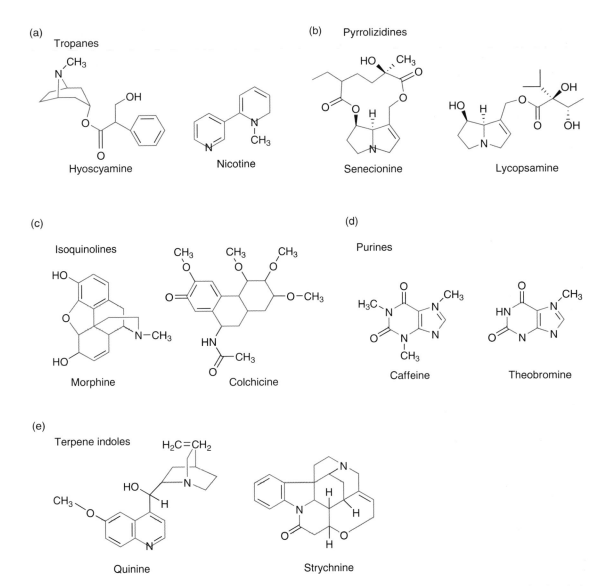

Fig. 4.13. Major groups of plant alkaloids. Examples of some of the principal families of plant alkaloids, showing their diverse nitrogenous multiple ring structures.

pyrrolizidine alkaloids for their own use. For example, some arctiid moths use plant-derived lycopsamines to synthesize pheromones used in courtship behaviour.

Isoquinoline alkaloids include the medically important painkillers morphine and codeine, which are produced commercially from the opium poppy *Papaver somniferum*. The toxic and addictive opiate, heroin, is produced from

the same plant. In general, isoquinoline alkaloids are toxic to humans due to their combined action on the gastrointestinal and central nervous systems. Other members of this group are used, in appropriate doses, for medicinal or other beneficial purposes. As discussed in Chapter 8, colchicine from *Colchicium autumnale* is used as a chromosome-doubling agent in breeding, as well as to treat gout in humans and as a cancer treatment in domestic pets. Berberine from *Coptis japonica* is an antimicrobial while emetine from *Uragoga ipecacuanha* is an anti-amoebic agent.

Purine alkaloids include the active ingredients in some of our most popular beverages but they can also be powerful insecticides. Caffeine is the major alkaloid component in leaves from the tea plant, *Camellia sinensis*, and in beans of the coffee plant *Coffea arabica*. The major purine alkaloid in beans of cocoa, *Theobroma cacao*, from which drinking and solid chocolate are made, is theobromine. Chocolate also contains small amounts of caffeine and anandamide, an endogenous cannabinoid also found in the human brain. Caffeine and theobromine are mild stimulants, although caffeine is also a diuretic that can cause heart palpitations in high doses and may be mildly addictive.

The function of these alkaloids in plants is largely insecticidal and they are particularly toxic to larvae that might otherwise browse on relatively soft young leaves, stems and seeds. A further function of caffeine is as an allelopathic agent that can readily leach from leaves into the soil where it inhibits the germination of other plants. This can sometimes cause problems if coffee is grown continuously in an area, as an excessive build-up of caffeine in the soil can eventually inhibit germination of new coffee plants.

Terpenoid indole alkaloids include well-known compounds such as strychnine, quinine and the anticancer agents vincristine and vinblastine. Strychnine, from *Strychnos nux-vomica*, is a popular rat poison but can also be highly toxic to humans. Quinine, from the bark of *Cinchona officinalis*, was a widely used antimalarial, but an increasing build-up of resistance in the target parasite has led to use of alternative plant compounds such as the sesquiterpenoid artemisinin. Quinine can also be used to treat irregular heartbeat, as can the related alkaloids, ajmaline and ajmalicine, both from *Rauvolfia serpentina*.

Terpenes and terpenoids

These non-polar compounds make up the largest class of natural products with over 25,000 forms described so far in higher plants. Terpenes and terpenoids are made up of one or more isoprene units (Fig. 4.14). Terpenoids or their derivatives include some of the primary compounds essential for plant function, such as carotenoids, the phytol side chain of chlorophyll, the electron transport intermediates plastoquinone and ubiquinone, and the major hormone groups abscisic acid, cytokinins and gibberellins (see Chapter 5). Volatile terpenoids, mixed with phenolics, are responsible for many distinctive floral aromas, while the monoterpenes, myrcene, limonene and α- and ß-pinene are the major volatiles in conifers. Among the many functions of terpenes and terpenoids are as photosynthetic pigments, redox intermediates, hormones, wound-healing resins, antimicrobial agents, pollinator attractants and herbivore repellents.

The basic unit of most secondary plant metabolites, including terpenes, is the simple hydrocarbon molecule isoprene. The term terpene usually refers to a hydrocarbon molecule while a terpenoid or isoprenoid refers to a terpene that has been modified, such as by the addition of oxygen. Isoprenoids are the building blocks of a wide range of metabolites including some plant hormones, sterols, carotenoids, rubber, the phytol tail of chlorophyll, and turpentine. The nomenclature of terpene compounds is as follows. The single five-carbon isoprene unit is termed a hemiterpene. Two isoprene units (C10) make up a monoterpene. Sesquiterpenes contain three isoprene units (C15), while diterpenes (C20) and triterpenes (C30) have two and three monoterpene units respectively. Tetraterpenes consist of four monoterpene units and polyterpenes contain more than four monoterpene units (i.e. more than eight isoprene units).

Terpenes tend to be relatively oily, hydrophobic compounds, as are many terpenoids, especially those with low oxygen contents. For this reason, such compounds are frequently found in specialized structures including resin ducts in conifers and epidermal glands on the leaves and petals of angiosperms. Terpenes have numerous functions in plants. Together with some low molecular weight phenolics, terpenes are major aroma components of flowers, where they attract potential pollinators. Terpenes are often synthesized as defence compounds in response to herbivory or other stresses. While many terpenes deter potential herbivores, pests or pathogens, others may attract predators of such undesirable organisms. For example, maize leaves that are damaged by the caterpillar *Spodoptera littoralis*

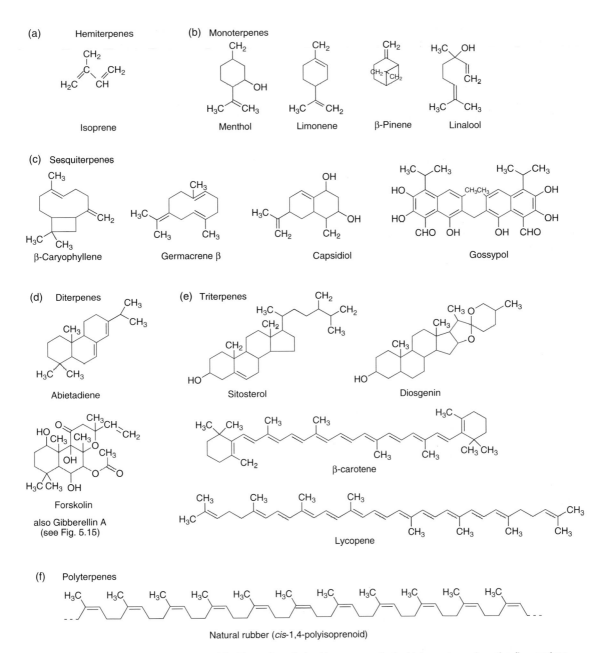

Fig. 4.14. Major groups of plant terpenoids. Examples of plant terpenes, all of which are based on the five-carbon isoprene group. Sizes range from simple monoterpenes such as menthol with two isoprene units to natural rubber, which is a polymer made up of thousands of isoprene units.

produce volatile terpenes that attract the parasitic wasp *Cotesia marginiventris*, which attacks the caterpillars. Also, maize roots damaged by the Coleopteran rootworm *Diabrotica virgifera* release the terpene (E)-β-caryophyllene,

which attracts nematodes that feed on *Diabrotica* larvae.

The pyrethrins are a class of irregular monoterpenes found in chrysanthemum that have especially strong insecticidal properties. Synthetic pyrethrins are still one of the most powerful groups of agricultural pesticides. Other terpenes have antimicrobial activities and can replace antibiotics in livestock feed, thereby reducing the likelihood that new strains of antibiotic-resistant bacteria might arise. Other undesirable microbes that are susceptible to terpenes include the yeast *Candida albicans* and the malarial parasite *Plasmodium falciparum*. The sesquiterpene lactone, artemisinin, which is obtained from sweet wormwood *Artemisia annua*, is one of the most important antimalarial agents now used in combinatorial treatment of malaria patients.

The diverse and constantly changing cocktail of terpenes, terpenoids and other defence compounds is one of the major defence mechanisms of plants. This means that many plants are unpalatable to potential herbivores. This in turn favours the evolution of highly specialized herbivores that are sometimes only able to forage on a single plant species to which they have become chemically adapted. Recent advances in DNA bar-coding technology have revealed that many apparently phenotypically identical herbivorous insect species that forage on different plants are actually made up of genetically distinct forms that constitute separate species. Despite looking identical (at least to the human eye), these so-called cryptic insect species preserve their reproductive isolation due to their unique ability to live on and feed off a single plant species.

Many terpenes and terpenoids can have profound effects on humans, both bad and good. Several diterpenes in euphorbias can damage the skin, while another group of diterpenes, the grayanotoxins, may cause fatal poisoning due to their effects on the central nervous tissue. In some cases, grayanotoxins in honey derived from rhododenron species can cause 'mad honey disease', which is an intoxication that is rarely fatal but can involve dizziness, weakness, excessive perspiration, nausea and vomiting. The major triterpenoid toxins are the saponins, which occur in species such as ivy, horse chestnut and yams. Diosgenin, which is the saponin from yam tubers, was the original precursor used to manufacture the contraceptive agent progesterone. Due to its ability to cause male infertility in mammals, the sesquiterpene gossypol, which is abundant in cotton plants, was an early candidate for a male contraceptive but its use was abandoned due to toxic side effects.

One of the best-known medically active terpenoids is the diterpene alkaloid compound taxol, which occurs in yew (*Taxus*) species. Taxol was found to inhibit the growth of cancerous cell cultures *in vitro* and is now used widely in cancer therapy where it is especially effective against ovarian, lung and breast cancers and Kaposi's sarcoma. The complex taxol molecule cannot be synthesized in the lab and is derived from needles or cuttings of yew trees. In addition to the beneficial compound taxol, all parts of the yew tree contain a highly toxic alkaloid, taxin. Some biotechnology targets for alkaloids and terpenes are discussed in Box 4.4.

4.6 Summary Notes

- Plants are capable of a much wider range of metabolic activities than animals. Products of plant metabolism are used for numerous applications, including foods, pharmaceuticals, clothing, industrial materials and fuels.
- The three main products of primary metabolism are carbohydrates, lipids and proteins, and these are also the three major macronutrients in the human diet.
- Important plant carbohydrates include sucrose, starch and cellulose. Sucrose is primarily used to transport carbon from source to sink tissues; starch is a carbon and energy store; and cellulose is the principal structural component of plant cell walls.
- Plant acyl lipids form the matrix of cellular membranes and can also serve as energy reserves in storage tissues such as fruits and seeds. The wide chemical diversity of plant storage lipids means they have many uses as renewable industrial feedstocks.
- Plants assimilate inorganic nitrogen compounds into amino acids and proteins either directly or in conjunction with symbionts such as soil-borne mycorrhizae or nitrogen-fixing Rhizobia.
- Plants synthesize a huge range of secondary compounds for purposes such as defence, strengthening, signalling, and as pigments and redox intermediates.
- The major groups' secondary compounds include phenolics, alkaloids and terpenoids, many of which are candidates for biotechnological manipulation.

Box 4.4. Biotechnology targets for alkaloids and terpenes

There is great interest in manipulating plant alkaloids for medical applications as well as to improve crop defences against pests and diseases in the field. Important targets include alkaloid-producing crops, such as tobacco (nicotine), coffee and tea (caffeine), plus high-value medically active compounds, such as quinine, morphine, codeine and colchicine and the potent anticancer agents vincristine and vinblastine.

In an effort to suppress production of morphinan alkaloids in opium poppy, RNAi has been used to silence the multigene family encoding codeinone reductase. This enzyme produces codeine from codeinone and morphine from morphinone and these two precursors were expected to accumulate in the transgenic plants. Surprisingly, however, the entire biosynthetic pathway leading to morphinal alkaloids was blocked a further seven enzymatic steps upstream from codeinone reductase. This indicated that codeine and/or morphine might normally regulate some of the earlier steps in their own biosynthetic pathway and their absence resulted in their complete shutdown. One interesting feature of these transgenic poppy plants was the accumulation of (S)-reticuline instead of codeine and morphine. This compound can stimulate growth of cultured hair cells in mice and is being assessed as a possible agent to combat baldness, which would provide an alternative, and perhaps more socially acceptable, market for opium poppies.

Biotechnological manipulation of terpenes is in its infancy and has been confined so far to research studies rather than commercial applications. Transgenesis is a popular option to manipulate plant terpene content. Several model species, including tobacco, tomato, lettuce and *Arabidopsis*, have been transformed by addition of one or more copies of a terpene synthase gene. In the case of tobacco, the addition of terpene synthase genes led to an increase in terpene emission from leaves of the transgenic plants. *Arabidopsis* terpene overproducers showed enhanced pest resistance, while tomatoes with enhanced linalool levels had enhanced flavour and aroma but, unfortunately, also had reduced levels of lycopene.

The high value and important medical uses of the diterpene taxol have led to several efforts to produce the compound in transgenic and non-transgenic cell cultures *in vitro*. This approach is promising because it avoids the need for harvesting needles or bark from slow-growing yew trees, while also providing a highly contained and secure production system that minimizes potential contamination risks. Finally, although terpenes are not generally thought of as possible candidates for biofuels, the tropical copaiba plant, *Copaifera langsdorfii*, produces an oil that can be used as a diesel fuel. This oil is largely made up of terpenes and the copaiba or similar tropical plants might be future sources of biofuels for local markets in tropical countries.

Further Reading

General

Bowsher, C., Steer, M. and Tobin, A. (2008) *Plant Biochemistry*. Garland, New York.

Jones, R.L., Buchanan, B.B. and Gruissem, W. (2000) *Biochemistry and Molecular Biology of Plants*. Wiley, New York.

Carbohydrates and lipids

Baud, S. and Lepiniec, L. (2010) Physiological and developmental regulation of seed oil production. *Progress in Lipid Research* 39, 235–249.

Haigler, C.H., Singh, B., Zhang, D., Hwang, S., Wu, C., Cai, X., Hozain, M., Kang, W., Kiedaisch, B., Strauss, R., Hequet, E., Wyatt, B., Jividen, G. and Holaday, S. (2007) Transgenic cotton over-producing spinach sucrose phosphate synthase showed enhanced leaf sucrose synthesis and improved fiber quality under controlled environmental conditions. *Plant Molecular Biology* 63, 815–832.

Murphy, D.J. (2010) Improvement of industrial oil crops. In: Singh, B.P. (ed.) *Industrial Crops and Uses*. CAB International, London, pp. 183–206.

Sanjaya, Durrett T.P., Weise, S.E. and Benning, C. (2011) Increasing the energy density of vegetative tissues by diverting carbon from starch to oil biosynthesis in transgenic *Arabidopsis*. *Plant Biotechnology Journal* 9, 1–10.

Saxena, I.M. and Brown, R.M. (2005) Cellulose biosynthesis: current views and evolving concepts. *Annals of Botany* 96, 9–21.

Nitrogen metabolism

Andrews, M., Lea, P.J., Raven, J.A. and Lindsey, K. (2004) Can genetic manipulation of plant nitrogen assimilation enzymes result in increased crop yield and greater N-use efficiency? *Annals of Applied Biology* 145, 25–40.

Igarashi, R.Y. and Seefeldt, L.C. (2003) Nitrogen fixation: the mechanism of the Mo-dependent nitrogenase. *Critical Reviews of Biochemistry and Molecular Biology* 38, 351–384.

Shewry, P.R. and Casey, R. (1999) *Seed Proteins.* Kluwer, Amsterdam.

Singh, B.K. (ed.) (1999) *Plant Amino Acids. Biochemistry and Biotechnology.* Decker, New York.

Secondary compounds

Aharoni, A., Jongsma, M.A. and Bouwmeester, H.J. (2005) Volatile science? Metabolic engineering of terpenoids in plants. *Trends in Plant Science* 10, 594–602.

Chen, F. and Dixon, R.A. (2007) Lignin modification improves fermentable sugar yields for biofuel production. *Nature Biotechnology* 25, 759–761.

Glazer, A.W. and Nikaido, H. (1995) *Microbial Biotechnology.* Freeman, New York.

Gould, K., Davies, K. and Winefield, C. (eds) (2009) *Anthocyanins: Biosynthesis, Functions, and Applications.* Springer, Berlin.

Zwenger, S. and Basu, C. (2008) Plant terpenoids: applications and future potentials. *Biotechnology and Molecular Biology Reviews* 3, 1–7.

Plant Organization and Development

5.1 Chapter Overview

In this chapter we will examine how the bodies of higher plants are organized and how they grow and develop in different and rather more flexible ways than those of higher animals. Unlike animals, plants are not limited to a narrow range of body forms, and the size and shape of adult plants often depends on environmental factors. This developmental plasticity is especially useful in agriculture. Although farmers have been manipulating plant bodies for millennia, our capacity for more radical types of redesign has been greatly enhanced by modern scientific discoveries and the advent of new biotechnologies. The middle sections of the chapter deal with plant reproduction and its manipulation. Finally, we examine the roles of hormones in regulating plant growth, differentiation, reproduction and stress responses. The elucidation of hormonal regulation and its manipulation via methods such as mutagenesis or transgenesis have profound implications for crop improvement.

5.2 Plant Organization and Growth

Higher plants have a relatively simple and flexible basic body organization with a highly developed capacity for indeterminate growth. This means that the final shape and size of an adult plant can often be significantly modulated by the environment. A major factor behind the indeterminate growth habit of plants is their sessile nature, which means that an individual plant is unable to move once it begins to develop. In order to exploit their local habitats most efficiently, therefore, plants require great flexibility in their growth form and their responses to the many biotic and abiotic opportunities and threats they face during their lifetimes.

The plant body

The body of a higher plant is made up of three basic types of tissue, namely an outer layer of *dermal* tissue enclosing a relatively undifferentiated mass of *ground* tissue through which runs the more highly differentiated *vascular* tissue (Fig. 5.1). Angiosperms also have three basic cell types, namely parenchyma, collenchyma and schlerenchyma. *Parenchyma* cells are most commonly spherical or elongated with a thin primary cell wall, although they may have a secondary wall that can occasionally be lignified. Such cells are found in all organs and carry out the key metabolic functions of photosynthesis and storage. Examples of parenchyma cells include the cortex and pith of roots and stems, and the mesophyll and palisade cells of leaves. *Collenchyma* cells are elongated with relatively thick but flexible cell walls that tend to occur in aggregates. They provide tough but elastic support for young growing tissues, such as stems and leaf petioles. *Schlerenchyma* cells, which are normally dead at maturity, have thick, lignified cell walls that are relatively rigid and help support the weight of mature plant organs.

The dermal tissue can be regarded as the 'skin' of the plant. It generally comprises a single layer of primarily parenchyma cells, often coated with a waxy cuticle, that provide a waterproof seal and protection from external attack. The ground tissue, which makes up the bulk of the plant, is predominantly parenchyma but collenchyma and schlerenchyma cells are found. Vascular tissue transports water, ions, minerals and photosynthetic assimilates such as sugars and amino acids throughout the plant body, which it also helps to support. It is made up of *xylem*, *phloem*, parenchyma and schlerenchyma cells.

Higher plants are also divided into an aerial shoot system, which is normally above ground, and a root system, which is normally below ground. The shoot consists of supporting stems, trunks and branches that hold the all-important photosynthetic organs, the leaves. The root system serves both to anchor the plant and to extract water and

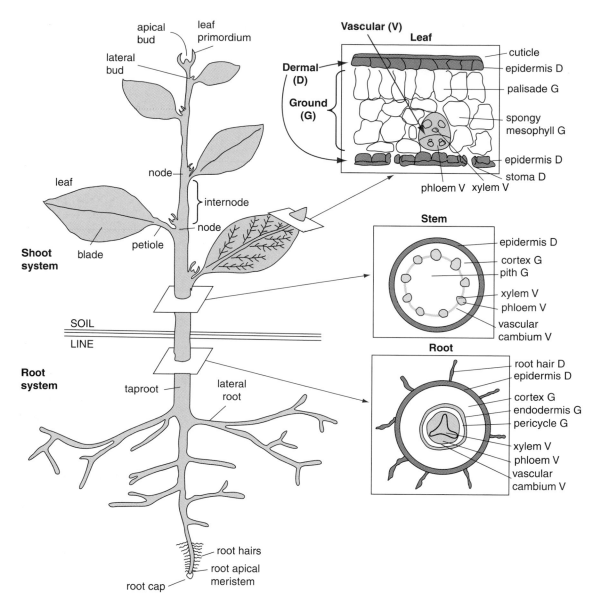

Fig. 5.1. Organization of the plant body. The plant body is made up of a branched aerial shoot system containing photosynthetic and reproductive structures, and a branched underground root system containing water- and mineral-extracting structures. Both systems consist of a major vertical stem axis off which a variable number of horizontal branches may be formed. Aerial branches produce the leafy foliage while root branches produce lateral roots that end in the fine root hairs through which most nutrients are absorbed. Major organs, such as leaves, stems and roots, are made up of the three basic types of tissue. Dermal tissue (D) forms a protective unicellular skin-like layer around the outside of each organ. Ground tissue (G) fills much of the interior of each organ and carries out many important metabolic functions, including photosynthesis. Vascular tissue (V) runs through each organ in pipe-like bundles and connects all parts of the plant with the root and shoot systems.

nutrients from the substrate (normally soil) in which it is located. Although it is sometimes neglected in comparison with the shoot, the root system of many plants can have a similar biomass to the shoot. In addition to its support and nutrient functions, roots often serve as food storage sites that enable plants to survive during periods of reduced photosynthesis, such as in overwintering perennial species. All plant tissues are derived from specialized regions of cell division known as *meristems* (Fig. 5.2). As discussed below, the activity of meristematic regions determines the overall shape and size of a given plant.

Growth patterns

Plants have highly flexible growth patterns. Meristematic regions in a plant contain rapidly dividing, undifferentiated cells that remain active throughout its lifetime. Although they tend to have characteristic adult forms, plants do not have anything like the pre-programmed body plan of animals. Differentiation in animals is almost always irreversible and is specified very early in development. This means that adult body shape and size is more or less fixed, with the result that mature members of the same animal species tend to look very similar to each other. In plants, some features might be relatively constant, such as leaf shape and branching patterns (opposite, alternate, etc.), but the exact shape of the mature organism can be highly unpredictable. Thus, a plant like a beech can vary from a small contorted shrub less than 1 m high if it grows on a rocky, windswept hillside to a magnificent 50 m tree growing in a more fertile and sheltered habitat.

One reason for this difference is that plants continue to grow throughout their life via multiple meristematic regions. This capacity for *indeterminate growth* in plants is widely exploited in practices such as pruning, coppicing, pollarding, and the use of cuttings in commercial and domestic plant husbandry. In contrast, animal growth tends to be *allometric*, whereby all parts of the body grow such that its overall proportions are maintained. Hence, highly active apical meristems located at the root or shoot apex tend to give rise to relatively tall and deeply rooted plants respectively, whereas more active lateral meristems will lead to a shorter, bushier type of plant. Nowadays biotechnologists can take even greater advantage of the plasticity of plant growth by producing novel forms that are both higher

yielding and more amenable to management and harvesting. A recent example is the development of new commercial apple varieties that are no more than 2 m high and grow like vines but produce far more fruits than traditional apple trees (Fig. 5.3).

Roots and shoots

The overall pattern of plant growth depends upon the activity of its major meristems. Apical meristems located at the tips of roots and shoots enable a plant to increase in length along the vertical axis, upwards for shoots and downwards for roots. Lateral meristems are found near the periphery of the plant, usually in a cylinder, and supply cells for the plant to increase its width by growing along the horizontal axis, i.e. by branching out sideways rather than up or down. Only dicots have lateral meristems, which is why monocots tend to grow as tall thin plants such as grasses and palm trees. In contrast, dicots tend to be broader and often have a bushier growth habit.

As shown in Fig. 5.2, the principal growth region of a root is located just behind the root cap, a thimble-like covering that protects the delicate root apical meristem. The root cap, which is constantly being sloughed off and replaced, is derived from meristem cells and secretes a polysaccharide slime that lubricates the soil through which the root is growing. The root apical meristem is a region of rapid cell division directed away from the root cap. Adjacent to the main meristem is the quiescent centre where cells divide much more slowly than other meristematic cells. This rather mysterious region is highly resistant to radiation and chemical damage and may be a reserve in the case of damage to the main apical meristem. Immediately behind the apical meristem are three primary meristematic regions that continue to divide to produce the three major tissue types.

The outermost root meristematic region is the *protoderm*, which produces cells destined to become dermal tissue. The central region is the *ground meristem*, which develops into the ground tissue. The innermost primary meristem is the *procambium*, which gives rise to vascular tissue. Behind these meristematic regions is the zone of elongation, where cells elongate up to ten times their original length and push the root further downwards into the soil. Finally, the zone of maturation is the region of the root where fully functional and differentiated cells develop.

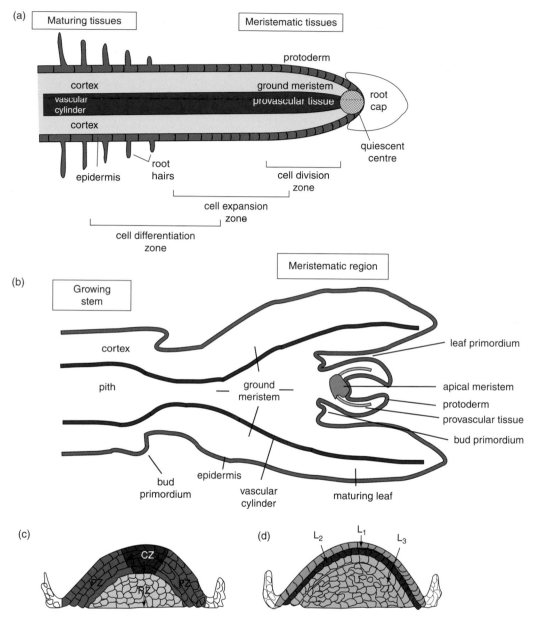

Fig. 5.2. Plant meristems. Plant cells tend to divide and differentiate at meristems. Apical meristems are responsible for growth along the principal vertical axis of the shoot and root while lateral meristems produce horizontal branching. (a) The root apical meristem produces partially differentiated protodermal, ground and provascular cells that first divide rapidly, then elongate, and finally differentiate into the respective specialized tissues. (b) Leaf apical meristems function in a similar way to those of roots except that they can also form leaves that grow directly from the stem. Lateral meristems, or bud primordia, produce branches that have their own meristems from which the plant can grow horizontally. (c) The tip of the shoot apical meristem consists of a central zone (CZ) containing relatively slowly dividing stem cells from which all other cell types are derived. To the side is the peripheral zone (PZ) where cells divide much more rapidly and give rise to leaf pimordia. Below is the rib zone (RZ), which forms the central tissues of the stem. (d) Cell layers in the meristem also contribute to tissue formation with the outermost (L_1) layer forming the epidermis and the inner (L_2 and L_3) layers forming internal structures such as ground and vascular tissues (c and d adapted from Bowman and Eshed, 2000).

The major upwards-directed region of shoot growth is the shoot apical meristem, which is a dome-shaped mass of dividing cells at the tip of the terminal bud. As in roots, this principal meristem gives rise to the three primary mersitems, protoderm, ground meristem and procambium, to form the three tissue types. On either side of the apical meristem, leaves arise as leaf primordia. Behind the apical meristem are axillary meristems, which may be dormant but can become activated in response to various stimuli to form branches off the main shoot. In this case, the axillary meristem becomes the apical meristem of the new branch.

The balance between growth of apical and lateral meristems will largely determine the overall shape of the plant (see Fig. 5.3). For example, some

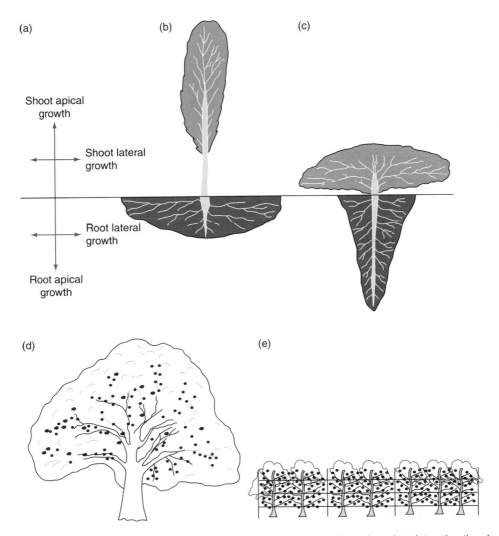

Fig. 5.3. Variation in adult body shapes in plants. (a) The primary meristematic regions that determine the shape and size of a typical higher plant. (b) Plants with dominant shoot apical meristems tend to have tall, slender bodies. Plants with dominant lateral root meristems tend to have shallow, widely growing roots. (c) Plants with dominant shoot lateral meristems tend to have short, bushy bodies. Plants with dominant apical root meristems tend to have deep, narrow growing roots. (d) Apple plants normally grow as medium-sized bushy trees that can reach over 20 m in height. (e) A combination of breeding and training by regular cutting of the shoot apical meristem produces a totally different body shape in many modern commercial apples, which grow as short (<2 m), vine-like plants that yield more fruit and are easier to manage than traditional varieties.

plants such as palms have dominant apical meristems that cause them to grow tall and thin. Other more bushy or shrubby plants have more active lateral meristems that favour extensive sideways growth at the expense of height. The grassy monocots have active apical meristems at the base of their shoots. This means that the tops of these shoots can be removed by grazing or mowing without damaging the meristem or inhibiting its growth. Thanks to this adaptation grassy plants are able to survive and even flourish in many of the herbivore-dominated landscapes created by pastoral agriculture.

Secondary growth involves lignification of cell walls and is found in both herbaceous and woody dicots. In herbaceous dicots, secondary xylem and phloem are in a single ring of discrete bundles whereas, in woody dicots, the secondary xylem forms a continuous cylinder. In contrast, monocot stems usually lack secondary growth and have no vascular cambium or cork cambium. Monocot stems tend to be uniform in diameter with scattered vascular bundles, rather than in a ring as found in dicot stems.

Leaves

Leaves are the major photosynthetic organs of plants and various forms of leaf organization evolved at least seven times in land plants. Small membranous structures called phyllidia arose separately in mosses and liverworts. Lycopods developed small, simple, single-veined structures called microphylls. Four lineages of more advanced plants, leptosporangiate ferns, eusporangiate ferns, sphenopsids and seed plants, independently developed more complex leaves, called euphylls or megaphylls, which are derived from whole shoot systems and are often multi-veined. In seed plants, leaf development involves a complex network of interacting gene families. Examples include the *KNOX* genes that help maintain indeterminacy; *HD-Zip* genes to maintain adaxial identity; and *YABBY* and *KANDAI* genes to maintain abaxial identity. These and other groups of homeotic and transcription factor genes act in concert to produce the characteristic dorsiventral asymmetry of many leaves that is required for them to develop into flat laminate organs capable of efficient photosynthesis.

Transport

Plants must transport water, minerals and nutrients from the soil and photosynthetic assimilates from the leaves, and distribute these substances throughout the plant body. Unlike animals, plants do not have a pumping mechanism to transport fluids around their bodies. Instead, they rely on the colligative and osmotic properties of aqueous solutions to transport dissolved materials as much as 50 m vertically. One of the most important innovations made by early land plants was the development of increasingly efficient vascular systems that enabled first the ferns and later the gymnosperms and angiosperms to colonize new habitats across the world. In higher plants, minerals, assimilates, hormones and other molecules can be transported in either of two networks within the vascular system, namely the xylem and phloem.

Water and minerals are absorbed from the soil, often via fungal mycorrhizae, and transported by the xylem to the rest of the plant. Xylem consists of water-transporting *vessels* and *tracheids*, supported by fibres and supplemented by parenchyma cells that enable materials to be loaded and unloaded from the vessels and tracheids (see Fig. 5.4). While most gymnosperms only have tracheids in their xylem, angiosperms also have vessels, which are more efficient at conducting water and are one of the factors behind the evolutionary success of angiosperms. Water movement is powered by the release of water vapour into the atmosphere due to transpiration through leaf stomata. The cohesion (stickiness) of water molecules in small tubes creates a suction effect, known as capillary action, that draws water up the xylem from the roots to replace the transpired water. Unlike the heart, which is an active pump in animals, the xylem is a passive system for fluid transport. However, most of the important soil minerals, including fertilizers such as nitrates and phosphates as well as many metal ions, are actively loaded into roots via ATP-dependent plasma membrane pumps and then loaded into the xylem to be carried in the water flow up to the rest of the plant.

The phloem transports assimilates, such as sucrose and amino acids, from actively photosynthesizing tissues to the rest of the plant. While the xylem carries water and minerals upwards from the soil, the phloem tends to carry assimilates downwards or laterally from the aerial parts of the plant to sink tissues such as flowers, fruits and roots. The phloem has two types of transport cell (see Fig. 5.5). First are the relatively wide, tubular *sieve tube members* responsible for long-range movement of

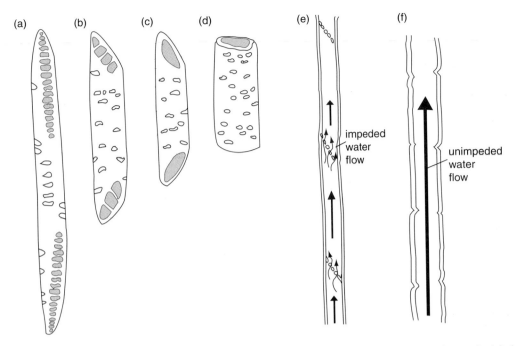

Fig. 5.4. Conducting cells of the xylem. The main conducting cells of the xylem are tracheids and vessels. (a)–(d) Progression from a relatively long, thin, tapering tracheid with a compound perforation plate to a wider, thicker vessel element with a simple perforation plate. (e) The compound perforation plates in tracheids and more primitive vessel elements impede water flow and impede the efficiency of nutrient transport. (f) The simple perforation plates in more advanced vessel elements, as found in most angiosperms, provide a much more efficient and rapid system for water flow from roots to shoots.

most assimilates. These cells lack nuclei and have very few organelles but are filled with a dense matrix containing long strands of P-protein (phloem protein) that facilitate assimilate movement. P-protein strands extend through holes in the sieve plate that connects adjacent sieve-tube members. Each sieve tube member is connected via plasmodesmata to a smaller, nucleated, organelle-rich *companion cell* that regulates its metabolism, provides energy (ATP) and proteins, and is involved in loading and unloading of assimilates.

The principal phloem contents are sucrose (0.3–1.0 M), the amino acids glutamate, glutamine, aspartate and asparagine (2–10 mM), hormones such as auxins, a few proteins, sRNAs, and a range of inorganic solutes including potassium, magnesium, phosphate and chloride. A few species transport alternative sugars to sucrose, such as stachyose, raffinose or polyols such as sorbitol or mannitol. According to the widely accepted 'mass flow' hypothesis, the active removal of assimilates at a sink tissue reduces the local concentration of solutes. This leads to the loading of more assimilates from source tissues and their continued transport through the phloem to sink sites due to osmosis.

In most crop plants, the regulation of phloem activity is especially important for the optimum allocation of assimilates to economically important sinks such as seeds, fruits and other storage organs. However, in some useful plants, such as pasture grasses, breeders are more concerned with ensuring assimilate allocation to new leaves, rather than storage organs, because the former are required for feeding livestock. In addition to its role in hormonal signalling, recent evidence suggests that the phloem is an important pathway for sRNA-, hormone- and protein-mediated signalling throughout the plant (see Figs 6.8 and 6.10). These signals may be involved in a wide range of processes, including the exchange of environmental information between different parts of the plant, regulation of source–sink dynamics, responses to pathogens, and the regulation of developmental processes such as flowering.

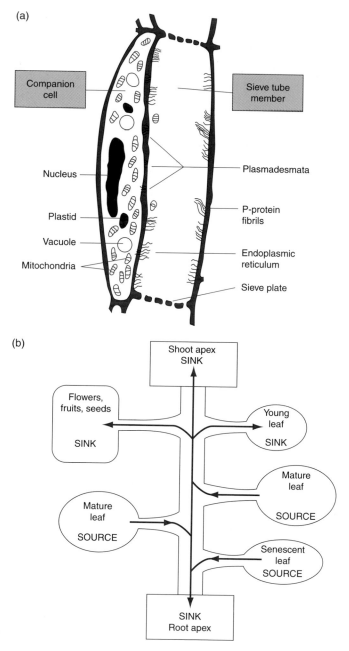

Fig. 5.5. Conducting cells of the phloem. (a) Assimilates flow through sieve tube members of the phloem. These enucleated cells are relatively wide with few organelles except for some peripheral ER and P-protein fibrils. Assimilates flow from one sieve tube member to another via sieve plates that contain protein channels that form pores that can be regulated to control the flow. Adjacent companion cells are densely packed with organelles, especially mitochondria, and are linked to sieve tube members by numerous plasmadesmata. (b) The phloem network transports assimilates and other compounds from source tissues such as photosynthetically active mature leaves to sink tissues such as growing meristems, roots, flowers, fruits and seeds, and immature leaves. Storage organs such as seeds, tubers and rhizomes, initially act as strong sinks when they are being formed but subsequently act as source tissues when their storage products are eventually used to support growth of new plant tissues.

5.3 Plant Reproduction

With the exception of relatively recently formed polyploids, most vascular plants have a dominant diploid adult generation. Although many higher plants are capable of asexual reproduction, most species can also reproduce sexually via a short-lived haploid stage. In flowering plants, the haploid male gamete is contained within the pollen grain and the haploid female gamete, the egg cell, is located in the embryo sac. Pollen grains are released from male floral organs and may fertilize eggs in female floral organs on plants as much as several kilometres away. For some crops, male sterility is a useful trait to minimize undesirable gene flow, e.g. from a transgenic variety, or to ensure seed sterility, e.g. for hybrid production. In most wild plants, self-fertilization (fertilization of an egg by pollen from the same plant) tends to be excluded by various physical or biological mechanisms although it can be a desirable attribute for breeders seeking to create genetically uniform crop varieties.

Following successful pollination, fertilization of an egg by a pollen nucleus gives rise to a zygote that develops into an embryo and then into a mature seed, eventually producing a new adult plant. However, unlike animals where there is just one fertilization between sperm and egg, flowering plants have another fertilization event whereby a second pollen nucleus fuses with two maternal polar nuclei in the embryo sac. This results in a triploid cell, which divides to produce the nutritive endosperm tissue that surrounds the growing embryo as the seed develops. The seed endosperm is a particularly important tissue for humans because it is the site of accumulation of our major source of dietary calories, namely the starch-rich storage reserves of cereal crops such as rice, wheat and maize. Globally speaking, endosperm-derived starch supplies over half of total human calorie intake. In most other grain crops, including seed legumes and oilseeds, the major seed reserves, such as starch, protein and oils, accumulate in the swollen cotyledonary tissues of the developing embryo rather than in the endosperm.

The processes of flowering, fertilization and seed development are major targets for manipulation by breeders seeking to improve crop yield and quality. However, because many plant species, including major crops such as potatoes and bananas, sometimes or always reproduce asexually, the mechanisms of vegetative or clonal propagation are also of paramount interest to plant biotechnologists. Finally, the development of technologies such as apomixis might eventually allow breeders to circumvent sexual reproduction altogether, which could considerably enhance their ability to manipulate plants for human use.

Sexual reproduction

Genetic variation is the raw material of evolution by natural selection and sexual reproduction is one of the key strategies used by most organisms to achieve sufficient levels of variation to enable them to cope with environmental changes. Other mechanisms for creating variation include horizontal gene transfer and mutation (see Chapter 3), although these processes operate over long-term evolutionary timescales and are more relevant to simpler organisms with relatively rapid life cycles than to complex, long-lived eukaryotes such as higher plants. A major constraint for sexually reproducing plants is their lack of mobility, which requires them to release their male gametes into the open environment rather than directly on to a recently laid egg mass or into the body of a receptive female. Many flowering plants have overcome this challenge by recruiting animal vectors to deliver their pollen. Indeed, the coevolution of insect pollinators with angiosperms may be one of the reasons for the evolutionary success of both groups of organisms (see Chapter 2).

Flowering

STRUCTURAL ORGANIZATION Most angiosperm flowers are made along similar lines with four concentric whorls – *calyx, corolla, androecium* and *gynoecium* – each comprising a separate floral organ (see Fig. 5.6). All four whorls are derived from modified leaves and intermediate stages of their transition from a laminate shape to a carpel or stamen can be seen in some of the more primitive angiosperms. The two outer whorls are not directly involved in gamete formation and are termed the perianth. In some species, especially basal angiosperms and monocots, the perianth consists of two whorls or a spiral of undifferentiated tepals that surround the stamens and carpels.

The outermost whorl (whorl 1) forms a calyx. The calyx is made up of several sepals, which are leaf-like structures that envelop and protect the flower bud during its early development. The second

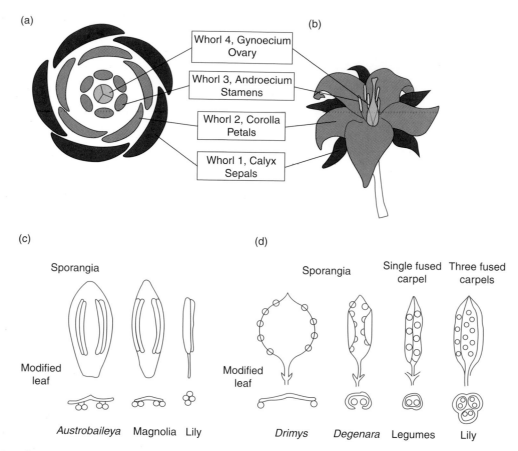

Fig. 5.6. Structure and evolution of the floral whorls. The four concentric floral whorls shown stylistically (a) and in terms of the respective organs (b). (c) Evolution of whorl 3, the androecium, from tubular male sporangia attached to a modified leaf in primitive angiosperms such as *Austrobaileya*, to the four fused chambers found in the anthers of lily. (d) Evolution of whorl 4, the gynoecium, from spherical female sporangia on a modified leaf in *Drimys* to the three fused carpels found in the lily.

whorl (whorl 2) forms a corolla, which is made up of the petals. Petals are also relatively leaf-like with a basic laminate structure, although they can assume a huge variety of shapes, sizes and colours. Showy, coloured petals that advertise the flower are found in plants that rely on animal vectors (especially insects) for pollination. In contrast, anemophilous, or wind-pollinated, flowers tend to have much-reduced or even absent petals. The number of sepals or petals in a flowering plant is characteristic of its class, with monocots having groups of three per whorl while dicots tend to have groups of four or five per whorl.

The two inner whorls are respectively the sites of male and female gametogenesis. Whorl 3 forms the androecium, which is made up of several stamens, each consisting of a stalk or filament that supports an *anther*. The anther is the site of male gamete formation in plants, the gametes being released as part of the pollen grains that also contain one or more additional nuclei that play a role in double fertilization. The innermost whorl (whorl 4 or carpel) forms the gynoecium, or female reproductive apparatus. Each flower normally contains a single gynoecium organized in three parts with a basal ovary, a tubular style and a terminal *stigma* that acts as the receptive surface for pollination.

GENETIC REGULATION Floral identity is specified by a small number of homeotic genes that act together to specify the various organs present in the whorls. The modes of action of floral homeotic genes are strikingly parallel to the major homeotic, or *Hox*, genes in animals that determine body organization in groups as diverse as insects, worms and vertebrates. Most of our knowledge about the molecular genetics of flowering comes from the dicots, *Arabidopsis* and *Antirrhinum*. While the original so-called ABC model derived from these species is still widely used, it does not necessarily apply to all angiosperms and has now been updated as the ABCE or 'quartet' model (see Fig. 5.7). At its simplest, the traditional ABC model specifies that in most angiosperms whorl 1, the sepals, is specified by A genes such as *apetala1* and *apetala2*. Whorl 2, the petals, is specified by B genes such as *apetala3* or *pistillata*. Whorl 3, the stamens, is specified by B genes plus the C gene group, *agamous*. Finally, whorl 4, the carpel, is specified by the C genes alone. According to the newer *ABCE model*, a further class called E genes, such as *sepallata1*, acts alongside A, B and C genes to specify all four floral organs although the E genes are functionally redundant.

The A, B, C and E gene products are proteins that act as transcriptional activators by 'switching on' large numbers of genes involved in specifying the nature of each of the floral organs. All of these proteins, except APETALA 2, contain the same DNA-binding domain, the MADS box, a motif also found in several other transcriptional activators involved in other aspects of plant development. In each whorl, dimers of the relevant MADS proteins are proposed to bind to specific DNA regions in the promoter regions of their target genes. The DNA regions that bind MADS box proteins have the consensus sequence CArG (i.e. $CC(A/T_6GG)$ and can either be adjacent to one another or some distance apart within the promoter. Tetramers form through protein–protein interactions between the MADS protein dimers, which generates a complex that is bound to two CArG-box binding sites. Formation of this complex leads to activation of gene transcription. The predicted composition of tetramers in the four whorls are: AP1–AP1–SEP–SEP in whorl 1 to specify sepals; AP1–SEP–AP3–PI in whorl 2 to specify petals; AG–SEP–AP3–PI in whorl 3 to specify stamens; and AG–AG–SEP–SEP in whorl 4 to specify carpels. Protein identities are as follows: AG, AGAMOUS; AP1, APETALA 1; AP3, APETALA 3; PI, PISTILLATA; SEP, SEPALLATA.

Advances in the understanding of the genetic regulation of floral development are helping to explain some of the phenotypes observed in crops such as brassicas. This research is also assisting in efforts to better control the flowering process during tissue culture-based propagation of some major crops. For example, as described in Box 5.1, it is helping breeders to avoid floral abnormalities that can result from attempts at mass clonal propagation of economically important tree species such as oil palm.

Gametogenesis

Gametogenesis is the process of producing haploid male and female gametes and occurs in the flowers of angiosperms. Male gametes are produced in the anthers, which are the terminal parts of the stamens in whorl 3, the androecium. Most anthers contain four chambers where male gametes are produced as part of the pollen grains. In each chamber, a diploid sporocyte cell divides asymmetrically to form one line of nutritive tapetal cells and a second line of microspore mother cells. The microspore mother cells undergo two meiotic divisions to form a tetrad of haploid cells, the microspores. The diploid tapetal cells form a ring around the developing microspores in each chamber. The tapetum then supports the further growth of the microspores as they undergo an asymmetric mitotic division to produce a pollen grain consisting of a larger vegetative cell and a smaller generative cell

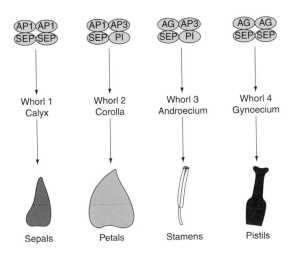

Fig. 5.7. The ABCE model of floral identity. The identity of each of the four floral whorls is specified by the combined action of four groups of homeotic genes. See text for details. AG, agamous; AP, apetala; PI, pistillata; SEP, sepallata.

Box 5.1. Floral abnormalities in oil palm

The importance of understanding floral initiation and its genetic regulation is illustrated by the problems faced by biotechnologists seeking to mass-propagate clonal lines of high-yielding oil palm plants. Oil palm is the major global edible oil crop and a valuable source of industrial products such as oleochemicals and biofuels.

Tissue samples from selected elite trees are cultured in the presence of nutrients and growth regulators in order to propagate millions of new seedlings. These seedlings are then planted on a vast scale in commercial plantations. Unfortunately, during the tissue culture process, the action of growth regulators can sometimes cause MADS box genes involved in the specification of floral identity to became abnormally methylated. This epigenetic phenomenon results in the stamens of both male and female flowers being transformed into carpels, lead-

ing to the development of misshapen flowers with a so-called *mantled* phenotype. These sterile flowers do not produce any fruits and the trees are economically worthless.

Due to the slow growth rate of the oil palm tree it took more than 5 years before the problem was recognized when mass propagation was first attempted for this crop. By this time, millions of sterile trees had been planted and these now had to be removed at a cost of millions of dollars. Research is now revealing how the floral identity genes operate normally in oil palm and how they can become inappropriately methylated during tissue culture. This is helping scientists to develop new methods for the early detection and minimization of such abnormalities, which is now enabling the industry to resume mass propagation of high yielding elite lines of oil palm.

(see Fig. 5.8A). It is important for the fate of the daughter cells that this mitotic division is asymmetric. Treatment of microspores with the microtubule-disrupting agent colchicine leads to a loss of asymmetry and the microspores often form haploid embryos rather than pollen grains. This technique is regularly used by breeders to create haploid plants, which, as we will see in Chapter 8, can be invaluable tools for crop improvement.

In normal pollen grains, each generative cell divides mitotically to form the two haploid 'sperm' nuclei responsible for the double fertilization process in angiosperm sexual reproduction. Pollen grains mature within the chamber, or locule, bounded by the tapetum. Towards the end of pollen development, the tapetum supplies precursors for the formation of the uniquely durable sporopollenin wall that surrounds mature pollen grains. The tapetum can also export special mixtures of lipids and proteins that form the sticky pollen coat found in some species where pollen is propagated by animal vectors. In such cases the pollen grains are often relatively large ($<300\,\mu m$ diameter) with internal storage reserves, such as lipids or carbohydrates that accumulate in a similar manner to seed storage reserves. In species that produce anemophilous (wind-borne) pollen, the grains tend to be smaller (20–$60\,\mu m$ diameter) and lighter, with fewer storage reserves and a relatively smooth coat.

In all angiosperms, as the anthers mature the tapetum and parts of the anther wall undergo apoptosis (programmed cell death). This causes the anther chambers to open and release the pollen grains. In plants that produce vector-borne pollen, the opening of the outer floral whorls, especially the often highly coloured petals, coincides with pollen release in order to attract appropriate vector species. Flowering in such species is programmed to coincide with the availability of vectors, which is why many insect-pollinated plants tend to flower later in the spring, when flying insects have hatched. In contrast, many wind-pollinated plants flower in the early spring when they can take advantage of windier conditions. Even if they are ready for floral opening and pollen release, many insect-pollinated plants are able to delay these processes during adverse weather conditions, such as rain or strong wind, that may limit insect availability.

Female gametes in angiosperms are produced within the carpel that is the innermost whorl 4, or gynoecium. Within the gynoecium is a series of ovules, each containing a female gametophyte or embryo sac where the egg, and later the embryo, develops. In the majority of angiosperms a megasporocyte undergoes two rounds of meiosis to produce four haploid cells, only one of which becomes a megaspore while the other three degenerate. The megaspore undergoes three sets of mitotic divisions

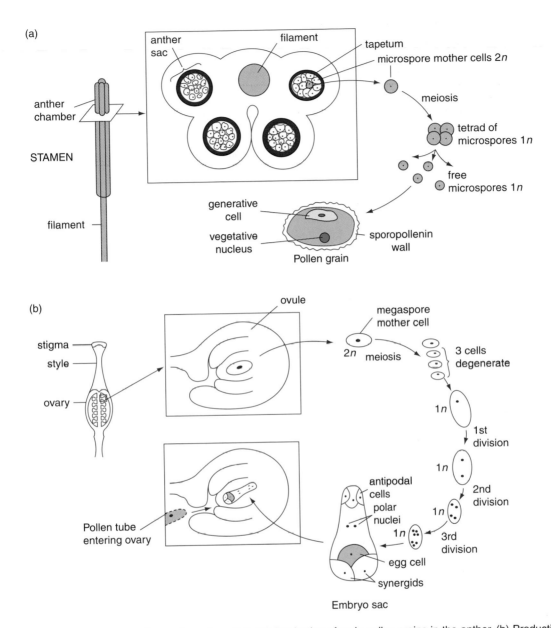

(a)

anther
sac

filament

tapetum

microspore mother cells 2n

anther
chamber

STAMEN

filament

meiosis

tetrad of
microspores 1n

free
microspores 1n

generative
cell

vegetative
nucleus

sporopollenin
wall

Pollen grain

(b)

ovule

megaspore
mother cell

stigma

style

ovary

2n   meiosis

3 cells
degenerate

1n

1st
division

1n

2nd
division

1n

Pollen tube
entering ovary

antipodal
cells

polar
nuclei

1n

1n

3rd
division

egg cell

synergids

Embryo sac

Fig. 5.8. Gametogenesis: pollen and egg formation. (a) Production of male pollen grains in the anther. (b) Production of female egg cells in the ovary. See text for details.

to produce a female gametophyte consisting of seven cells containing eight nuclei with the egg cell at the basal end. Within the embryo sac, the egg cell is flanked by two synergids that help guide the pollen tube into the ovary where one of the sperm cells will fuse with the egg nucleus to form the zygote. Meanwhile the large central cell contains the two polar nuclei that will fuse with a second sperm cell to form the triploid endosperm (see Fig. 5.8b).

Fertilization

In most species, pollen grains are transported from one plant to another either by wind or via biological vectors. By far the most common pollen vectors are insects such as beetles, flies, butterflies and moths. Insect pollination, or entomophily, with beetles as the earliest pollinators was probably the original method of pollen dispersal in angiosperms. This theory is supported by the presence of petals as the floral whorl 2 from the very beginning of angiosperm evolution. Other animal vectors of pollen include birds, bats, reptiles and occasionally small ground-dwelling mammals. Petals are important for attracting potential pollinators but have become secondarily reduced or even lost in wind-pollinated plants, such as grasses. For example, there is a small structure termed a lodicule at the base of grass flowers that swells up and initiates opening of the flowers prior to pollen release. Although it looks nothing like a petal, the lodicule expresses typical whorl 2 genes and is evidently a former petal or complete corolla that became redundant once grasses developed wind pollination. In such plants, the lodicule has now acquired a new function in assisting pollen release.

In order to germinate, a pollen grain must land on the surface of a receptive stigma, which is the uppermost part of the female reproductive organ, the carpel. Most plants favour outbreeding and pollen is normally prevented from germinating on a stigma from the same plant. This can be achieved by the physical or temporal separation of pollination and stigma development, or via biochemical signals that identify self-pollen. In many species where pollen and receptive stigmas are produced at the same time, small peptides on the pollen coat ensure the self-incompatibility of pollen and stigma from the same plant. If a self-pollen grain lands on a stigma, these peptides bind to receptors on the stigmatic surface causing the immediate vicinity of the stigma to lose water, which renders it incapable of hydrating the pollen grain. This is a highly localized response and adjacent parts of the stigmatic surface are still able to support germination of compatible non-self pollen grains.

During a compatible pollination response, the stigma facilitates pollen grain hydration, resulting in the emergence of a tube from one of several apertures in the pollen wall. The hydrated pollen tube rapidly grows through the surface of the stigma and enters a nutritive transmitting tissue from which it grows towards the ovary. The stigma is separated from the ovary by a tubular style of varying length according to the species (see Fig. 5.9). One function of the style may be to impose a form

of selection on the dozens of pollen grains that may germinate almost simultaneously on the stigma, so that only the fastest and fittest will reach the ovary and fertilize an egg. This process is analogous to the competition between the thousands of animal sperm typically deposited in a female vagina as they race to fertilize egg cells in the uterus.

Embryogenesis and seed development

A developing seed is a chimeric entity consisting of several tissues derived from a mature ovule. It normally contains a diploid *embryo* as well as the triploid *endosperm* and maternally derived diploid cells that form the seedcoat. Of the various seed tissues, only the embryo survives into the next generation. The seedcoat is discarded after germination and the endosperm is simply a food reserve for the embryo that is used up either during the latter part of seed development or shortly after germination. In species with a transient endosperm, the mature seed consists almost exclusively of an embryo that stores its food reserves in enlarged cotyledons. Examples include most grain legumes such as peas and beans, and oilseeds such as sunflower and rapeseed. In species with a persistent endosperm, the mature seed consists of a relatively small embryo surrounded by an extensive endosperm that acts as a store of food reserves such as starch. Examples include cereal grains, castor beans and most umbelliferous seeds such as parsley and carrots (Fig. 5.10).

Many seeds develop within fruits, which are exclusively maternal organs. Some fruits are relatively large and may contain numerous small seeds that are released as the fruits dehisce. Often, fruits are fleshy and nutritious and may be coloured and/ or scented in order to attract animal vectors. These animals eat the fruits but tend to pass out the undigested seeds in their droppings, hence both dispersing and fertilizing them. In the most important group of crop plants, the cereals, what looks like a mature seed is actually a dry fruit called a *caryopsis*. This structure consists of a seed, containing a starchy endosperm and a much smaller embryo or germ, with an outer thin layer of pericarp or fruit skin that is tightly attached to the seed.

Following fertilization, the *zygote*, or fertilized egg, becomes located at the base of the embryo sac, establishing a defining polarity that is the context for subsequent embryo development. Cells arising from divisions in the basal region of the zygote

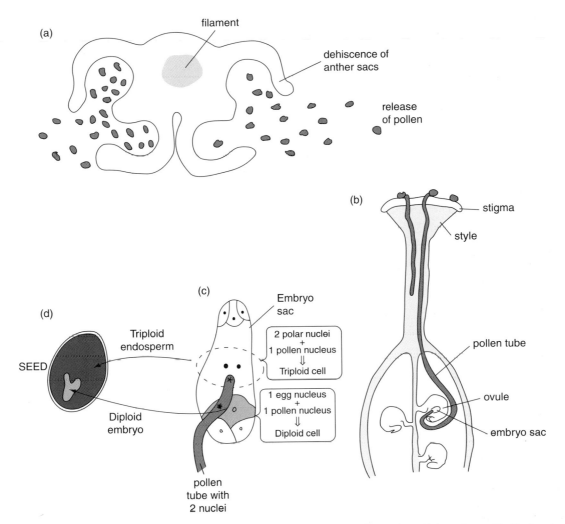

Fig. 5.9. Fertilization. (a) Anther sacs open following dehiscence and release pollen grains, which are transmitted to female stigmas via either wind (anemophily) or animal vectors, of which the most important group are insects (entomophily). (b) Pollen grains landing on a stigma tend to be rejected (but not in all cases) if they are from the same plant or a different species. Following a compatible response, a pollen grain will germinate and produce a tube that transports the haploid vegetative and generative nuclei to an ovule. (c) The ovule contains a haploid egg cell that fuses with the pollen generative nucleus to produce a diploid zygote. The ovule also contains two haploid polar nuclei that fuse with the pollen vegetative nucleus to produce a triploid cell. (d) The diploid zygote divides and differentiates to produce the embryo, or future plant, while the triploid cell divides to form the nutritive endosperm tissue found in angiosperm seeds.

differentiate into elongated cells that form the suspensor, a stem-like structure that links the embryo with the maternal tissue. The suspensor is a conduit for delivery of nutrients to the embryo, especially during early stages of seed development, although it degenerates later in seed maturation. Above the suspensor, a series of rapid cell divisions forms a globular mass of small cells that gradually begin to differentiate into the various embryo tissues. Embryogenesis *per se* is the phase of rapid cell division and differentiation that occurs very early in seed development. During the remainder of seed development, cell division no longer occurs. Instead there is a phase of cell elongation and embryo growth followed by distinct periods of storage product accumulation, dehydration and

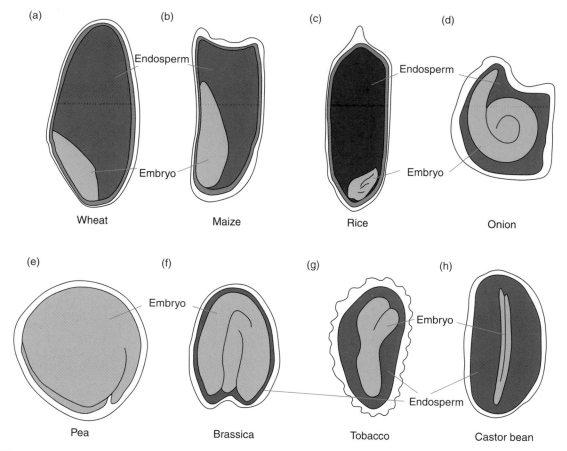

Fig. 5.10. Embryos and endosperm in angiosperm grains and seeds. Mature cereal grains such as wheat, maize and rice (a)–(c) contain relatively small embryos surrounded by a much larger starchy endosperm. In monocots like onion (d), a relatively larger embryo occupies over half the seed volume. In dicot seeds there is a full range of endosperm contents ranging from zero in peas (e), a single cell layer in brassicas (f), almost half the seed volume in tobacco (g) to most of the volume in castor bean (h).

final maturation, all of which are governed by different homeotic genes and associated patterns of hormonal regulation.

The course of seed development can vary considerably in angiosperms but broadly characteristic patterns for dicots and monocots, as exemplified by *Arabidopsis*, are shown in Fig. 5.11. Briefly, the early 8- or 16-cell embryo consists of a globular mass of relatively undifferentiated cells. The globular mass develops into a heart-shaped embryo due to accelerated growth in two regions that make up the lobes of the 'heart'. These lobes are the precursors of the cotyledons, which have a mainly storage function. At this stage, other cell types also differentiate with the outer layer forming a protoderm that develops into the epidermis, while the lower region

includes the hypocotyl, which forms the stem of the seedling after germination, and the radicle, which forms the new root. By 4 days after fertilization, embryo differentiation is complete. For the next 6 days, the embryo expands due to cell enlargement, and storage reserves, mainly oil and protein, are laid down in the swollen cotyledons. The latter stages of embryo growth are at the expense of the endosperm, which eventually shrinks until it is only one or two cells thick in a mature seed.

In maize, the pattern formation stage of embryogenesis takes about 10 days and results in a single-lobed embryo resembling a tiny blade of grass attached to an even smaller radicle and hypocotyl. Over the next 50 days the embryo expands considerably via cell enlargement and some storage

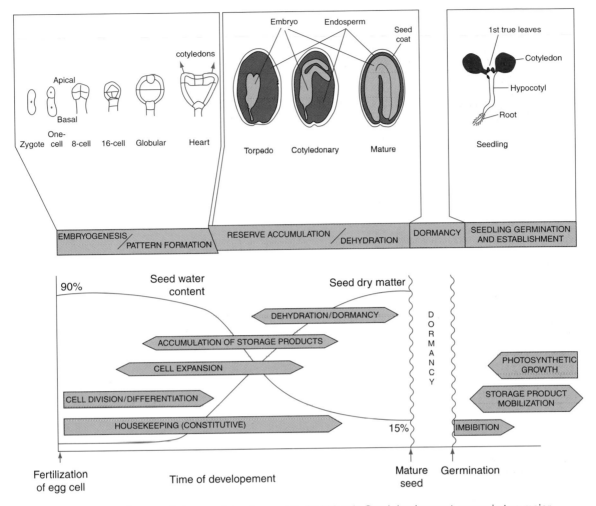

Fig. 5.11. Sequence of events during seed development in *Arabidopsis*. Seed development occurs in two major phases, embryogenesis/pattern formation and reserve accumulation followed by dehydration. Mature seeds may then experience a prolonged period of dormancy before they finally germinate to produce a new plant. See text for additional details.

reserves are accumulated, especially in the outer layer or scutellum. However, the vast majority of the reserves, including nearly all of the starch, accumulate in the endosperm, which continues to expand until it forms the bulk of the grain volume. The mature maize grain consists mainly of a large starchy endosperm plus a much smaller embryo or germ that contains oil and protein reserves.

The accumulation of important storage reserves, such as starch, oil and protein, in seeds, fruits and modified roots is regulated by interactions of hormones such as ABA with a large number of regulatory genes such as *pickle* and *wrinkled1*.

The *pickle* gene encodes the chromatin-remodelling factor CHD-3, which results in the suppression of metabolic pathways normally expressed during seed development. Hence *pickle* genes are active in non-seed tissues such as leaves and stems. Mutant *Arabidopsis* plants with a non-functional *pickle* gene accumulate storage oil and protein in ectopic locations such as leaves. In the future, disruption of *pickle* genes via RNAi or induced mutagenesis may be useful for the engineering of biomass crops to accumulate high-calorie oils as well as lignocellulosic fuels in their leaves and stems (see Box 4.2).

Asexual reproduction

Sexual reproduction is an important strategy for maintaining desirable variation in most eukaryotic populations. However, asexual reproduction can also be effective in enabling already well-adapted individuals to produce clonal offspring that can rapidly exploit an existing ecological niche in the short to medium term. It is also one of the most effective ways for humans to propagate economically useful plants. One of the major differences between complex multicellular animals and the higher plants is the much higher incidence of asexual strategies for reproduction in plants. Although parthenogenesis occasionally occurs in animals (as an analogy to one of the forms of plant apomixis), there are no analogous mechanisms to the numerous forms of vegetative propagation found in all plant groups.

As discussed previously, cell fate during animal development is irreversibly programmed in a manner that normally precludes differentiated tissues from being reprogrammed to form new clonal tissues or individuals. In contrast, plants can readily use vegetative propagation to form new clonal tissues, organs and complete individuals, even from fully differentiated structures such as stems and roots. This ability has been important in facilitating the spread of many plant species as they colonized the earth. As shown in Fig. 5.12, although sexual reproduction has become more common during plant evolution, almost 45% of plants can still propagate themselves clonally. By exploiting vegetative propagation, people have been able to manipulate many useful plants, including crops, ornamentals and undomesticated species, for millennia. Today, biotechnologists are developing new forms of asexual propagation of plants in order to create uniform, true-breeding hybrids or transgenic varieties for human use.

Vegetative reproduction

Many plant species are able to propagate clonal copies from vegetative organs such as shoots, roots and stems, or from modified versions of such organs such as tubers, bulbs, corms, rhizomes and stolons (see Fig. 5.13). However, vegetative propagation does not always result in the separation of a new clonal individual from the parent plant. In many cases, what appears to be a widely dispersed group of separate plants of the same species is actually a single super-plant in which most of the shoots are linked underground by a common root system. The ability to remain connected to an existing mature root system can enable a newly formed shoot to exploit a large volume of fertile soil, plus the potentially extensive food reserves stored in many modified roots or stems. This can be a powerful competitive advantage for a new vegetatively propagated shoot when compared with a sexually propagated seedling germinated from a seed with limited food reserves that must form a complete root system *de novo*. The ability to produce new shoots supported by a pre-existing root system is one of the factors behind the ability of several

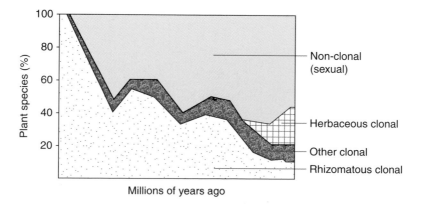

Fig. 5.12. Clonal propagation in the fossil record. The earliest land plants were propagated clonally, i.e. asexually. Non-clonal, sexual propagation became more common as vascular plants diversified but, even today, as many as 45% of extant plant species, including many crops, use some form of clonal propagation. Data from Tiffney and Niklas (1985).

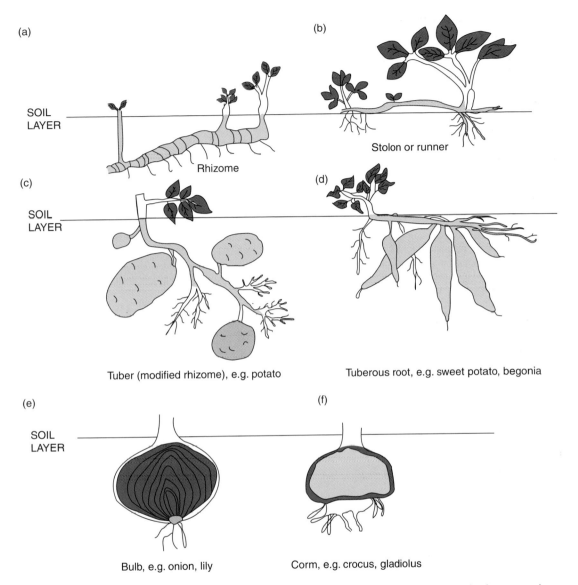

Fig. 5.13. Mechanisms of vegetative reproduction. The most common forms of vegetative reproduction occur via modified stems such as rhizomes (a), stolons (b) or tubers (c) or swollen roots (d). Bulbs (e) are modified underground stems with fleshy swollen leaves while corms (f) are modified stems.

invasive plant species to colonize new habitats and displace the native flora (see Box 5.2).

Around the world, many staple food or cash crops are vegetatively propagated, including potatoes, sweet potatoes, yams and sugarcane. Some of our oldest tree crops, such as olives, were traditionally propagated by grafting new stem cuttings on to an existing rootstock, once the old stem had reached the end of its useful life. Thanks to the old rootstock, with its extensive mycorrhizal network, the new stem will grow much more rapidly and eventually yield a lot more fruit than would a newly planted seedling. These traditional forms of propagation were used successfully by farmers (see Chapter 7) and nowadays they are still used by breeders alongside newer techniques of vegetative propagation. For example, in the 1950s, Indonesian farmers discovered

Box 5.2. Invasive plant species and vegetative propagation

Many of the most troublesome invasive plant species owe a great deal of their success to their ability to withstand pruning and to propagate themselves vegetatively. Across Northern Europe, one of the most economically damaging of these plants is Japanese knotweed, *Fallopia japonica*. This plant tends to displace indigenous flora and is very difficult to eradicate, especially as it has few if any enemies outside its home range in Japan. Introduced *Fallopia japonica* is male-sterile and therefore can only propagate itself vegetatively. This means that knotweed plants tend to be all genetically similar to one another. Indeed, DNA analysis of knotweed plants in Britain has revealed that they all comprise a single genetically identical clonal population. These plants make up for their lack of genetic diversity by their ability to store assimilates produced during the growing season in fleshy rhizomes. The rhizomes can spread underground and re-sprout some distance from the parent plant. Even if the parent plant is cut down or treated with herbicide, new shoots can quickly emerge from undamaged rhizomes.

Because it grows and spreads faster than most native vegetation, knotweed forms a monoculture that sustains far lower levels of animal life, reducing local biodiversity. Knotweed can also grow through walls, tarmac and even concrete, which makes it a particular menace in building sites. For these reasons, knotweed is now classified as a controlled species in Britain. It is illegal to transport or plant knotweed and all parts of the plant, including rhizomes, must be removed from any construction site before building work is permitted. Despite these measures, knotweed continues to spread across Europe and North America and a great deal of research is now focused on finding improved chemical treatments and/or new pest or disease agents for more effective biocontrol of knotweed.

---

that yields of the edible roots of cassava, *Manihot esculenta*, can be greatly increased by grafting stalks of the vigorous wild relative *M. glaziovii* on to crop rootstock. More recently, breeders have produced several similar cassava grafting hybrids that have as much as sevenfold increases in edible root yield.

In commercial agriculture, vegetative propagation is often used to create uniform populations of superior varieties of highly heterozygous plants, such as recent hybrids or relatively undomesticated, highly outcrossing species. This can be difficult to achieve via seed propagation where seed-derived progeny are often less uniform and of lower quality than their parents. It is especially true of tree and vine crops, such as nuts, apples and berries when the parent variety is a high-yielding hybrid that does not breed true. A potential drawback of vegetative propagation is the possibility of pathogen transmission from an infected parent to its progeny. In several vegetatively propagated crops in developing countries, including cassava, sweet potatoes and cooking bananas, this mechanism of pathogen transmission can seriously affect food production. One of the aims of breeders is to produce high-yielding and pathogen-free cuttings for farmers, as this can have a huge impact on crop health and productivity.

Apomixis

Apomixis in flowering plants is a particular form of asexual plant reproduction that involves clonal propagation from seeds, or agamospermy, rather than from vegetative structures such as roots or shoots. Apomixis is widespread in higher plants and is most common in the largest monocot family, the Poaceae. In major dicot families, such as the Asteraceae and Rosaceae, apomictic genera are also widespread. Given its widespread distribution, apomixis probably evolved independently several times in angiosperms and it is therefore puzzling that it appears to be absent in the large legume family, the Leguminoseae.

As shown in Fig. 5.14, there are several naturally occurring mechanisms of apomixis in angiosperms. In the commonest forms of heritable apomixis (diplospory and apospory), diploid cells derived from megasporocytes or other embryo sac cells fail to undergo meiosis, but nevertheless acquire the ability to enter into embryogenesis and produce fertile seeds. Alternatively, as in several *Citrus* and *Garcinia* species, diploid somatic (i.e. non-reproductive) cells from around the embryo sac become adventitiously embryogenic and produce viable seeds. In either case, embryogenesis and seed development occur inside otherwise normal

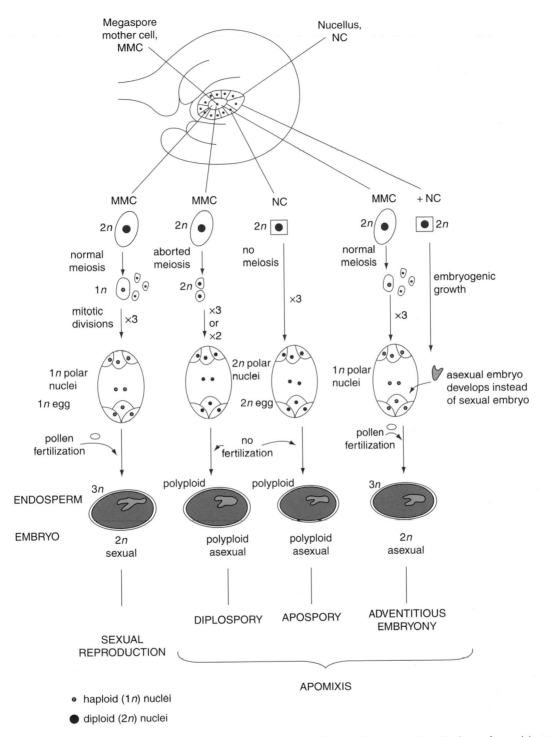

Fig. 5.14. Mechanisms of apomixis compared with sexual reproduction. Three of the principal mechanisms of apomixis are diplospory, apospory and adventitious embryony. In terms of commercial use, adventitious embryony has the advantage of producing seeds with a diploid somatic embryo and a normal triploid endosperm. Adapted from Koltunow (1993).

reproductive structures without either meiosis or fusion of gametes. The resulting progeny are genetic clones of the parent plant and will have the same ploidy level. In another form of non-heritable apomixis, as found in some *Solanum*, *Nicotiana* and *Lilium* species, a haploid embryo can develop from an unfertilized egg or from another haploid cell within the gametophyte. The progeny are sterile haploids that are normally of limited value, although they can become useful to breeders if they can be converted into doubled haploids via chromosome doubling technologies (see Chapter 8).

Because reproductive or germline cells are more protected from the accumulation of mutations than vegetative or somatic cells, apomixis is more likely to maintain genetic fidelity than vegetative reproduction. For example, Bermuda grass, *Cynodon dactylon*, which propagates vegetatively, has up to ten somatic mutations per genome per generation. In contrast, the rate of mutations in germline cells is typically less then one per genome per generation. A further advantage of apomixis is that the presence of a seed phase facilitates dispersal from the mother plant with the additional possibility of the seed remaining dormant during adverse environmental conditions. It is unlikely that apomixis is completely obligate in any particular plant species. However, facultative apomixis can enable the rapid spread of advantageous variants in a particular environment without necessarily compromising the long-term adaptability of a species.

Apomixis has several advantages for plant breeders. Unfortunately none of the major crops or other useful plants are naturally apomictic and it may require gene transfer to get the trait into such plants. Recent efforts to develop apomixis technology for crop plants and the possible benefits of such varieties are discussed in Chapter 10.

5.4 Hormones and Signal Transduction

Plant hormones are analogous in some respects to animal hormones. Hormones are small organic molecules normally produced remotely from their sites of action. They regulate often-complex physiological and/or developmental processes, acting either individually or in combination with other hormones. Many hormones act directly on genes, or via signal transduction pathways, to initiate or block transcription. Often the target of a hormone is a regulatory gene that itself affects the activity of many other genes. Hormones such as gibberellins,

ABA, auxins and jasmonates interact with *cis*-acting DNA sequences in the promoter regions of their target genes (see Chapter 2). Often these *cis*-acting sequences are highly conserved between plant species. The structures of the major classes of plant hormone are shown in Fig. 5.15.

In a few cases plant and animal hormones are chemically related compounds. For example, the cholesterol-derived steroid hormones, which are one of the commonest and most important classes in animals, are structurally similar to the brassinosteroid hormones of plants. The simple gaseous hydrocarbon, ethylene, which is classified as a hormone in plants, can affect neighbouring plants as well as its host. The hormones jasmonic and salicylic acid, which often act in concert in plant defence responses, can be methylated *in vivo*, making them volatile and potentially able to influence neighbouring plants and even animals. The ability of a volatile plant hormone to affect other plants is analogous to the pheromones produced by most animals. However, not all plant hormones are directly comparable to animal hormones. For example, some plant hormones affect the cells in which they are synthesized, as well as more remote tissues.

Plant hormones are sometimes referred to as plant growth regulators, but in many instances it is not just growth or development that is regulated by these molecules. For example, they also regulate physiological processes such as pathogen responses. Here, we will use the term hormone because this most accurately describes their major functions. The term 'plant growth regulator' will be restricted to describing those naturally occurring or synthetic analogues of plant hormones used commercially in the cultivation and postharvest management of plants, such as fruits, vegetables and cut flowers.

Major classes of plant hormone

There are hundreds of plant compounds that could be described as hormones. Eight of the most important groups of plant hormone are described below. With the exception of ethylene, each of these groups contains structurally related members, ranging from only two for jasmonates to more than 100 in the case of the gibberellins. The biological roles and commercial uses of the major plant hormones and growth regulators are listed in Table 5.1. In addition to these major hormone classes, there are probably numerous other compounds that affect growth, development and environmental response

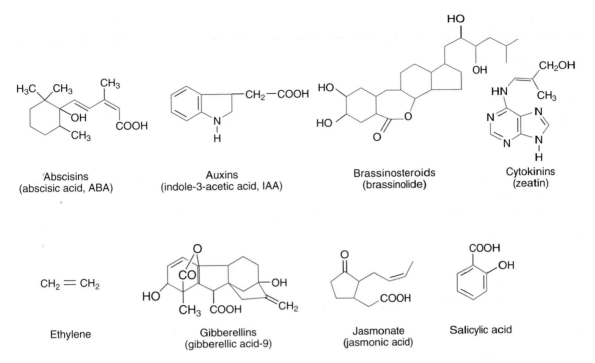

Fig. 5.15. Structures of the major classes of hormones in plants. See text for details.

processes in plants. For example, several peptides with putative hormone-like functions have been reported. Also, as outlined above, sRNA molecules have been identified in the phloem and may act as transportable signals that regulate gene expression in remote parts of the plant in processes such as leaf development and pathogen responses. The manipulation of plant hormones and their target genes are important aspects of academic and commercial applications of plant biotechnology.

Abscisic acid (ABA)

As its name implies, this terpenoid-derived compound was originally believed to be involved in the abscission of fruits and leaves. Although ABA is now believed to have little involvement in controlling abscission its name has persisted. The ABA biosynthetic pathway shares its early stages with the plastid-located carotenoid pathway until the long-chain C-39 neoxanthin molecule is cleaved to produce the much shorter chain C-14 compound xanthoxal, which gives rise to ABA. While ABA is the best-known and commonest member of this hormone class, it includes several other abscisins.

In general, abscisins tend to inhibit growth, for example they induce dormancy in seeds and buds and prevent their premature germination.

During seed development and germination, ABA and gibberellins (GA) act antagonistically to achieve appropriate growth. Early in seed development a high GA:ABA ratio ensures rapid cell division and expansion, but as the seed matures GA levels fall and ABA levels rise, causing cessation in growth, onset of desiccation tolerance and promotion of dormancy. Following germination, GA levels rise rapidly and the new seedling begins to grow into an adult plant. The most common practical use of ABA is for tissue culture and micropropagation where it can help initiate developmental pathways such as somatic embryogenesis. Such techniques are used commercially on a massive scale, e.g. for clonal propagation of millions of seedlings in some plantation crops, and also for the regeneration of some transgenic crops.

Auxins

Auxins were the first group of plant hormones to be studied and one of their original investigators

Table 5.1. Roles and uses of the major classes of plant hormones and growth regulators.

Hormone class	Chemical forms	Developmental processes affected	Commercial uses
Abscisins	Derivatives of ABA, abscisic acid, a simplified 15-carbon product of carotenoid metabolism	Leaf senescence, bud and seed dormancy; seed desiccation; stomatal closure; seed storage product accumulation	Regulation of fruit ripening in tree crops
Auxin	Aromatic amino acid derivative: IAA, indole-3-acetic acid Synthetic analogues: 2,4 D, 2,4-dichlorophen-oxyacetic acid; NAA, 1-naphthaleneacetic acid; IBA, indole-3-butyric acid	Cell elongation, apical dominance; photo- and gravi-tropism; ethylene synthesis; adventitious root formation	Synthetic analogues used as defoliants; to increase yield in fruits; and for root initiation in culture
Brassinosteroids	Variations on classical steroid ring structure: Over 125 forms	Stem and cell elongation; pollen tube growth; leaf bending and unrolling; xylem differentiation	Promoting plant growth
Cytokinins	Derivatives of adenine nucleotides with isopentanyl side chains Synthetic version: Kinetin	Cell division, leaf expansion and senescence; retardation or inhibition of apical dominance	Induction of cell division in tissue culture; improve-ment of seed germination; delaying leaf and flower senescence
Ethylene	Simple 2-carbon hydrocarbon monoene	Fruit ripening, lateral cell expansion; abscission; senescence	Stimulation of fruit ripening; promoting latex flow in rubber trees
Gibberellins	5-ring gibberellane skeleton with a wide variety of side chains	Stem elongation in shoots and buds; seed dormancy; fruit growth; leaf expansion; flowering	Production of large seedless fruits; promoting malting in barley
Jasmonates	Oxidized derivatives of α-linolenic acid: Methyl jasmonate; Jasmonic acid	Insect and wound responses; many plant developmental processes	Protection from arthropod pests and fungal patho-gens; reduction of chilling damage
Salicylic acid	Simple phenolic compounds derived from benzoic acid: Methyl salicylate; Salicylic acid	Antifungal and antibacterial responses	Major ingredient in the painkiller aspirin

was Charles Darwin. In general, all auxins have a carboxylic acid group linked to an indole ring. The most common auxin, indole-3-acetic acid (IAA), can be derived from several precursors, the best known of which is the aromatic amino acid tryptophan. Auxins are transported in their native form or as inactive conjugates linked to sugars or proteins. Auxin conjugates can be stored and then activated by deconjugation when required. As with most soluble hormones, auxins can be transported over long distances via the phloem network or directly from cell to cell.

Auxins, cytokinins and gibberellins all tend to promote the growth of plant tissues and several man-made auxin analogues are used in plant manipulation as discussed below. As well as stimulating growth, auxins affect many other processes such as apical dominance, lateral root formation, development of the vascular system, and parthenocarpy. Auxins have roles in these diverse and apparently unconnected processes via interactions with other plant hormones, and it is the nature of these interactions rather than the identity of the hormones themselves that determines the developmental response in the tissue concerned.

Due to the instability of the major naturally occurring auxin, IAA, synthetic analogues tend to be used in the many practical applications of auxins in agriculture, horticulture and tissue culture (Table 5.1). One of the most important commercial and scientific uses of auxins is in tissue culture, as shown in Fig. 5.16. This method involves the sequential addition of auxins and cytokinins to undifferentiated callus tissue to promote differentiation of roots and photosynthetic shoots, to regenerate new plants from cultured cells. Because they disrupt the normal balance of endogenous hormones in plant cells, some synthetic

auxin analogues can be used as herbicides (see Chapters 9 and 10). The herbicidal synthetic auxins cause cell elongation by acidification of the cell wall, leading to disorganized cell growth. Tissue proliferation results in stem swelling and disruption of the phloem, preventing movement of assimilates from leaves to roots and ultimately resulting in plant death.

Brassinosteroids

At least 50 brassinosteroid molecules have been described in plants and algae. The best studied group is the C23 brassinosteroids such as brassinolide, which is derived from the steroid precursor campesterol. Other brassinosteroids are synthesized from the phytosterols, sitosterol, cholesterol and isofucosterol. They play roles in many processes

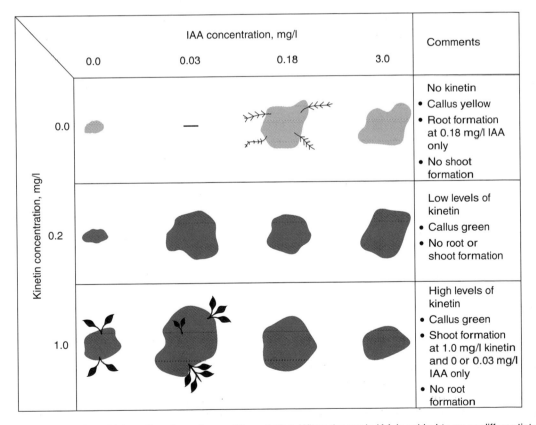

Fig. 5.16. Effects of cytokinin and auxin on tissue differentiation. When the auxin IAA is added to an undifferentiated callus culture, it eventually results in root formation but this is suppressed at the highest concentration. In contrast, the cytokinin kinetin produces greening and shoot formation at higher concentrations. When added together, IAA and kinetin tend to stimulate greening but shoot and root formation are suppressed. Adapted from Mauseth (1991).

including cell elongation, differentiation of tracheary elements, pollen tube growth and ethylene production. Brassinosteroid-deficient mutants tend to be dwarf and exhibit particular growth abnormalities in the dark. Although brassinosteroids have no direct commercial applications at present, their powerful effects on many aspects of plant development make them potential targets for biotechnological manipulation. In particular, there are indications that, while they may not have large effects where crops are already growing well, brassinosteroids may be useful in improving crop performance in suboptimal growing environments.

Cytokinins

These terpenoid-derived compounds include the zeatin-type cytokinins found in most plants. Similar compounds are also found in fungi and bacteria. The first stage in their biosynthesis is the transfer of an isopentanyl group from dimethylallyl diphosphate to ATP or ADP in plants or to AMP in fungi and bacteria. Cytokinins promote cell division in plants and act together with auxins to regulate cell differentiation. They may also help delay senescence or ageing of some tissues and mediate auxin transport throughout the plant. Cytokinins tend to act synergistically with auxins during most of the major growth and developmental stages in the life cycle of a plant.

Ethylene

Ethylene is an extremely simple two-carbon hydrocarbon gas of formula $CH_2=CH_2$. It is involved in processes such as fruit ripening, abscission, senescence and lateral cell expansion. Because of the key role of ethylene in fruit ripening in particular, the manipulation of its biosynthesis and function in plants is a primary target for biotechnologists seeking to delay ripening and spoilage in fruit crops that are picked early before they are ripe and often transported unripe for long distances. Ethylene biosynthesis involves a short pathway from the amino acid methionine with 1-aminocyclopropane-1-carboxylate (ACC) as a key intermediate.

The enzymes ACC synthase (ACS) and ACC oxidase (ACO) are encoded by large multigene families in most plants, and ACS in particular has an important role in the regulation of ethylene production in response to various stimuli. This has led to the manipulation of ACS using antisense-mediated down-regulation of gene expression. For example, during the 1990s the company Zeneca produced transgenic tomatoes with reduced ACS activity leading to lower ethylene production in their fruits and a delayed ripening and senescence. As discussed in Chapter 11, these tomatoes were on sale in the UK from 1996 to 1999 but were discontinued after the *Pusztai affair*. The ethylene signal transduction pathway is discussed below (see also Fig. 5.17).

Gibberellins

Although over 125 different gibberellins have been detected in plants, many are inactive in bioassays and may instead be precursors, breakdown products, or have as-yet undiscovered roles. All gibberellins are based on a five-ring gibberellane skeleton with a variety of side chains. Like several other hormones, gibberellins are derived from terpenoid precursors such as geranyl geranyl pyrophosphate via complex metabolic pathways that are not yet fully characterized. Gibberellins are of historical interest because one of the traits examined in Mendel's experiments on heredity, namely dwarf versus tall pea plants, resulted from a lesion in gibberellin biosynthesis. Gibberellins are also of great agricultural interest because the semi-dwarf phenotypes in the cereal crops, wheat and rice, which formed the basis of the *Green Revolution* of the 1960s and 1970s, are caused by various mutations that affect the formation or perception of gibberellins during plant growth (see Chapter 9).

The most important role of gibberellins during normal plant development is to promote cell growth and expansion, especially in young shoots. Gibberellins are also involved in delaying leaf senescence and abscission, dormancy breaking, mobilization of reserves in seeds, stimulation of flowering in long-day plants and in promoting parthenocarpy in some species. As with other hormone classes, gibberellins rarely act alone. The nature of the response elicited by gibberellins is governed by combination of the nature of the target tissue or organ, and by a series of interactions with one or more additional plant hormones. DELLA proteins are repressors of gibberellin signalling whose functions are conserved in different plant species. Gibberellins promote stem growth by causing degradation of DELLA proteins via the

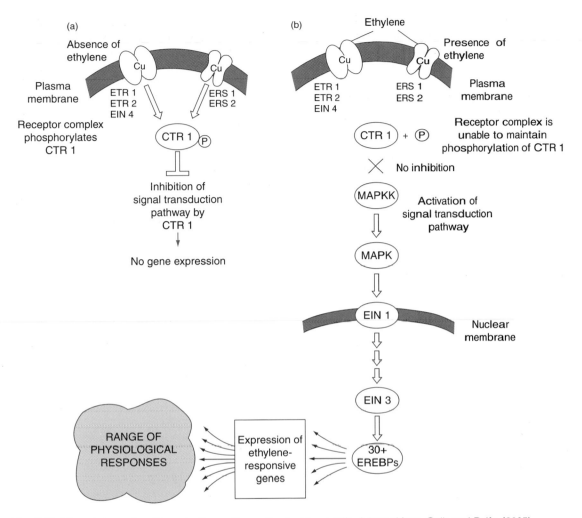

Fig. 5.17. The ethylene signal transduction pathway. See text for details. Adapted from Opik and Rolfe (2005).

ubiquitin–proteasome pathway. The most widely utilized dwarfing alleles in wheat encode gibberellin-resistant forms of a DELLA protein that act as constitutive repressors of stem growth. Dwarfism can also be caused by mutations in genes controlling the biosynthesis or signalling pathway of gibberellins (see Fig. 9.9).

Jasmonates and salicylic acid

These two hormones often act in concert, particularly in the regulation of plant defence responses, although they can also act individually as in the role of jasmonic acid in wound responses (see Chapter 6). Jasmonic acid is an oxidized and cyclized C-12 derivative of the principal fatty acyl component of the thylakoid membranes of chloroplasts, α-linolenic acid. It is synthesized from the oxidation of α-linolenic acid via lipoxygenase and the allene oxide synthase pathway. Salicylic acid is a relatively simple carboxyl derivative of benzoic acid. Although the details of the pathway have yet to be resolved, data from *Arabidopsis* mutants deficient in isochorismate synthase suggest that it is formed from shikimic acid via chorismic acid in a similar manner to bacteria. The *Arabidopsis* mutants were unable to produce salicylic acid in response to pathogen attack, suggesting that the normal plant defence response requires an induction of salicylic acid biosynthesis *de novo*.

Although neither of these hormones has any direct use in commercial agriculture, manipulation of their signalling pathways is of great interest for the investigation of plant defence and stress responses. Finally, the acetylated form of salicylic acid, commonly known as aspirin, is one of the most widely used human medications with many functions, including as a painkiller, an antipyretic and in prevention of blood clots. Aspirin is an ancient folk remedy that was originally derived from willow trees but it is now synthesized chemically.

Hormones and signal transduction in plants

Plant hormones are signalling molecules that frequently act via a cascade of linked processes known as a signal transduction pathway. Often a hormone will bind to a receptor that initiates a series of responses leading to downstream effects such as cell elongation or senescence. Signal transduction pathways are frequently branched and may be linked with one or more parallel signal transduction pathways initiated by different hormones.

In plants, one of the better-characterized signal transduction pathways is that of ethylene as summarized in Fig. 5.17. Full details of this pathway are still being worked out but a brief overview is as follows. Ethylene receptors isolated from plants including *Arabidopsis* and tomato are copper-binding transmembrane proteins located in the endoplasmic reticulum. These receptors interact with kinases or phosphatases in the next stage of the signalling pathway. Addition or removal of phosphate groups by these proteins can either activate or repress the function of downstream components in the pathway. In the absence of ethylene binding, the receptor complex causes the protein CTR1 to become phosphorylated and activated. The activated form of CTR1 effectively shuts down the pathway by inhibiting the downstream kinases.

Ethylene binding to one of its receptor proteins blocks the phosphorylation of CTR1, thereby inactivating it and releasing the downstream kinases from inhibition. These kinases activate further nuclear-located proteins, such as EIN2 and EIN3. Eventually, a series of about 30 ethylene-response-element binding proteins can become activated. Each activated protein can bind to specific DNA motifs in the promoter regions of numerous ethylene-responsive genes. The end result might be induction of the expression of many dozens of genes with numerous divergent functions whose only common feature is to have an ethylene-response element in their promoters.

The action of any given hormone might have more than one set of physiological outcomes, and any given physiological outcome might result from the action of several hormones. This mechanism is known as cross-talk and such cross-talking networks are especially important in the perception of and response to biotic and abiotic stresses, as discussed further in Chapter 6.

5.5   Summary Notes

- The plant body has a relatively simple organization, consisting of outer dermal tissues enclosing a mass of ground tissue through which pass the vascular tissues, xylem and phloem, which transport water, minerals, assimilates and hormones.
- Plant growth is modular and highly flexible with new tissues and organs able to grow to replace those that are lost or damaged.
- Most higher plants can undergo sexual reproduction via a short-lived haploid stage. Pollen and egg cells fuse to produce a diploid zygote while a second fertilization produces the triploid endosperm tissue that nourishes the growing embryo in the seed.
- Almost half of angiosperm species are capable of asexual reproduction, which occurs most frequently via vegetative propagation and less so via apomixis.
- The eight major classes of plant hormone regulate most aspects of body organization, development and response to the environment. Plant hormones tend to act in concert with each other, often regulating gene expression via signal transduction pathways.

Further Reading

General

Mauseth, J.D. (1991) *Botany: An Introduction to Plant Biology*. Saunders College Publishing, Philadelphia.

Opik, H. and Rolfe, S. (2005) *The Physiology of Flowering Plants*. Cambridge University Press, UK.

Scott, P. (2008) *Physiology and Behaviour of Plants*. Wiley, New York.

Taiz, L. and Zeiger, E. (2010) *Plant Physiology*, 5th edn. Sinauer, New York.

Plant organization and growth

Bowman, J.L. and Eshed, Y. (2000) Formation and maintenance of shoot apical meristems. *Trends in Plant Science* 5, 110–115.

Ogas, J., Kaufmann, S., Henderson, J. and Somerville, C. (1999) PICKLE is a CHD3 chromatin-remodeling factor that regulates the transition from embryonic to vegetative development in Arabidopsis. *Proceedings of the National Academy of Sciences USA* 96, 13839–13844.

Zhang, S., Sun, L. and Kragler, F. (2009) The phloem-delivered RNA pool contains small noncoding RNAs and interferes with translation. *Plant Physiology* 150, 378–387.

Plant reproduction

Adam, H., Jouannic, S., Orieux, Y., Morcillo, F., Richaud, F., Duval, Y. and Tregear, J.W. (2007) Functional characterization of MADS box genes involved in the determination of oil palm flower structure. *Journal of Experimental Botany* 58, 1245–1259.

Koltunow, A.M. (1993) Apomixis: embryo sacs and embryos formed without meiosis or fertilization in ovules. *Plant Cell* 5, 1425–1437.

Soltis, D.E., Chanderbali, A.S., Kim, S., Buzgo, M. and Soltis, P.S. (2007) The ABC model and its applicability to basal angiosperms. *Annals of Botany* 100, 155–163.

Tiffney, B.H. and Niklas, K.J. (1985) Clonal growth in land plants: a paleobotanical perspective. In: Jackson, J.B.C., Buss, L.W. and Cook, R.E. (eds) *Population Biology and Evolution of Clonal Organisms.* Yale University Press, New Haven, pp. 35–66.

van Dijk, P. (2009) Apomixis: basics for non-botanists. In: Schön, I., Martens, K. and van Dijk, P. (eds) *Lost Sex: the Evolutionary Biology of Parthenogenesis.* Springer, Berlin, pp. 47–62.

Hormones and signal transduction

Khripach, V., Zhabinskii, V. and de Groot, A. (2000) Twenty years of brassinosteroids: steroidal hormones warrant better crops for the XXI century. *Annals of Botany* 86, 441–447.

Muangprom, A., Thomas, S.G., Sun, T.-P. and Osborn, T.C. (2005) A novel dwarfing mutation in a Green Revolution gene from *Brassica rapa. Plant Physiology* 137, 931–938.

6 Plant Responses to the Environment

6.1 Chapter Overview

Like all other living organisms, plants are surrounded by an external environment that is often hostile. Plants must successfully interact with the biotic and abiotic components of this environment in order to stay alive and reproduce. Unlike motile organisms such as animals and most microbes, plants are unable to move away from environmental threats and must therefore develop various strategies to deal with such stresses. The theme of this chapter is the various forms of environmental stress that are encountered by plants, the mechanisms used to cope with such stresses and how such mechanisms can be manipulated to improve plant performance.

6.2 Importance of Stress

Environmental and biological stresses impose major limitations on plant growth and productivity and are therefore of great interest to biotechnologists. Stress can be defined as 'any change in environmental conditions that might reduce or adversely affect the normal growth and/or development of an organism'. Each individual organism (even members of the same species) may have different ranges of optimal environmental conditions. A given organism can also respond differently to the same form of stress, depending on its age, developmental stage and/or previous experience of the stress. Moreover, what one organism perceives as a stress may be another's opportunity. For example, the high temperatures in a brush fire are an obvious and rather extreme form of stress that can often be lethal to plants. However, some seeds are adapted not just to withstand wildfires, but actually require exposure to fire in order to germinate into what is likely to be a landscape largely cleared of competitors.

Although plants are unable to move away from environmental stresses they can take limited forms of evasive action to avoid localized stress. For example, a plant might avoid an area of nutrient-poor or polluted soil by growing in a different direction. Plants can also avoid long-term stresses by special adaptive mechanisms. Xerophytes avoid water loss by closing their stomata during the day. Some plants avoid dry ground by growing a single deep taproot through layers of dry soil until they reach the water table. Here, their roots can branch out to exploit the wetter soil. Plants can often adapt to stresses such as cold or high light intensity if they are exposed gradually, but may be badly damaged by the same stress if exposed suddenly. A tolerant plant is one that is able to withstand full exposure to an environmental stress with minimal adverse impact. Tolerance is normally a long-term adaptation to stable or frequently recurring environmental conditions. Such a stress need not be chronic (i.e. long lasting); indeed, most stresses tend to be sporadic in their occurrence.

Examples of chronic stresses include constant exposure to sub-zero temperatures (in polar regions), extreme desiccation (in a desert), or long-term inundation in salty water (along a sea coast or estuary). Examples of sporadic abiotic stresses include periodic flooding during rainy seasons, the often severe diurnal heat/cold cycle in deserts, and occasional droughts or bush fires. Sporadic biotic stresses include disease outbreaks, intermittent herbivory and seasonal pests. Because such stresses are sporadic, plant tolerance mechanisms do not always need to be 'switched on', but they should always be available for relatively rapid deployment. A key attribute of effective stress tolerance is therefore the early detection of a potential threat and the timely induction of an appropriate adaptive or defensive response.

Although stresses are often studied in isolation, it is much more common in the real world for organisms to experience simultaneous and interconnected multiple stresses. For example, in the daytime desert plants may face extreme heat, lack of moisture and high irradiation levels. In contrast, at night they may still face dehydration stress but this is now combined with cold stress (desert temperatures can fall below zero overnight). Fungi tend to attack plants that have been previously damaged by wind or by herbivores. They also attack plants that are not well nourished and therefore may be easier to penetrate via their waxy cuticle if this is not well maintained. In each case, the plant is facing a combination of stresses, each of which might require the deployment of quite different response mechanisms.

The principal biotic (living) stresses come from pests, diseases, other plants and herbivores. Important abiotic (non-living) stresses include nutrient limitations, drought, heat, cold, salinity,

flooding and various forms of pollution. Many stresses that have different environmental origins can produce similar effects at the molecular level in a plant. Two common molecular stresses that arise from many different environmental factors are intracellular dehydration and increased concentrations of oxidative free radicals. For this reason cross-talking networks are especially important in the perception and response to what are often linked forms of biotic and abiotic stresses (see Fig. 6.1).

As shown in Table 6.1, the combined effects of biotic and abiotic stresses on the major temperate and tropical crops cause serious yield losses. The importance of plant stress responses as targets for biotechnology is indicated by the fact that over

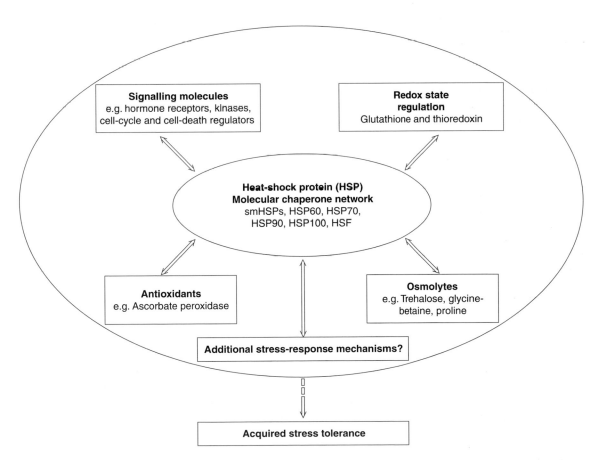

Fig. 6.1. Networks of abiotic stress perception and response pathways in plants. The various stress response pathways in plants are highly interconnected. A central role is played by the heat-shock protein (HSP)/molecular chaperone network but this also interacts with hormones and other signalling molecules, redox regulators, osmolytes and antioxidants. Acquired stress tolerance in plants is often a result of such coordinate interaction of stress response mechanisms acting to prevent cellular damage and to re-establish homeostasis. HSF, heat shock factor. Adapted from Wang *et al.* (2004).

Table 6.1. Effects of combined stresses on yields of some major crops.

Crop	Record yield (t/ha)	Average yield (t/ha)	Average yield (% of record yield)	Average loss (% of record yield)	
				Biotic	Abiotic
Soybean	7.4	1.6	21.8	9.0	69.3
Barley	11.4	2.1	18.0	6.7	75.4
Wheat	14.5	1.9	13.0	5.0	82.1
Maize	19.3	4.6	23.8	10.1	65.8
Potato	94.1	28.3	30.1	18.9	54.1
Sugarbeet	121.0	42.6	35.2	14.1	50.7

Data from Buchanan *et al.* (2000).

99% of first-generation transgenic crops, in terms of land area occupied, were designed to cope with two particular forms of biotic stress, namely competition from other plants (i.e. weeds) and attack by insect pests. As we will see later in Chapter 9, these early transgenic stress-related traits tend to be regulated by single genes. However, most of the agronomically important stress-related traits in crops are physiologically and genetically complex and are regulated by large numbers of genes. Understanding the regulation and manipulation of these complex traits is therefore among the most important targets for crop biotechnologists.

6.3 Abiotic Stresses

Any aspect of the physical environment can potentially act as an abiotic stress to an individual plant. Also, a particular stress might be relatively innocuous under some circumstances but may have more severe effects in other conditions. For example, a sudden cold snap may be more damaging to non-adapted plants than gradual cooling, even if the latter eventually results in cooler temperatures. For plants adapted to particular stresses, such as drought-tolerant desert cacti or salt-tolerant mangroves, drought or salinity may not be perceived as stressful but rather as part of their normal growth conditions. For the purposes of biotechnology, we can recognize several types of abiotic stress that can have severe effects on the growth and development of useful plants, especially crops. These include thermal conditions (heat and cold), water availability (drought and flood), mineral and nutrient availability (e.g. lack of nitrate or excess of salt) and man-made pollutants (e.g. heavy metals, ozone or acid rain).

At the molecular level, different abiotic stresses often have similar effects on cellular systems, such as membranes, or on macromolecules, such as proteins and DNA. A frequent common factor is an increase in levels of oxidative free radicals that damage many biomolecules. Reactive oxygen species (ROS) are oxygen radicals and non-radical derivatives of oxygen that are formed during normal metabolic processes such as photosynthesis as well as during many forms of abiotic stress. ROS damage can lead to injury, premature senescence or death. Examples of ROS include superoxide (O_2^-), hydroxyl (OH), peroxyl (ROO) and alkoxyl (RO^-) radicals; as well as the non-radical intermediates, singlet oxygen ($^1O_2$), hydrogen peroxide (H_2O_2) and ozone (O_3). Because abiotic stresses can affect any part of a plant cell, each cellular compartment deploys specific sets of stress-related genes and proteins (see Fig. 6.2).

One of the most common molecular features of abiotic stresses is protein dysfunction resulting from conformational change. Following thermal or oxidative damage, proteins become denatured and may aggregate and/or lose function. Molecular chaperones have roles in protein folding, assembly, translocation and degradation in normal cellular processes, but also stabilize proteins and membranes and assist protein refolding under stress conditions. One of the most important families of stress-related chaperones is the heat-shock proteins or HSPs (see Table 6.2). A second common feature of many abiotic stresses is some sort of membrane abnormality, sometimes due to protein damage but often due to direct effects on the lipid bilayer itself.

The perception of abiotic stress involves multiple signalling pathways, each responding to different forms of stress while also communicating with other pathways to generate a coordinated response by the plant. At the molecular level, abiotic stress signalling involves: receptor-coupled phospho-relay,

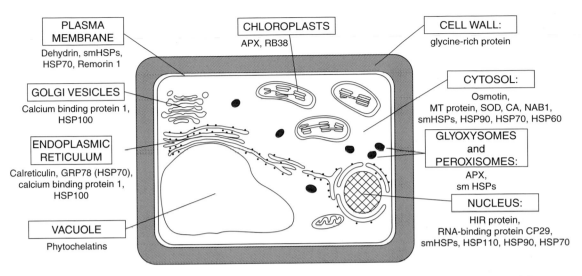

Fig. 6.2. Localization of important stress-related proteins in the plant cell. Abiotic stresses affect all parts of the plant cell, with membrane systems and proteins being especially susceptible to damage. For this reason, each of the major compartments in a plant cell contains specific classes of stress-response proteins and signalling pathways. APX, ascorbate peroxidase; CA, calcium activated proteins; CP, chlorophyll a/b-binding protein; GRP, glucose related protein; HIR, hypersensitive-induced protein; MT, metallothionein; NAB, nucleic acid binding protein; RB, plant retinoblastoma protein; SOD, superoxide dismutase. Adapted from Timperio *et al.* (2008).

Table 6.2. Abiotic stress-induced proteins in plants.

Stress	Specific proteins induced	Non-specific proteins induced	
		ROS scavengers	Heat shock proteins
Heat	Galactinol synthase, choline kinase, glutaredoxin, thaumutin, peptidyl prolyl isomerase	Ascorbate peroxidase, glutathione S-transferase, catalase, superoxide dismutase	HSP110, HSP100, HSP90, HSP70, smHSPs, CPN60, BiP
Chilling	Osmotin, dehydrin, glycine-rich protein, RNA-binding protein	Ascorbate peroxidase, glutathione S-transferase, methionine synthase, thioredoxin	HSP90, HSP70, smHSPs, CPN60, CPN20
Freezing	Glycine-rich protein, antifreeze protein	Ascorbate peroxidase, CA	HSP70, smHSPs
Drought	Osmotin, dehydrin, aquaporin, LEA proteins	Aldolase/aldehyde reductase, methionine synthase	HSP70, smHSPs, HSC70
Light	NAB1, RB38	Ascorbate peroxidase	HSP70B, smHSPs, CPN60, CPN23, CPN20
Salt	Osmotin, dehydrin, remorin1, HIR protein, GF14a, GF14b, ABP	COX6b-1, triosephosphate isomerase, enolase, UGPase	HSP90, HSP70, smHSPs
Ozone	Cysteine synthase, isoflavone reductase, calcium binding protein1, calreticulum	Glutathione S-transferase, ascorbate peroxidase, GPX, methyltransferase	PR5, PR10, HSP70, HSP60, smHSPs, ClpA
Heavy metals	Metallothioneins, PC proteins, phytochelatins	GSH-derived proteins	HSP70

Data based on Timperio *et al.*, 2008.

phosphoinositol-induced Ca²⁺ changes; a mitogen-activated protein kinase (MAPK) cascade; and transcriptional activation of stress-responsive genes (see Fig. 6.3). Several signalling components are associated with the plant response to high temperature, freezing, drought and anaerobic stresses. Because some signalling molecules regulate a range of different downstream responses, it is sometimes possible to manipulate multiple processes in a complex stress-response network by targeting one of these key signalling factors, as outlined in Box 6.1. Some of the transgenes used to engineer oxidative stress tolerance in experimental plants are listed in Table 6.3, although none of them have so far resulted in commercial crop releases.

Thermal stresses

Most plants are adapted to grow within a specific temperature range and will experience stress outside this range. Susceptibility to thermal stress can vary considerably in different tissues or developmental stages of a plant. A common form of thermal stress that affects crops and ornamental plants is late frost, which can be especially damaging to flowers and can drastically limit seed set and hence yield. Tolerance to late frost was the key to enabling crops such as maize and sunflower to be grown in many cooler temperate latitudes. The three

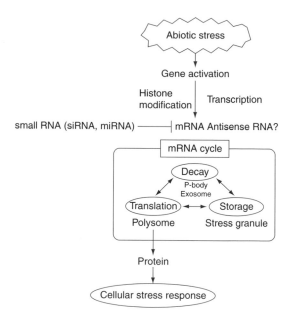

Fig. 6.3. Transcriptional regulation of abiotic stress responses. Abiotic stresses result in the transcriptional activation of numerous genes with the production of mRNA that may be subject to epigenetic regulation via histone modification. Post-transcriptional gene silencing via small RNA or antisense RNA may then occur. In other cases, the mRNA enters a cycle of storage, translation and/or degradation during the stress response. Adapted from Urano *et al.* (2010).

Box 6.1. Manipulating multiple abiotic stress pathways in transgenic plants

There are numerous examples of improvements in plant stress responses following the overexpression of conserved stress-related genes from other organisms in laboratory studies. For example, overexpression of a homologue of the yeast stress-related *DBF2* kinase genes resulted in multiple stress tolerance in transgenic *Arabidopsis* plants. Some tolerance to salt stress was also achieved in transgenic tobacco plants overexpressing calcineurin, a Ca²⁺/calmodulin-dependent protein phosphatase known to be involved in salt-stress signal transduction in yeast.

Stress-related kinases have also been targets for efforts to obtain multiple stress tolerance in transgenic plants. For example, overexpression of an osmotic-stress-activated protein kinase, SRK2C, resulted in a higher drought tolerance in *Arabidopsis*. In another experiment, a truncated tobacco mitogen-activated protein kinase kinase kinase (MAPKKK), called NPK1, activated a multiple oxidative signalling cascade resulting in cold, heat, salinity and drought tolerance in transgenic plants. In other cases, down-regulation of signalling factors can enhance abiotic stress tolerance. For example, the a and b subunits of farnesyltransferase ERA1 are negative regulators of ABA signalling. Antisense-mediated down-regulation of these two ERA1 subunits resulted in enhanced drought tolerance of *Arabidopsis* and rapeseed plants.

While these results are highly encouraging, it is less certain if such single-gene manipulations, particularly using constitutive gene promoters, will necessarily lead to transgenic crops in the field that have enhanced stress tolerance without any adverse side effects, such as reduced yield.

Table 6.3. Examples of transgenes used to engineer oxidative stress tolerance.

Transgene	Transgenic host plant	Type of stress tolerance
Alfalfa ferretin	Tobacco	Tolerance to iron-induced oxidative damage
APX1	Arabidopsis	Heat tolerance
APX3	Tobacco	Oxidative stress protection
Arabidopsis Fe-SOD	Tobacco	Ozone damage protection
Cytosolic Cu/Zn-SOD	Tobacco	Ozone damage protection
GST/GPX	Tobacco	General stress tolerance
GST (Nt107)	Tobacco	Cold and salinity tolerance
GST (NtPox) and GST (ParB)	Arabidopsis	Aluminium and oxidative stress protection
E. coli GR	Tobacco plastid	Cold and salinity stress protection
E. coli GR	Tobacco cytosol	SO_2 and paraquat tolerance
Mn-SOD	Rapeseed	Aluminium stress protection
Pea Cu/Zn-SOD	Tobacco	High light and chilling tolerance
Tobacco Mn-SOD	Alfalfa plastid	Drought and freezing stress protection
Tobacco Mn-SOD	Tobacco plastid	Ozone damage protection

APX, ascorbate peroxidase; GPX, glutathione peroxidase; GR, glutathione reductase; GST, glutathione S-transferase; SOD, superoxide dismutase.
Data partially adapted from Slater *et al.* (2008).

major forms of thermal stress are excessive heat, chilling and freezing.

Heat

The vast majority of multicellular eukaryotes have a maximum temperature threshold of about 50°C. For most plants the thermal death threshold is 45–50°C, although this depends on the duration of the stress, tissue age, water content and adaptability of the plant. Some exceptional plants, such as agave and cacti, can withstand temperatures up to 60°C thanks to their ability to re-radiate excess incident light from their leaves rather than converting it to heat. Excessive heat causes damage to most cellular structures, principally due to its effects on membrane proteins and lipids. Many proteins are easily heat-denatured, including some key enzymes and pigment/protein complexes involved in photosynthesis. Heat stress may also exacerbate water stress and the two often occur together.

Heat stress syndrome is a series of metabolic dysfunctions and physical constraints that accumulate in plants broadly as follows. The first stage is a decrease in photosynthesis and increase in respiration caused by stomatal closure. This leads to decreased transpiration-mediated cooling and an increased internal temperature. At the cellular level, cell membranes are damaged and begin to leak, which exacerbates water loss. Plant growth is inhibited and food reserves are metabolized, but these cannot be allocated to all tissues due to failure of the phloem network. Membrane damage leads to generation of free radicals that eventually result in loss of membrane integrity, irreversible protein breakdown and cell death.

PHOTOSYSTEM II AND WARM/COOL ADAPTED CROPS The most sensitive parts of a plant to heat stress are its membranes and especially the photosynthetic thylakoid membranes (see Fig. 6.4). Membrane lipids containing highly unsaturated fatty acids are more susceptible to malfunction at high temperatures. Some plants adapt to elevated temperatures by increasing the saturated fatty acid content of their membrane lipids, which effectively makes them more 'solid'. Another strategy is to increase the proportion of some structural lipids that help retain membrane integrity. The high sensitivity of photosynthesis to heat is probably largely the result of damage to labile components of the oxygen-evolving complex of photosystem II, which is a peripheral thylakoid membrane protein. This is a potential target for manipulation to extend the growth range of some cool-season crops that show reduced emergence or yield at high temperatures (see Box 6.2).

MOLECULAR RESPONSES – THE ROLE OF HSPS At the molecular level, the major response of plants to thermal stress is the induction of genes encoding heat-shock proteins, which can protect or repair heat-sensitive proteins by acting as molecular

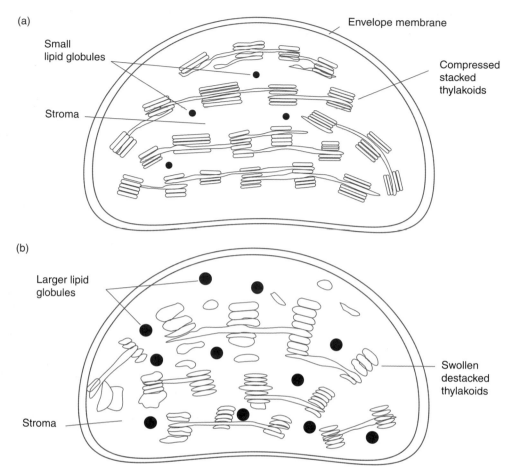

Fig. 6.4. Thylakoid membrane damage due to abiotic stress. (a) During many forms of abiotic stress, one of the most labile parts of the plant cell is the thylakoid membrane system of choloroplasts, and in particular some components of the photosynthetic pigment–protein complexes. (b) Damage to these proteins results in swelling of the thylakoids and a reduction in their ability to form the characteristic granal stacks. This results in a reduction or cessation of photosynthesis and may lead to complete breakdown of the thylakoid membranes and the formation of lipid globules in the stroma, as frequently occurs after ozone damage to leaves.

chaperones. Five classes of heat-shock protein (HSP) genes have been discovered, four of which are highly conserved between eukaryotes and prokaryotes. The four conserved classes are categorized according to size as HSP60 (the chaperonin/GroEL group), HSP70 (DnaK group), HSP90 and HSP100 (Cpl group). The fifth group is the small HSPs, several of which have important roles in plants. As shown in Fig. 6.5, HSP genes are highly regulated and may be induced in response to other stresses as well as heat.

Under normal conditions, a rapid response to heat is regulated by a heat-shock factor (HSF) protein that is normally bound to an HSP70 protein but dissociates upon heat stress, and reassembles into a trimer. This trimer binds to the promoter region of several heat-shock protein genes, resulting in their immediate expression. Plants are unusual in typically having more than 20 genes encoding heat-shock factor proteins, whereas other eukaryotes only contain 1–3 such genes. Another regulatory mechanism involves transcription factor proteins that are inactive when tethered to membranes via a transmembrane domain. Heat stress results in proteolytic release of the transcription factor, which then travels to

Box 6.2. Crop adaptations for different thermal environments

Many of our major annual crops are specifically adapted for growth in relatively warm or cool seasons. Warm-season annuals include common bean, cotton, cowpea, cucurbits, finger millet, grain amaranth, lima bean, maize, mung bean, pearl millet, pepper, pigeon pea, rice, sesame, sorghum, soybean, sunflower, sweet potato, tobacco and tomato. Cool-season annuals include barley, brassicas, rapeseed, faba bean, linseed, garbanzo bean, Irish potato, lentil, lettuce, lupin, mustard, oat, pea, radish, rye, spinach, triticale, turnip, vetch and wheat.

One of the most common thermal adaptations involves membrane thermostability. This can be evaluated by measuring electrolyte leakage from leaf discs subjected to a range of temperatures. Results showed that the photosystem II of the cool-season species, wheat, is more heat sensitive than the photosystem II of rice and pearl millet, which are warm-season species. High soil temperatures can also reduce plant emergence. The maximum threshold temperatures for germination and emergence are higher for warm-season than for cool-season annuals. For example, the threshold maximum seed zone temperature for emergence of cowpea is about 37°C compared with 25°C to 33°C for lettuce. During the vegetative stage, high daytime temperatures can damage components of leaf photosynthesis, reducing carbon dioxide assimilation rates and depressing crop yields.

Among the cool-season annuals, pea is very sensitive to high daytime temperatures, with plant death occurring when air temperatures exceed about 35°C for sufficient duration. In contrast, barley is very heat tolerant, especially during grain filling. Cowpea, which is a warm-season annual, can produce substantial biomass even when growing in daytime air temperatures peaking at about 50°C, although its vegetative development may exhibit abnormalities such as leaf fasciations (abnormal bands in the growing tips). Greater heat tolerance is defined as being where a specific plant process is less damaged by high tissue temperatures; it can involve constitutive effects or may require acclimation. Tolerance to high soil temperatures during seed germination requires constitutive genetic effects, although the maternal plant environment during seed development and maturation also can influence the heat tolerance of seed during germination. Tolerance to high tissue temperatures during plant emergence and early seedling growth therefore appears to be a complex multigenic process that involves both constitutive and acclimatization effects.

---

the nucleus where it induces expression of a variety of heat-responsive genes.

Heat-shock proteins are not the only mechanism of heat tolerance in plants. For example, some cowpea genotypes vary greatly in their heat tolerance during reproductive development, but still produce the same set of heat-shock proteins in their leaves when subjected to mild heat stress. Equally, some heat-responsive genes may also be involved in regulating other stress responses. For example, an *Arabidopsis* gene involved in salt and pathogen defence responses (and similar to a human transcription X-box binding gene) is also heat inducible and involved in heat tolerance. This is evidence of the sort of cross-talk between different stress-responsive signalling pathways outlined in Fig. 6.1. The prospect of increased temperatures in many temperate crop-growing regions of the world has led to efforts to enhance thermal stress tolerance in some of the major cool-season crops discussed in Box 6.2.

Chilling

Depending on the species and extent of acclimatization, chilling damage to plants can potentially occur between 20°C and 0°C. The resultant injuries may include disruption to germination, flower and fruit development, yield and storage life. Minor chilling stress at non-lethal temperatures is normally reversible. Also, exposure to gradually decreasing temperatures above the critical range can result in hardening of plants that may reduce or eliminate injury during subsequent cold exposure. Ultrastructural symptoms of chilling injury become evident before obvious physical symptoms are visible. Such symptoms include: swelling and disorganization of chloroplasts and mitochondria; reduced size and number of starch granules; dilation of thylakoids and granal destacking (see Fig. 6.4); formation of small vesicles of chloroplast peripheral reticulum; lipid droplet accumulation in chloroplasts; and condensation of chromatin in the nucleus. As shown in Fig. 6.6, adaptive responses

---

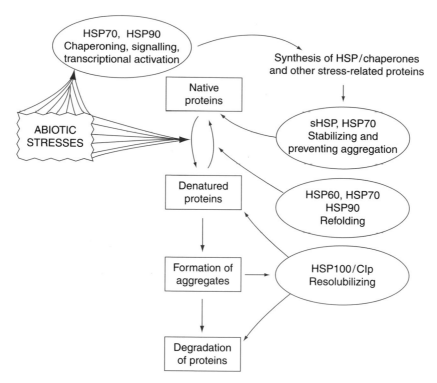

Fig. 6.5. Role of HSP/chaperone network in abiotic stress responses. The many different types of HSPs and chaperones have a variety of roles in stress responses. A key role is the prevention of protein denaturation and loss of function that frequently accompanies abiotic stresses. In addition to stabilizing native protein conformations, HSPs can prevent or reverse aggregation and subsequent degradation of denaturated proteins as well as assisting with their refolding into a functional native conformation. Adapted from Wang *et al.* (2004).

include: decreasing membrane lipid unsaturation; increasing the levels of osmotically active solutes; and an overall decrease in metabolic rate.

Freezing

Freezing injury may occur when the external temperature drops below the freezing point of water and ice crystals are formed. The more susceptible plant varieties can be killed by the first touch of frost. At the other extreme, native plants in cold climates may survive extremely low temperatures without injury. The key to avoiding freezing damage is to prevent or limit the formation and growth of intracellular ice crystals. Such ice crystals can puncture membranes, resulting in cell death after thawing. Extracellular freezing can also be hazardous because cytosolic water tends to be drawn towards ice crystals forming in intercellular spaces. This results in elevated intracellular ion and solute

concentrations that can eventually be lethal. In some cases, water can remain liquid as low as −47°C without nucleating and forming ice. When nucleation of this supercooled water does occur, intracellular ice forms very rapidly, resulting in death of the plant.

Plants can adapt to withstand freezing by: cessation of metabolic activity at low temperatures; avoiding nucleation sites so that intracellular ice crystals cannot form; and by accumulating cryoprotectants. The latter include some sugars and polyhydric alcohols that allow membranes and cytosol to lose much of their water while remaining intact, and specific proteins, such as dehydrins. Because the major stress encountered during extracellular freezing is dehydration, one way for cold-adapted plants to survive is by becoming tolerant to low intracellular moisture levels. This can be achieved by up-regulating HSP/chaperones to protect vulnerable proteins and to accumulate solutes to protect membranes and other structures such as chromosomes.

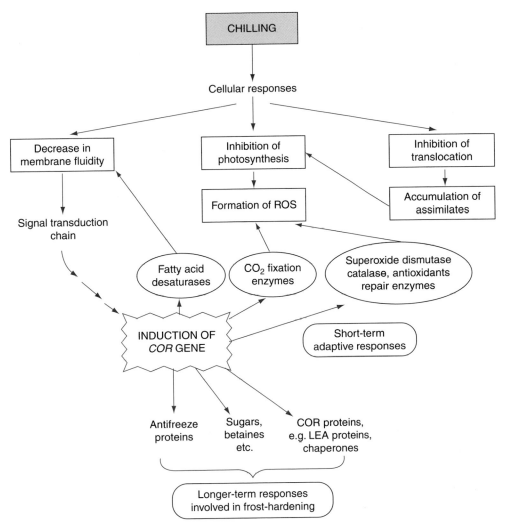

Chapter 6

Fig. 6.6. Cellular responses to chilling stress. In chilling-resistant plants, chilling stress causes a decrease in membrane fluidity. This triggers a signalling cascade leading to activation of *COR* (cold responsive/regulated) genes leading to a series of short-term adaptive responses. It also sets up a series of longer-term responses, such as formation of antifreeze proteins and sugars, that eventually result in the acquisition of frost hardening by the plant. LEA, late embryogenesis abundant. Adapted from Opik and Rolfe (2005).

Water stresses

Some form of water stress is commonplace in the life cycle of most plants. Summer droughts can occur even in relatively damp temperate climates and plants can generally cope for several weeks without rain. In contrast, plants might be inundated by rainfall and roots could be flooded for several days. Therefore, most plants are relatively well adapted to a range of water stresses as long as

they are not too long lasting. However, prolonged exposure to water deficit can only be tolerated by a few specialized xerophytic species, none of which is a crop. For staple crops, by far the most serious form of water stress is an extended drought. Over the past 10 millennia, long-term droughts have repeatedly resulted in famine and abandonment of agriculture around the world. Droughts have directly contributed to the collapse of complex and sophisticated civilizations in the Americas, China,

Near East, Africa and India. Given the prospect of more frequent and longer lasting droughts in some crop-growing regions, improved water-use efficiency is a particularly important trait to incorporate into crop-breeding programmes.

Drought and water deficit

Plants have numerous structural and physiological adaptations to cope with a lack of water. Some species minimize water loss via adaptations such as sunken stomata, deciduous leaves and elevated levels of intracellular solutes. In other cases, plants seek to escape or avoid drought, such as by producing seeds that only germinate after significant rainfall. Many plants in the dry heart of Australia escape drought by remaining dormant as seeds for decades, but can rapidly germinate, flower and set seed during one of the brief wet spells in the region. These plants combine short life cycles with high rates of growth and gas exchange, using maximum available resources while moisture in the soil lasts. Drought avoidance involves minimizing water loss by closing stomata, reducing light absorbance through rolled leaves and decreasing canopy leaf area. It is also necessary to maximize water uptake by allocating more resources for root growth, recovering nutrients stored in older senescent leaves and by higher rates of photosynthesis.

In addition to structural adaptations, drought tolerance involves coordination of physiological and biochemical responses at cellular and molecular levels (see Fig. 6.7). These responses include osmotic adjustments, more rigid cell walls and

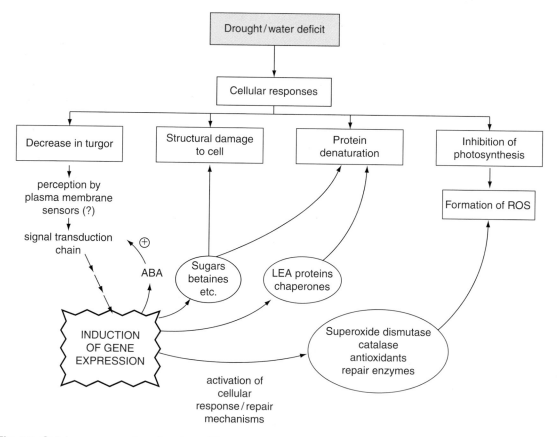

Fig. 6.7. Cellular responses to water stress. Water stress leads to a variety of cellular responses, some of which are also seen in chilling stress (see Fig. 6.6). As with chilling stress, the primary perception mechanism is via membrane sensors that trigger a signalling cascade resulting in the expression of specific sets of genes. Most of the responses involve cellular repair mechanisms and antioxidant systems to mitigate damage by reactive oxygen species (ROS). LEA, late embryogenesis abundant. Adapted from Opik and Rolfe (2005).

smaller cells. True dehydration tolerance as found in euxerophytes is relatively uncommon. Examples of euxerophytes include the creosote bush (*Larrea tridentata*), resurrection plant (*Craterostigma plantagineum*) and some lichens. Such plants tend to have very slow growth rates and hyper-accumulate solutes such as trehalose and proteins such as dehydrin. While these adaptations enable euxerophytes to cope with prolonged drought stress, they are also metabolically expensive and are therefore unlikely to be suitable for transfer to crop plants without incurring a substantial yield loss. So far, development of drought tolerance via direct selection in crop breeding programmes has been hampered by low trait heritability and large genotype × environment interactions. As discussed in Chapter 10, new biotechnological approaches are now being developed as drought tolerance is increasingly recognized as one of the most important future threats to global food security.

Flooding and water surplus

Excess water is less of a problem for most plants than water deficit. Many plants have adapted to aquatic habitats while most others are able to tolerate flooding providing it is transient. Most of these adaptations are structural features to cope with water movement and anoxia. A major problem associated with prolonged immersion in water is a lack of oxygen. Like animals, plants require oxygen for respiration and waterlogged roots will eventually die and rot as they become anaerobic. Many aquatic plants, or hydrophytes, have adapted by developing an extensive system of air-filled holes or spaces called lacunae in leaves and aerenchyma in other tissues. In such plants, up to 60% of leaf volume may be air-filled to allow for gas exchange throughout the plant. Aquatic plants often have thin, strap-like and supple leaves to withstand tidal currents and wave action. Many species have rhizomes that keep them attached to the soft bottom and enable rapid vegetative growth, which is often their primary mode of reproduction. When released, their pollen grains are often in gelatinous strands that are carried by water currents to fertilize other plants.

Salt stress

After drought, the most severe form of abiotic stress is salinity, which affects at least 20% of irrigated land worldwide and severely affects yields of many crops (Table 6.4). Like many other stresses, salt stress is multifactorial and involves additional stresses, especially dehydration. High intracellular levels of Na^+ and Cl^- ions can directly affect macromolecules such as proteins or DNA. But high

Table 6.4. Salt susceptibility of some major crops.

Crop	Threshold salinity (dS/m)*	Yield decrease (%)
Bean (*Phaseolus vulgaris*)	1.0	19.0
Eggplant (*Solanum melongena*)	1.1	6.9
Onion (*Allium cepa*)	1.2	16.0
Pepper (*Capsicum annuum*)	1.5	14.0
Maize (*Zea mays*)	1.7	12.0
Sugarcane (*Saccharum officinarum*)	1.7	5.9
Potato (*Solanum tuberosum*)	1.7	12.0
Cabbage (*Brassica oleracea*)	1.8	9.7
Tomato (*Lycopersicon esculentum*)	2.5	9.9
Rice (*Oryza sativa*)	3.0	12.0
Groundnut (*Arachis hypogaea*)	3.2	29.0
Soybean (*Glycine max*)	5.0	20.0
Wheat (*Triticum aestivum*)	6.0	7.1
Sugarbeet (*Beta vulgaris*)	7.0	5.9
Cotton (*Gossypium hirsutum*)	7.7	5.2
Barley (*Hordeum vulgare*)	8.0	5.0

*According to the USDA, saline soil can be defined as having an electrical conductivity in excess of 4 dS/m (about 40 mM NaCl). However, most grain and vegetable crops are highly susceptible to soil salinity at levels well below 4 dS/m.
Data from Chinnusamy *et al.* (2005).

extracellular or vacuolar levels of the same ions can also lead to water loss from the cytosol (analogous to the effect of extracellular ice crystals discussed above), leading to a form of dehydration stress. Therefore, mechanisms of salt tolerance normally involve the exclusion of excess salt from the cytosol. This can be achieved by a variety of means including: sequestration of Na^+ and Cl^- in vacuoles or in specialized salt glands; blocking Na^+ entry into the cell; Na^+ exclusion from the transpiration stream; or excretion of salt on to leaf surfaces.

Salt stress is one of the major challenges facing world agriculture with more and more land becoming hypersaline due to misuse of irrigation and/or climate change. This causes falling yields and food shortages. One way of addressing salt stress is to study existing mechanisms of salt tolerance in halophytic species that are already adapted to growing in saline environments. Halophytes are found directly on coasts and in estuarine and salt-marsh habitats, but can also occur inland on soils containing high levels of salts. Such plants have to deal with potentially toxic levels of sodium, carbonate and chloride ions that might enter via their roots. In many cases water is selectively absorbed and the salt is excluded. Succulents like *Mesembryanthemum crystallinum* store water in swollen leaves. In other plants, excess salt is deposited on the leaf surface (e.g. tamarisk) or in special salt glands (e.g. the saltbush, *Atriplex*).

Many halophytes have a highly concentrated leaf sap with an osmotic pressure in the range of −17 MPa, compared with the normal range of −1 to −3 MPa. This means that such plants can absorb water osmotically even from soil that contains relatively high levels of salt. With their less negative water potentials, normal plants are unable to extract water from such soils and quickly die from dehydration. Mangroves are important halophytes in tropical estuarine habitats that exhibit several different mechanisms for coping with saline conditions. The black mangrove and the white mangrove can take up seawater through their roots, but excrete excess salt through pores or salt glands on the leaf surface. In contrast, the red mangrove, *Rhizophora mangle*, excludes the salt from seawater at the root–substratum interface. Tidal mangroves are also noted for their pneumatophores, erect roots that are exposed at least part of the day enabling submerged parts of the plant to obtain oxygen while remaining attached to an anaerobic substrate (sea mud).

Although many halophytes can tolerate high levels of intracellular salt, this must be sequestered in vacuoles as their cytosolic enzymes and other components are still susceptible to salt poisoning. Such halophytes tend to maintain a high ratio of K^+/Na^+ in the cytosol. Compartmentation of sodium occurs via a vacuolar transporter protein that moves potentially harmful Na^+ ions from cytosol into the large central vacuole. These ions, in turn, act as an osmoticum within the vacuole to maintain water flow into the cell. In *Arabidopsis*, three classes of low-affinity K^+ channels have been identified: these are K^+-inward rectifying channels; K^+-outward rectifying channels; and voltage-independent cation channels. The K^+-outward rectifying channels could play a role in mediating the influx of Na^+ into plant cells. These channels, which open during the depolarization of the plasma membrane, could mediate the efflux of K^+ and the influx of Na^+ ions. Na^+ competes with K^+ uptake through Na^+/K^+ co-transporters and may block K^+-specific transporters of root cells under saline conditions. This can result in toxic levels of Na^+ plus insufficient K^+ for enzymatic reactions and osmotic adjustment. Na^+ influx is controlled by the low-affinity Na^+ transporter, AtHKT1.

When a plant is exposed to high salinity, various genes are up-regulated, the products of which are either directly or indirectly involved in plant protection. Susceptibility or tolerance to salt stress in plants is conferred by the coordinated action of many stress-responsive genes that are also involved in cross-talk with components of other stress-related signalling pathways. At the molecular level, salt tolerance involves a complex interplay between components such as: ion pumps like the Salt Overly Sensitive (SOS) plasma membrane Na^+/H^+ antiporter; SOS signalling pathways; ABA; calcium; transcription factors; mitogen-activated protein kinases; glycine betaine; proline; reactive oxygen species; and DEAD-box helicases, to name but a few.

Initially, transgenic strategies for engineering salinity stress tolerance targeted single genes in biochemical pathways that are end points of response cascades. Common targets are vacuolar Na^+ transporter genes as outlined in Box 6.3. More recently, upstream master switches that regulate the activity of many downstream genes and proteins are also being targeted. Candidates for such master switches can be identified by combining data from large-scale functional screening programmes with the increasing genomic data now

Box 6.3. Biotechnology targets for salt-stress tolerance

The most popular target genes for enhancing salt-stress tolerance are those encoding vacuolar transporters. Such transporters are able to remove potentially toxic Na^+ ions from the cytosol for sequestration in vacuoles. The *AtNHX* gene, encoding a vacuolar transporter from *Arabidopsis thaliana*, was overexpressed in transgenic *Arabidopsis*, tomato, rapeseed, rice, tobacco, maize and tall fescue plants. The transgenic plants grew normally under low-salt conditions but also grew, flowered and produced seeds in the presence of 200 mM NaCl. This salt concentration is 40% of that of seawater and will normally inhibit the growth of almost all crop plants (see Table 6.4). In another example, transgenic plants overexpressing AVP1, encoding the vacuolar H^+-pyrophosphatase, showed enhanced salt tolerance.

Transgenic melon and tomato plants expressing the *HAL1* gene, encoding a vacuolar antiporter protein, showed increased salt tolerance as a result of retaining more K^+ than the control plants under salinity stress. A vacuolar chloride channel gene involved in cation detoxification, *AtCLCd*, has been overexpressed in *Arabidopsis* to confer salt tolerance by compartmentalizing Na^+ ions in the vacuoles. The gene from the Salt Overly Sensitive I (SOSI) locus in *Arabidopsis*, which is similar to plasma membrane Na^+/H^+ antiporter genes from bacteria and fungi, was overexpressed using the CaMV 35S promoter. The extent of up-regulation of *SOSI* gene in transgenic plants was correlated with the increase in Na^+ tolerance.

Another target for manipulation is to restrict the entry of salt into root cells or the xylem in order to preserve an optimum cytosolic Na^+/K^+ ratio. *Arabidopsis* mutants in which the *hkt1* gene was knocked out had much reduced uptake of Na^+ and the plants exhibited enhanced growth in saline soils. This suggested that the gene product, AtHKT1, was a salt tolerance determinant and this was confirmed when transgenic wheat plants were produced in which the equivalent *HKT1* gene was knocked out via antisense technology. The resulting wheat plants showed reduced Na^+ uptake and grew better than controls under moderate salinity. However, as with other forms of stress tolerance manipulation discussed here, the use of single-gene transgenic approaches has yet to be convincingly demonstrated to be applicable in the highly complex conditions experienced by crops in the field, as discussed in Chapter 12.

---

available for many plant species. Genes encoding osmolytes, ion channels, receptors, calcium signalling components and other signalling factors or enzymes may confer salinity tolerance when transferred to sensitive plants. While these results are encouraging, the complexity and diversity of salt stress response mechanisms in plants mean that this remains a highly challenging trait to engineer into crops via current transgenic approaches. Moreover, as with many other desirable crop traits, even if it can be achieved, the metabolic cost of transgenic salt tolerance may come at an unacceptable cost in terms of yield reduction.

Pollutants

Pollutants are chemicals, most commonly resulting from human activity, that may be deleterious to plant growth. Examples include industrial residues such as heavy metals, gases such as ozone, nitrogen oxides (NO_x), and sulfur dioxide (SO_2), and agrochemicals such as some fertilizers and biocides.

Among the most important metallic toxins are copper, nickel, zinc, cadmium and selenium. Otherwise essential metals may also be toxic at high concentrations. For example, excess iron results in release of ROS and consequent oxidative damage. As with salt tolerance, some plants can tolerate toxic metals to some extent either by excluding their entry via the root system or by sequestering them in compartments such as vacuoles. A common method is to chelate the metal by binding it to compounds such as phytochelatins and transporting the phytochelatin–metal complex to the vacuole for storage. Phytochelatins are low-molecular-weight thiols formed from the amino acids glutamate, cysteine and glycine, with the general formula $(\gamma\text{-Glu-Cys})_n\text{Gly}$. Exposure of many plants to cadmium induces phytochelatin synthase gene expression in leaves, leading to sequestration of the metal.

Metallothioneins are metal-binding proteins that are especially effective at sequestering copper and zinc ions as well as more toxic metals such as cadmium and mercury. Although metallothioneins may play a role in the homeostasis of essential metal ions and in detoxification of heavy metals, they have also been implicated in many other roles in plants. Examples include oxidative stress, pathogen responses, embryogenesis, leaf senescence and fruit ripening. There is considerable interest in manipulating metal-binding compounds in plants, for example in order to increase tolerance of crops growing in regions subject to metal pollution. In other cases, metal-accumulating plants could be used either to clear polluted land of metals or to sequester valuable metals, such as gold, in a form that can be more readily extracted for economic benefit.

The most serious phytotoxic gaseous pollutant is tropospheric ozone, which is produced as a result of photochemical reactions in the atmosphere involving NO_x and volatile organic compounds. By far the most important sources of these compounds are vehicle exhaust fumes, fossil-fuel power plants and other industrial processes. Unlike stratospheric ozone, which shields the earth from harmful cosmic radiation, tropospheric ozone is a major pollutant that can severely affect plant growth and crop yield. Since the mid-20th century, global levels of tropospheric ozone have steadily increased with measurable effects on crop productivity in some regions. As early as the 1960s, increasing ozone pollution was one of the reasons for the demise of citrus production in southern California. More recently, increasing ozone release due to industrialization in many developing countries, especially in Asia, is threatening to reduce local crop yields. Ozone readily enters leaves via the stomata and one characteristic effect is the breakdown of lipids in thylakoid membranes to form triacylglycerol-rich globules. Loss of pigments results in chlorosis and photosynthesis may be much reduced. Some plants are relatively ozone tolerant due to their ability to up-regulate ROS scavenging systems at elevated ozone concentrations. Research is now focusing on identifying the genes involved and their possible transfer to ozone-susceptible crop species.

6.4 Biotic Stresses

The principal biotic stresses affecting plants come from pests, pathogens, other plants and herbivores. *Pests* are generally regarded as smaller animals, such as insects, mites or worms, that might eat parts of a plant and/or burrow into its tissues for shelter or to lay eggs, often causing huge damage in the process. Some rodents and birds have become pests that can especially threaten grain crops both before and after harvest. *Pathogens* are microorganisms such as fungi, viruses and bacteria that might only cause minor damage to part of a single leaf right up to the death of the entire plant. *Other plants* principally exert stress by competing for resources such as light and nutrients. Sometimes this might involve physical contact between plants. Such contacts can range from relatively harmless epiphytes that use a host plant for support to the much more serious effects of parasitic plants that remove nutrients from and often kill their hosts. Finally, *herbivores* can be defined as animals that eat plants, ranging from phytophagous insects (that may be also classified as pests) to large grazing vertebrates such as buffalo or deer.

The combination of multiple biotic stresses can be a major limitation on plant productivity both in the wild and in agriculture. It is estimated that global crop losses due to biotic factors average at least 30–40% of potential yield, and such losses can be much larger during severe outbreaks. During the 20th century, the major focus of plant breeders was understandably (in view of the human population explosion) focused on yield and quality traits. To some extent this has led to loss of some endogenous biotic resistance in commercial crop varieties. As discussed in Chapter 8, this deficiency was partially compensated for by development of chemical agents, or biocides, to control biotic stress. Examples include insecticides, fungicides, nematicides and herbicides. Today, chemical biocides are essential tools that enable intensive forms of conventional agriculture to produce sufficient food for the growing global population.

However, the use of some agrochemicals has been increasingly questioned on the grounds of financial cost, safety and environmental impact. Some chemical biocides have been banned altogether but in other cases more target-specific and sustainable alternatives have been developed by the industry. An attractive alternative to external chemical agents is to breed crops with a greater endogenous capacity for resistance to some of the major biotic stresses. Research into biotic stress manipulation is therefore one of the most active areas of plant biotechnology. As early as the 1980s, some biotic stress-related traits, including several forms of herbicide tolerance and insect resistance, were discovered to be amenable to manipulation by single-gene insertions. It is therefore no surprise that such traits were present in almost all the first-generation transgenic crops grown since the mid-1990s (see Chapter 9).

Pests

Some of the most serious animal pests of major crops are listed in Table 6.5. Of these, insects and mites are by far the most damaging. Even in advanced farming systems, insect and mite damage is responsible for average annual yield losses of between 10 and 20%. Although some adult insects such as locusts can occasionally cause great damage to crops, insect larvae are by far the major threat. Some of the most common insect pests are the larvae of beetles, butterflies and moths. Flowering plants have coevolved with such insect pests for tens of millions of years and have developed numerous physical and chemical defence mechanisms. For example, some plants produce thickened external coverings or secrete chemical deterrents, while others mount localized defence responses to insect damage as shown in Fig. 6.8.

Plants produce hundreds of potential insecticidal compounds, some of which are proteins, such as proteases and amylases that interfere with digestion in larval pests. Lectins are insecticidal proteins that cause agglutination of cells, while chitinases can damage the cuticle of larvae. Genes encoding many insecticidal proteins have been transferred to various crop plants, as listed in Table 6.6. In some cases, the resulting transgenic varieties have shown significant resistance to insect pests in the laboratory. However, very few have been tested in field trials and none of them are currently in commercial production. A more successful approach to creating insect resistance in transgenic plants has been the use of genes encoding a class of protein toxins from the Gram-positive soil bacterium *Bacillus thuringensis* (Bt). The use of transgenic crops expressing Bt genes is now one of the most widespread commercial applications of plant biotechnology with such crops being grown on over

Table 6.5. Examples of animal pests affecting major crops.

Pest organism	Order	Crops affected	Effects
Insects			
Locust	Orthoptera	Grasses	Eats all aerial parts of plants
Bollworms	Lepidoptera	Cotton, maize	Feed inside developing cotton boll and maize ear
European corn borer	Lepidoptera	Maize	Feeds inside stems leading to plant damage or death
Colorado beetle	Coleoptera	Potato	Eats foliage, greatly reducing yield
Armyworm, Leafworm	Lepidoptera	Cotton, maize, rice, tobacco	Eats young leaves, greatly reducing yield
Brown plant hopper	Homoptera	Rice	Sucks sap (phloem)
Tobacco budworm	Lepidoptera	Cotton, tobacco	Eats seeds and damages fibres
Mediterranean fruit fly	Diptera	Citrus, peach, guava, apricot, fig, apple	Eats or damages soft fruits
Sawflies	Hymenoptera	Wheat, orchard and forest crops	Eat foliage, greatly reducing yield
Nematodes			
Globodera spp.		Potatoes, sugarbeet, rice	Rot tubers and roots
Anguina spp.		Cereals	Attack flowers and leaves
Trichodorus spp.		Pea, lucerne, potato, sugarbeet, tobacco	Root damage, carry rattle and browning viruses
Slugs			
Helix spp.		*Citrus* spp., ornamentals	Foliage and fruit loss
Grey field slug		Cereals	Eats newly sown seed
Rodents			
Field mice		All grain crops	Eats field and stored grain
Birds			
Many seed-eating species		All grain crops	Eats newly sown and ripe grain

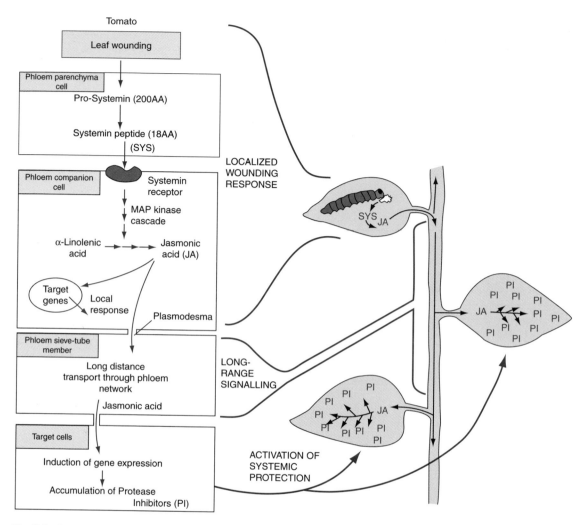

Fig. 6.8. Systemic responses to wounding. In this example, wounding of a plant tissue, for example by a browsing caterpillar, results in both localized and systemic responses. Physical damage to the leaf triggers the conversion of pro-systemin to the systemin peptide in phloem parenchyma cells. The systemin peptide moves to phloem companion cells where it triggers a MAP kinase cascade resulting in formation of jasmonic acid (JA). JA then stimulates a local response to the immediate threat but is also transported through the phloem network to generate a wider systemic response. For example, JA can induce gene expression in remote target tissues leading to the accumulation of protease inhibitors (PI) that act as protective insecticidal agents in the event of future attack. Adapted from Taiz and Zeiger (2010).

50 Mha worldwide in 2010. The origins and impact of Bt technology are examined in more detail in Chapter 9.

After insects, nematodes are the most destructive crop pests/parasites, causing over $100 billion in annual losses to world agriculture. Nematodes inflict direct damage by eating leaves, stems, flowers or roots, as well as considerable indirect damage via transmission of viral diseases during feeding.

Nematodes have acquired several genes enabling them to digest plant cell walls (e.g. cellulases, pectate lyases and xylanases) via horizontal gene transfer from bacteria. As with insects, many wild plants that have coevolved with nematodes carry partial or complete resistance to these pests. However, such resistance is much less common in crop species and nematodes are becoming an increasing problem, especially in developing countries.

Table 6.6. Some insecticidal plant compounds and new transgenic host plants.

Active compound	Gene	Plant of origin	Target insect group	Transgenic plant species
Agglutinin	WGA	Wheat germ	Lepidoptera, Coleoptera	Maize
Chitinase	BCH	Bean	Homoptera, Lepidoptera	Rapeseed
Trypsin	CMe	Barley	Lepidoptera	Tobacco
	CMTI	Squash	Lepidoptera	Tobacco
	CpTI	Cowpea	Coleoptera, Lepidoptera	Apple, lettuce, wheat, rice, tobacco, tomato, rapeseed, sunflower
Proteinase	Pot PI-I	Potato	Lepidoptera, Orthoptera	Petunia, tobacco
	Pot PI-II	Potato	Lepidoptera, Orthoptera	Birch, lettuce, rice, tobacco
	PI-I	Tomato	Lepidoptera	Lucerne, tobacco, tomato
	PI-II	Tomato	Lepidoptera	Tobacco, tomato
Serine protease	C-II	Soybean	Coleoptera, Lepidoptera	Rapeseed, potato, poplar, tobacco
	PI-IV	Soybean	Lepidoptera, Orthoptera	Potato, tobacco
	MTI-2	Mustard	Lepidoptera	Arabidopsis, tobacco
α-Amylase inhibitor	α-AI-Pv	Common bean	Coleoptera	Pea, tobacco, azuki bean
	WMAI-I	Cereals	Lepidoptera	Tobacco
Tryptophan decarboxylase	TDC	*Catharanthus roseus*	Homoptera	Tobacco
Lectin	GNA	Snowdrop	Homoptera, Lepidoptera	Rapeseed, potato, rice, sugarcane, sweet potato, tobacco, tomato
	p-lec	Pea	Homoptera, Lepidoptera	Potato, tobacco
	rice lectin	Rice	Lepidoptera, Coleoptera	Maize
	jacalin	Jack bean	Lepidoptera, Coleoptera	Maize
Peroxidase	Anionic POX	Tobacco	Lepidoptera, Coleoptera, Homoptera	Tobacco, tomato, sweet gum

Data from Slater *et al.* (2008).

Nematode infestation may be promoted by intensive cropping, infrequent rotations and poor phytosanitary practices. Nematodes can be controlled to some extent by management methods, such as chemical nematicides, crop rotation, soil sterilization and chemical agents. Some of these chemicals, such as carbamates, are among the most toxic pesticides in widespread use.

Therefore, development of transgenic nematode-resistant crops could have great economic and environmental benefits. Research approaches include the induction of so-called 'suicide genes' in plant cells that are infected with a nematode, or the expression in plants of protease inhibitors that inhibit nematode growth. To date these studies are still some way from commercial application, but they remain a promising option for the future. An alternative method would be to breed non-transgenic crop varieties with endogenous resistance. There has been some success using wide crosses to transfer genes for resistance to potato cyst eelworm (*Globodera rostochiensis*) from a wild potato into modern cultivars. Unfortunately, the wider use of this approach is daunting due to the large number of crop varieties with different soil requirements and the range of race-specific resistance traits required for incorporation into breeding programmes. However, it is possible that transgenic breeding approaches, e.g. using nematicidal Bt genes such as some forms of Cry5 or Cry6 (see Table 9.3), might be useful in the future.

Other common pests include slugs, mites, rodents and birds. These are normally controlled by a combination of management and chemical methods. In a limited number of cases, biological control has been used, especially to combat introduced pests that may have few or no natural enemies in their new habitat. For example, oil palm is originally a Central African plant that is now widely grown in the tropics. In the Far East, a serious pest is the rhinoceros beetle, *Oryctes rhinoceros*, which eats the developing shoot apical meristems, effectively halting further growth. To combat this threat, the phytophagous fungus *Metarhizium anisopliae* and the *Oryctes* baculovirus have been introduced as biological control agents. On the same plantations, rats that attack the palm fruits can be controlled by introducing barn owls, *Tyto alba*, that are provided with nest boxes high in the trees.

Although, as detailed above, biological control has sometimes been highly effective, there have been some dramatic failures. One of the most notorious of these was the introduction of cane toad into Australia in 1935 to control the cane beetle (*Dermolepida albohirtum*). This insect is a serious pest of sugarcane crops where the greatest damage is done by larvae that hatch underground and eat the roots, which either kills or stunts the growth of the plant. The introduced cane toads were relatively ineffective at controlling the cane beetle. Even worse, they escaped from sugarcane plantations and have seriously affected native wildlife, becoming a more serious biological menace than the cane beetle that they were originally meant to control.

In addressing pest control it is invariably better to adopt several different approaches than to rely on a single strategy such as use of one particular pesticide or biocontrol agent to which the pest could eventually become resistant. This has led to the establishment of integrated pest management systems in several major crops. To take oil palm as an example again, some of the major pests of this crop in Malaysia and Indonesia come from the bagworm group of insects including *Mahasena corbetti*, *Metisa plana* and *Cremastopsyche pendula*. Several methods are being used to control bagworm as follows: the insects are lured into sex pheromone traps; other host plants are grown in the vicinity to encourage populations of parasitoids,

such as wasps, that prey on bagworms; and live Bt sprays are applied to affected palms to kill bagworms. Finally, transgenic approaches to bagworm control are also being tried. For example, in laboratory trials, transgenic oil palms expressing the cowpea trypsin inhibitor gene were found to be resistant to bagworm larvae.

Pathogens

Pathogenesis is a form of plant–microorganism interaction in which the host plant derives no benefit and some harm from the association. Harmful effects can range from mild symptoms to a fatal disease. Plant diseases are caused by pathogenic microorganisms including fungi, oomycetes, bacteria and viruses (see Table 6.7). Pathogens can be further divided into necrotrophs, which kill their host and feed on its body, and biotrophs, which require a living host to complete their life cycle. Fungi, oomycetes and bacteria can be either necrotrophs or biotrophs (depending on the pathogen species) whereas viruses are obligate biotrophs. Some pathogens may be highly specific for a single species or race of plant while others can infect several species. Some of the molecular responses to pathogen attack are shown in Fig. 6.9.

Mechanisms of pathogen resistance vary between 'non-host' and 'host' with many intermediate states. Non-host resistance occurs when an entire plant species is resistant to a pathogen via a complex multigenic mechanism. Host resistance occurs when a single plant variety is resistant to a specific pathogen to which other members of the same species may still be susceptible. This form of resistance is due to the interaction of a protein in the host with a protein in the pathogen, often called a gene–gene interaction. Host resistance normally leads to a hypersensitive response, often involving localized cell death and the formation of blocking structures that prevent further spread of the pathogen. Because it relies on a single gene, it is more amenable to transfer into elite crop varieties via breeding, but it is also more likely to break down in a relatively short period. As with pest-induced wounding, pathogens may sometimes elicit specific systemic responses in adapted plants as shown in Fig. 6.10.

Fungi and oomycetes

Fungi and oomycetes are by far the most widespread and economically damaging groups of

Table 6.7. Examples of plant pathogens affecting major crops.

Pathogenic organism	Crops affected	Disease	Annual losses ($)
Fungi and oomycetes			
Puccinia graminis	Wheat	Black stem rust	Multibillion
Fusarium spp.	Wheat, barley	Scab	Multibillion
	Cotton	Wilt	
Phytophthora spp.	Potato	Late blight	Multibillion
	Cocoa	Black pod	450 million
	Soybean	Root rot	120 million
Verticillium dahliae	Potato	Verticillium wilt	Multimillion
Bipolaris maydis	Maize	Southern corn leaf blight	Multimillion
Viruses			
Rice tungro virus complex	Rice	Rice tungro disease	Multibillion
Maize streak virus	Maize	Maize streak	Multibillion
Grassy stunt virus	Rice	Grassy stunt	Multibillion
Cassava African mosaic virus	Cassava	Mosaic disease	>2 billion
Bacteria			
Pseudomonas spp.	Beans	Leaf spot, blight, brown spot, canker	Multimillion
Xamthomonas spp.	Rice, tomato, pepper, beans, citrus, strawberries	Bacterial spot, wilt, leaf spot	Multimillion
Erwinia spp.	Potato, apple, pear	Soft rot, wilt, necrosis, fireblight	Multimillion
Clavibacter spp.	Potato, tomato	Wilt, canker, leaf spot, stunt	Multimillion

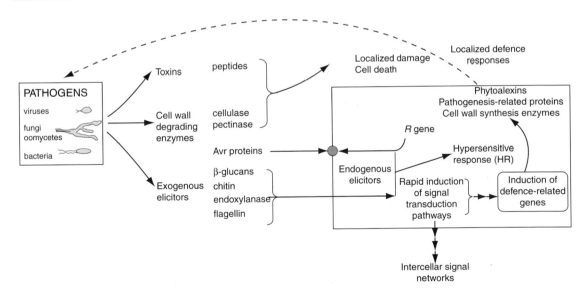

Fig. 6.9. Molecular responses to pathogen attack. Pathogens can cause localized damage to plant tissues by releasing toxins or by attacking cell walls as they attempt to enter the plant. In many cases, a plant is able to recognize an attacking pathogen via various elicitors that can cause it to mount various forms of defensive response. In some cases, the response may be localized cell death in the affected region and the sealing of the wound site to prevent further spread of the pathogen. In other cases, antimicrobial compounds such as phytoalexins or pathogenesis-related (PR) proteins may be produced. These defensive responses are caused via a series of signal transduction pathways involving mediators such as jasmonic acid and reactive oxygen species leading to expression and/or activation of transcription factors followed by activation of target genes.

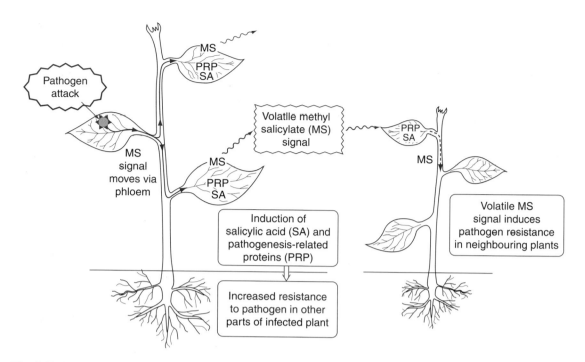

Fig. 6.10. Systemic responses to pathogen attack. In this example, pathogen attack on a leaf triggers production of methyl salicylate (MS), which moves through the phloem network to other leaves throughout the plant. MS and salicylic acid (SA) then stimulate the formation of pathogenesis-related proteins (PRP) in these leaves, resulting in an increased level of resistance to any further pathogen attack. In addition, MS may be released from a leaf as a volatile signal that can trigger pathogen resistance in neighbouring plants. Adapted from Taiz and Zeiger (2010).

pathogens affecting plants. The fungus-like oomycetes, also called water moulds, include some of the most virulent plant pathogens. Despite appearances, oomycetes are unrelated to fungi as they have cellulose, rather than chitin, cell walls. Instead, they are believed to be descended from algae that lost their plastids after becoming heterotrophs (Chapter 2). Their plant-like cell walls, which make them difficult to detect or respond to, may be one reason for their evolutionary success as pathogens. Notable examples include several *Phytophthora* spp., including the causes of: late potato blight, *P. infestans*; sudden oak-death, *P. ramorum*; soy root rot *P. sojae*; and fruit rot leaf spot of cucumbers and squash, *P. capsici*.

Fungi and oomycetes usually grow on or within plants as microscopic 1–2 µm diameter filaments known as hyphae. Masses of hyphae are produced, creating a huge surface area over which the pathogen can absorb nutrients from its host. In general, most healthy plants tend to resist attack by most fungal and oomycete pathogens although a few

pathogens have evolved mechanisms to overcome plant defences. However, as with animals, unhealthy plants that are damaged, wounded, or under another form of biotic or abiotic attack can often be infected by opportunistic fungi to which they would normally be resistant. This is an example of multiple stresses occurring together. For example, crops already debilitated by drought or nutrient stress are much more likely to suffer from fungal attack.

Bacteria

Bacteria can only enter a plant through natural openings such as stomata or through damaged tissue. By far the most common entry point for pathogenic bacteria is via wound sites created by mechanical damage (e.g. wind or pruning) or by the action of pests and herbivores. A few groups, such as some members of *Pseudomonas*, *Xanthomonas* and *Erwinia*, can also enter plants via stomata, nectaries and lenticels. By detecting compounds such as phenolics produced by damaged plant tissue some bacteria, such as the soil-dwelling *Agrobacterium tumefaciens*, can move to wound sites via chemotaxis. In other cases bacteria may be carried by an insect vector that transmits it to the plant tissue during feeding. For example, specialist leafhoppers that feed on xylem often carry the xylem-specific bacterium *Xylella fastidiosa*. Once these bacteria enter the xylem, they reproduce and are taken up by other feeding leafhoppers and are spread to other plants.

Many bacterial species can cause disease in a wide range of plant hosts, but particular strains within a species might only have a limited host range. Therefore some species are divided into several 'pathovars', each with a different host range. Sometimes certain strains (called 'races') within a pathovar can only cause disease in particular cultivars of a crop. A compatible race–cultivar interaction results in establishment of infection and resulting disease. In such cases the bacterium is said to be virulent and the plant is susceptible. If large numbers of bacteria that do not normally cause disease are injected into a plant, there is a so-called hypersensitive response involving localized necrosis of the affected tissue. This response by the plant involves induction of genes that lead to programmed cell death followed by desiccation and collapse of the necrotic tissue and the formation of a callose seal that normally blocks any further spread of the pathogen within the plant (see Fig. 6.9).

Viruses

The vast majority of plant viruses require a biological vector. The commonest vectors are pest species such as insects or nematodes, although some fungi may also transmit viruses. Viruses can also be transmitted to plant progeny via asexual reproduction and, much less frequently, via sexual reproduction if they infect seeds. Virus–vector associations tend to be highly specific and are often limited to a single species. For example, some insect-transmitted viruses attach themselves to specific sites in the mouthparts or salivary glands of their vector via a coat protein or via an additional helper or transmission protein. Once the vector has pierced the plant, such viruses dissociate from their binding sites and enter the damaged plant tissues. Some viruses are able to replicate in their vector without disabling it and such vectors can remain constantly infective towards plants for the remainder of their lives.

There are four main strategies for controlling viral infection of plants. The first is to avoid sources of infection, such as diseased material. This is especially important in vegetative propagation as many viruses are transmitted to clonal progeny from diseased parental stock. This has been a perennial problem that still affects many crops. Solutions include: (i) tissue culture or heat treatment to ensure virus-free stock; (ii) to control vectors, especially pest species, using methods such as insecticides or by planting in areas where vectors are absent; (iii) the plants themselves can be protected from systemic infection by inoculation with an attenuated form of the virus. This is analogous to vaccination as practised for animals and humans in order to confer resistance to viral infection; and (iv) transgenic plants can be created that express part or all of the virus coat protein.

Vaccination or transgenesis is necessary for virus control in crops because there are very few resistance genes available in related species for crossing into elite varieties to create durable endogenous resistance. Another promising approach is to create transgenic plants that express parts of viral RNA or DNA sequences. In some cases viral coat proteins, or fragments thereof, have been expressed in plants, resulting in interference with the early stages of viral replication. This strategy has been highly effective in many lab studies and, as described in Chapter 9, it has been used successfully on a commercial scale to control the ringspot virus in papaya crops. In other cases, the expression of short sequences of viral RNA or DNA in a plant results in silencing of viral gene expression and avoidance of infection. Such approaches have great potential to combat serious viral disease but their long-term durability has yet to be fully assessed.

Other plants

One of the major constraints on the growth of plants both in the wild and under cultivation is competition for resources such as light, water and nutrients from other plants, including members of the same species.

Conspecifics

Conspecifics are members of the same species. Competition for resources between conspecifics is much less common in wild plants than in cultivated species. Seed dispersal mechanisms tend to distribute seedlings away from each other and their mother plant. Dispersal is further reinforced by the ability of many seeds to remain dormant for months or years, thus ensuring temporal, as well as spatial, dispersal before germination. When plants do occur in large tightly packed groups of conspecifics, such as bluebells, Japanese knotweed or daffodils, such groups are often made up of genetically identical clones that originate from a single mother plant via asexual reproduction. Because they are clones, these plants are not really in competition with each other but are more like the different branches of a tree.

A man-made form of competition between conspecific plants occurs when crops are grown at high densities. This is one reason why crop growth tends to be limited by nutrient availability and why crops respond so well to supplementary fertilizer and minerals. Over the years, farmers have learned by trial and error how to achieve optimal spacing for each crop plant. This spacing can vary greatly between cultivars and may differ from year to year depending on climatic conditions. In some cases it may be necessary to cull surplus plants within a crop stand to enable the remainder to reach their optimal yield. One way to increase crop density is to maximize the ability of each plant to acquire nutrients and to photosynthesize within the smallest possible footprint. This has led to interest in redesigning the canopy architecture of some crops, e.g. by manipulating recently discovered genes that regulate branching or leaf design (see Chapter 12).

Weeds

Weeds are plants that compete for resources or otherwise impair the performance of a more desirable plant, such as a crop or ornamental. A weed is simply a plant that is growing in the 'wrong' place as defined by humans, whether they are gardeners, farmers or conservationists. Most weeds are regarded as undesirable because they compete in some way with man-made floral assemblages such as field crops, tree plantations, pastureland, decorative lawns or annual borders. Other plants may be defined as weeds because they are relatively new entrants into ecosystems, whether 'natural' or man-made. Examples include the many so-called 'invasive' plant species that have displaced native flora after being introduced (normally, but not always, by humans) into a new habitat (see Box 5.2).

Even crop plants can be regarded as weeds if they are growing in the 'wrong' place. Such crop weeds are known as volunteers and examples might include uninvited barley plants growing in a wheat field, rapeseed plants growing in a pea field, or vice versa.

Globally, weeds account for about 25% of crop losses due to biotic factors, although local losses can be much higher. As discussed in Chapter 9, after the mid-20th century, manual weed control was increasingly replaced by chemical herbicides, commonly known as weedkillers. By definition, herbicides are toxic to plants and can be either selective (affecting some plants) or non-selective (affecting most or all plants). Many hundreds of herbicides are available to farmers. Some of the more effective herbicides are the non-selective (or broad-spectrum) agents, which kill most or all weeds. The drawback of broad-spectrum herbicides is that they can also affect growing crops, especially at critical stages of development such as seedling establishment or grain filling. This has led to the development of crop varieties that are tolerant to one or more broad-spectrum herbicides. Such varieties can be treated with the herbicide in question at any time during the growing season to control weeds without adversely affecting development of the crop itself. One of the problems with the constant use of a limited repertoire of the more popular herbicides is that it facilitates the evolution of resistant weed biotypes. As shown in Table 6.8, this is becoming an increasingly serious challenge and one that may be exacerbated by the increasing use of a small range of transgenic herbicide-tolerance traits in some major crops, as discussed in more detail in Chapter 9.

Parasitic plants

Most weeds merely compete with useful plants and thereby limit their development and productivity. However, a potentially more serious threat is posed by plants that directly parasitize a useful plant. It has been estimated that 1% of angiosperm species are parasitic, including some that affect crops. One of the most destructive plant parasites is the witchweed, *Striga hermonthica*, which can severely limit the productivity of staple crops such as maize, sorghum, millet and upland rice in sub-Saharan Africa. Striga plants attach themselves to the roots of their host from which they withdraw photosynthetic

Table 6.8. Herbicide-resistant weeds.

Herbicide group	Mode of action	Example herbicide	Resistant biotypes
ALS inhibitors	Inhibition of acetolactate synthase (ALS)	Chlorsulfuron	108
Photosystem II inhibitors	Inhibition of photosynthesis at photosystem II	Atrazine	68
ACCase inhibitors	Inhibition of acetyl-CoA carboxylase (ACCase) and fatty acid synthesis	Diclofop-methyl	38
Synthetic auxins	Disruption of auxin-mediated processes	2,4-D	28
Bipyridiliums	Disruption of electron transport at photosystem I	Paraquat	24
Ureas and amides	Inhibition of photosynthesis at photosystem II	Chlortoluron	21
Glycines	Inhibition of EPSP synthase and disruption of protein synthesis	Glyphosate	18
Dinitroanilines	Microtubule assembly initiation	Trifluralin	10
Thiocarbamates	Inhibition of lipid synthesis	Triallate	8
PPO inhibitors	Inhibition of protoporphyrinogen oxidase (PPO)	Oxyflurofen	5
Tirazoles, ureas, isoxazolidiones	Bleaching: inhibition of carotenoid biosynthesis	Amitrole	4
Chloroacetamides	Inhibition of very long chain fatty acid formation and cell division	Butachlor	4
Nitriles	Inhibition of photosynthesis at photosystem II	Bromoxynil	3
Six other herbicide groups	Various	Various	8
Total number of unique herbicide-resistant weed biotypes			347

Data from http://www.WeedScience.org (2010).

assimilates, minerals and water. This results in the characteristic 'witch' appearance of the stunted and withered host crop. Striga can also cause serious damage to its host plant by producing phytotoxins. Striga infests as much as 50 Mha of farmland in Africa, causing yield losses from 20 to 80% or even total crop failure after severe infestations.

Striga seeds remain dormant and viable in the soil for up to 20 years. With every planting season, some dormant seeds, stimulated by crop exudates, germinate and infest the host crop. The number of Striga seeds in the soil therefore increases and compounds the problem for future years. The cost of the damage caused annually by Striga is estimated at $10 billion, affecting the livelihoods of more than 100 million people. One of the most common crop targets attacked by Striga is the cereal sorghum, which is one of the most important subsistence crops in Africa. For this reason, biotechnological solutions to the Striga problem were sought in the 1990s, with a good degree of success. Breeders used a broad-based research approach involving molecular genetics, biochemistry and agronomy to identify genes for Striga resistance in a wide range of sorghum germplasm. Plants carrying these genes were introgressed into locally adapted and more modern sorghum varieties to create Striga-resistant hybrids. Following successful field trials, these hybrids were widely disseminated in Africa in 2003.

Other economically important parasitic higher plant genera include dodder *Cuscuta* spp., which infect clover, potato and lucerne crops; and broomrapes, *Orobanche* spp., which infect a wide range of crops including tomato, aubergine, potato, cabbage, bell pepper, sunflower and beans. Broomrape infestations of crops are particularly serious in the Mediterranean and Near East regions where they affect 16 Mha and can cause 20–100% loss of crop yield.

Herbivores

Herbivory can be a serious threat to plants as shown by the immense damage often caused by livestock overgrazing. Overstocking of domesticated grazers such as cattle and sheep can lead to degradation of pastureland, including removal of plant cover and erosion of the remaining soil. Grazing by domesticated herbivores is largely responsible for the open, almost treeless landscape of many upland areas of Europe that were originally forested. Where they are present, large wild herbivores can sometimes be problematic for crops, but they are relatively easily controlled by fencing. However, other smaller burrowing plant-eaters such as rabbits and rodents can still cause serious damage to the foliage and seed of crops. In general, undesirable herbivores are best controlled by excluding them from crop areas or by culling.

Domesticated grazing herbivores require plants for food and this can come either directly from forage or from feed rations. Almost 70% of global agricultural land is used as forage for livestock grazing, and some of it is potentially suitable for arable crop production instead. In addition, a growing proportion of grain crop production is used to generate animal feed. Examples include feed rations based on whole grains, such as maize, oats and soybean, and protein-rich seed cakes produced as a by-product of crops such as sunflower or rapeseed. There is an increasing trend towards intensive methods of livestock management whereby animals such as pigs and cattle are kept in enclosed feedlots and fed grain rations rather than grazing in the open. This also increasingly applies to smaller livestock species, such as poultry, that are fed on grains such as maize and soybean.

Tropical and temperate grasslands containing mainly forage grasses and legumes occupy about 2.4 billion ha that generate some 19 billion t of green dry matter per year. In contrast, arable crops occupy about 1.4 billion ha and generate roughly 5 billion t dry matter equivalent of useful grain, fruit and root products. A list of the most economically important forage grass and legume species is given in Table 6.9. The manipulation of forage plants by modern breeding methods is an area of growing interest that is benefiting from increased knowledge of plant genetics and development gained from studies of model and arable crop species.

6.5 Summary Notes

- Due to their sessile nature, plants are particularly subject to a wide range of stresses from non-living (abiotic) and living (biotic) parts of their environment.
- Many stresses tend to occur together and plant stress responses involve multiple linked mechanisms that often involve several types of stress.
- Major categories of abiotic stress include thermal (heat/cold), water (drought/flooding), salt and man-made pollutants (ozone, heavy metals).
- A common feature in many stresses is oxidation and plants have numerous mechanisms to prevent or repair oxidative damage.

Table 6.9. Forage grasses and legumes for domesticated livestock.

Temperate – cool season	Tropical – warm season
Forage grasses	
Perennial ryegrass (*Lolium perenne*)	Bermudagrass (*Cynodon dactylon*)
Italian/annual ryegrass (*Lolium multiflorum*)	Manila grass (*Zoysia* spp.)
Tall fescue (*Festuca arundinacea*)	*Paspalum* spp.
Meadow fescue (*Festuca pratensis*)	*Pennisetum* spp.
Orchardgrass, cocksfoot (*Dactylis glomerata*)	Buffelgrass (*Cenchrus ciliaris*)
Timothy (*Phleum* spp.)	*Brachiaria* spp.
Bromegrass/prairie grass (*Bromus* spp.)	*Panicum* spp.
Wheatgrass (*Agropyron* spp.)	Forage maize (*Zea mays*)
Bentgrasses (*Agrostis* spp.)	Bamboos
Wild rye (*Elymus* spp.)	*Dichanthium* spp.
Forage legumes	
White clover (*Trifolium repens*)	Egyptian clover, berseem (*Trifolium alexandrium*)
Red clover (*Trifolium pratense*)	*Leucaena* spp.
Subterranean clover (*Trifolium subterraneum*)	*Sylosanthes* spp.
Sweet clovers (*Melilotus* spp.)	*Macroptilium* spp.
Vetches (*Vicia* spp.)	*Cetrosema* spp.
Lupins (*Lupinus* spp.)	*Desmodium* spp.
Soybean (*Glycine max*)	

- The major forms of biotic stress involve animal pests (especially insects), microscopic pathogens and other plants.
- All forms of stress tolerance are important targets for crop improvement and several biotechnological approaches have been used in this area.

Further Reading

Abiotic stress

Baniwal, S.K., Bharti, K., Chan, K.Y., Fauth, M., Ganguli, A., Kotak, S., Mishra, S.K., Nover, L., Port, M., Scharf, K.D., Tripp, J., Weber, C., Zielinski, D. and von Koskull-Döring, P. (2004) Heat stress response in plants: a complex game with chaperones and more than twenty heat stress transcription factors. *Journal of Bioscience* 29, 471– 487.

Buchanan, B., Gruissem, W. and Jones, R. (eds) (2000) *Biochemistry and Molecular Biology of Plants*. American Society of Plant Physiologists, Rockville, Maryland.

Chinnusamy, V., Jagendorf, A. and Jian-Kang, Z. (2005) Understanding and improving salt tolerance in plants. *Crop Science* 45, 437–448.

Opik, H. and Rolfe, S. (2005) *The Physiology of Flowering Plants*. Cambridge University Press, UK.

Pardo, J.M. (2010) Biotechnology of water and salinity stress tolerance. *Current Opinion in Biotechnology* 21, 185–196.

Slater, A., Scott, N.W. and Fowler, M.R. (2008) *Plant Biotechnology*. Oxford University Press, Oxford.

Timperio, A.M., Egidi, M.G. and Zolla, L. (2008) Proteomics applied on plant abiotic stresses: role of heat shock proteins (HSP). *Journal of Proteomics* 71, 391–411.

Urano, K., Kurihara, Y., Seki, M. and Shinozaki, K. (2010) 'Omics' analyses of regulatory networks in plant abiotic stress responses. *Current Opinion in Plant Biology* 13, 132–138.

Wang, W., Vinocur, B., Shoseyov, O. and Altman, A. (2004) Role of plant heat-shock proteins and molecular chaperones in the abiotic stress response. *Trends in Plant Science* 9, 244–252.

Biotic stress

Danchin, E.G.J., Rosso, M.-N., Vieira, P., de Almeida-Engler, J., Coutinho, P.M., Henrissat, B. and Abad, P. (2010) Multiple lateral gene transfers and duplications have promoted plant parasitism ability in nematodes. *Proceedings of the National Academy of Sciences USA* 107, 17651–17656.

Ejeta, G. (2007) Breeding for Striga resistance in sorghum: Exploitation of an intricate host–parasite biology. *Crop Science* 47, S-216–S-227.

Taiz, L. and Zeiger, E. (2010) *Plant Physiology*, 5th edn. Sinauer, New York.

WeedScience.org (2010) Available at: www.weedscience.org/In.asp.

Westwood, J.H., Yoder, J.I., Timko, M.P. and dePamphilis, C.W. (2010) The evolution of parasitism in plants. *Trends in Plant Science* 5, 227–235.

7 Domestication and the Empirical Exploitation of Plants

7.1 Chapter Overview

This chapter will focus on empirical plant exploitation from the initial domestication of crops until the beginnings of modern scientific plant breeding and agronomy in the late 19th century. We will begin by examining the mechanisms responsible for the domestication of the relatively few wild plants that became transformed into crops. In the second half of the chapter, we will review the various empirical methods of breeding and management used to improve and exploit crops. Crop improvement involves management techniques (e.g. crop rotation, irrigation and drainage) plus the use of biological (e.g. breeding and cloning) and chemical (e.g. pesticides and fertilizers) tools. We will see that early advances in crop production were mostly due to improved management whereas, with a few exceptions, chemical and biological methods had rather modest impacts on crop productivity until modern times.

7.2 Domestication of Crop Plants

Crop and livestock domestication occurred during an unusually stable and relatively warm/moist climatic period that started about 11,000 years ago and is still with us today. Genomic studies show that for most crops domestication involved a gradual accumulation of mutations and that, initially at least, it was not a deliberate decision by hunter-gatherers to become farmers. Rather it was a largely non-intentional process that can be envisaged as a form of mutually beneficial coevolution between humans and certain plants and animals. About 12,000 years ago, during the adverse climatic conditions of the Younger Dryas Interval, people in a few regions of the world responded to food shortages by more intensive forms of plant management. Only a few plants in a few geographical locations responded to human manipulation by developing domestication related traits, such as large, non-shattering seeds.

Plants that developed such traits as a result of random mutations were recognized as improvements and selectively propagated by people. Over many generations, this led to the imposition of human selection (or breeding) in favour of domestication traits, many of which would be highly disadvantageous for plants growing in the wild. In some cases, such as rice and breadwheat, the domesticated versions became new species that now rely on humans for their existence, as they are no longer capable of survival in the open environment. For some crops, the domestication process is still under way. In such 'recalcitrant' crops, some domestication traits have proved difficult to improve. However, new biotechnological tools are proving of great value in this area and in the future it may also be possible to use genomic and molecular technologies to convert completely wild plants into new crops.

Centres of origin

A 'Centre of Origin' is a region in which several wild plants began the process of domestication to crops suitable for cultivation. The 'Centres of Origin' concept was developed during the 19th and 20th centuries by de Candolle, Vavilov and Harlan. They noticed that a few regions of the world were especially rich in crop species that had been grown there for many millennia (see Fig. 7.1). Such areas also tended to have many wild relatives of locally cultivated crops and were obvious candidates as the places where these crops were first domesticated. In contrast, in many other parts of the world people grew fewer staple crops and no wild relatives were present. In such crop-poor regions, it is likely that the crops had been imported from the primary centres of origin some time after their initial domestication.

For example, Mesoamerica (Mexico and Central America) has a huge diversity of maize and squash

Fig. 7.1. Centres of origin of major crops. The seven most important centres of origin are shown. There are several additional more localized areas of crop origin, such as New Guinea, where people domesticated crops such as taro (*Colocasia esculenta*) and sugarcane (*Saccharum sinense*), but the vast majority of globally significant crops come from the seven major centres listed below.

1. Western Eurasia – barley, wheats, rye, fig (*Ficus carica*), pea (*Pisum sativum*), lentil (*Lens esculenta*), flax/linseed (*Linum usitatissimum*), olive (*Olea europea*), sugarbeet (*Beta vulgaris*), sesame (*Sesamum indicum*).
2. East Asia – rice, millets (*Panicum* spp.), hemp (*Cannabis sativa*), mulberry (*Morus alba*), onion (*Allium chinense*), tea (*Camellia sinensis*), soybean (*Glycine max*), orange (*Citrus sinensis*), Asian yam (*Dioscorea* spp.) grapefruit (*Citrus* spp.), cabbage (*Brassica oleracea*).
3. South Asia – rice, banana and plantain (*Musa* spp.), breadfruit (*Artocarpus communis*), coconut (*Cocos nucifera*), mango (*Mangifera indica*), mungbean (*Vigna radiata*), pigeon pea (*Cajanus cajan*).
4. Ethiopia – coffee (*Coffea* spp.), castor bean (*Ricinus communis*), cowpea (*Vigna sinensis*), sesame (*Sesamum indicum*), tef (*Eragrostis tef*).
5. Sahara/West Africa – pearl millet (*Pennisetum glaucum*), African yams (*Dioscorea cayenensis/rotunda*), oil palm (*Elaeis guineensis*), African rice (*Oryza glaberrima*), okra (*Abelmoschus esculentus*).
6. Mesoamerica – squash (*Cucurbita pepo*), avocado (*Persea americana*), common bean (*Phaseolus vulgaris*), cocoa (*Theobroma cacao*), maize, tetraploid cotton (*Gossypium hirsutum*), pepper (*Capsicum annuum*), sweet potato (*Ipomoea batatas*).
7. Andes – manioc/cassava (*Manihot utilissima*), groundnut (*Arachis hypogea*), tobacco (*Nicotiana tabacum*), potato (*Solanum tuberosum*), tomato (*Lycopersicon esculentum*), pineapple (*Ananas comosus*).

Adapted from Murphy (2007).

varieties, plus numerous wild relatives. The early cultivation of these crops in Mesoamerica is supported by historical and archaeological evidence. Further afield in adjacent regions of North and South America, far fewer maize and squash varieties are grown, wild relatives are absent, and the crops appear to be much later introductions. The same is true for wheat and barley, which have a well-defined centre of origin plus many wild relatives in the Near East, especially in the so-called 'fertile crescent' that extends in an arc along the Upper Tigris and Euphrates basins to the Taurus foothills of Anatolia and down to the Jordan Valley. Today, wheat and barley are grown as introduced species around the world but all of these cultivated varieties can be traced back to their origins in the Near East.

The Centres of Origin concept is significant in two ways. First, it shows that crop domestication happened independently in different parts of the world. Second, it demonstrates that such domestications were relatively rare events – hence the small number of primary centres of diversity. People outside these primary centres would have constantly experimented with non-agricultural uses of food plants and possibly tried to domesticate some of the best wild species (see Chapter 1). However, it seems that the types of plants available to such cultures were simply not amenable to domestication. For example, before the 19th century, hunter-gatherers in southern Africa, Australia and California partially lived off wild grasses for many millennia. During this time, they employed many techniques to encourage growth and improve the yield of their edible grasses. And yet they never produced domesticated versions of these grasses, while cultures in other regions appear to have domesticated other grassy species, namely the cereals, with relative ease.

This implies that it is not necessarily the activity of the humans that is the primary determinant of crop domestication, but rather the availability of the 'right sort' of plant. In other words, the initial stages of crop domestication may have been determined largely by a combination of plant genetics and environmental factors, such as climate, rather than being due to conscious human intervention. Therefore, if edible plants with the genetic potential for domestication happened to be present in an area, the chances for the development of agriculture would have been much higher. On the other hand, if no edible plants in a region possessed such genetic attributes, agriculture based on local plants could not have developed, no matter how clever or resourceful the local human population. The only way that such people, including most Europeans and North Americans, could develop agriculture would be to import already-domesticated crops, from elsewhere.

The domestications of the major ancient crops were highly variable in duration, and localization. For example, some crops, such as potatoes, barley, emmer and einkorn wheat, cassava, maize and bottle gourd, may have been domesticated on a very few occasions, and sometimes only once. In contrast, squash, cotton, millet and common beans were domesticated several times. In each centre of origin, several different types of plant were domesticated. These tend to be species that are complementary in terms of nutrition, such as starch-rich cereals and protein-rich pulses. Globally speaking,

in addition to the 'big three' cereals (rice, wheat and maize), other major groups included starchy root/tuber species such as potatoes, squash and yams, and pulses such as lentils, peas and beans. Although these plants are from very diverse families, they responded in broadly similar ways to domestication. In contrast, many close relatives of cereals, pulses and root crops were not domesticated, despite a great deal of effort by farmers. One genetic feature shared by most crop species is that control of the group of traits, often called the 'domestication syndrome', resides in a small number of physically linked genes.

Biology of crop domestication

Instances of crop and livestock domestication were comparatively rare and in many cases single key mutations played a vital part in such processes. For example, all the cultivated maize grown today is probably descended from a single teosinte plant with a specific mutation to its *tb1* gene (see Box 7.1). This mutation resulted in an altered growth pattern from a small bushy plant with small cobs to a tall, erect, apically dominant plant with enlarged cobs, as shown in Fig. 7.2. In the past decade, genomic analysis has greatly elucidated the ancient crop domestications and is suggesting new ways in which such crops might be genetically manipulated in the future. A list of some of the genomic changes, i.e. mutations, associated with domestication-related genes, is shown in Fig. 7.3. In each case, the random disruption of a few genes had far-reaching consequences for the plant. This also shows that mutations do not always result in loss of biological function: sometimes they create valuable new functions such as sweet-tasting maize cobs or enlarged tomato fruits.

Major domestication-related traits in plants

As listed in Table 7.1, major domestication-related traits include non-shattering seeds, large seeds, high yield, synchronous flowering and seed set, loss of seed dormancy, and traits related to harvesting and food preparation. In Table 7.2 there is a list of the domestication-related genes of rice that have now been isolated and in some cases characterized. Many non-crop plants also contain genes that could potentially result in domestication-friendly phenotypes like seed retention and lack of dormancy.

Box 7.1. Role of the *tb1* and *tga* mutations in maize domestication

The *tb1* (*teosinte branched 1*) gene encodes a transcription factor protein responsible for regulation of a set of genes involved in several processes, including formation of lateral branches. In domesticated maize this gene has mutated to an inactive form, resulting in suppression of main-stem branching so that the plant now has a single main stem and much larger cobs. The changes caused by mutation of this gene are visually dramatic in terms of overall architecture of maize but, more importantly, they also result in a plant that produces an increased food yield due to the larger size of its seed cobs (Fig. 7.2). The *tb1* gene acts with another gene, called *barren stalk1*, in regulating vegetative lateral branching in maize. Both genes were co-selected early in the transition of wild teosinte to cultivated maize.

The *tga1* (*teosinte glume architecture*) gene encodes a putative transcription factor and controls the formation of a hard grain case, or glume, around the seed kernels. This gene played a vital role in the domestication of maize. In teosinte, a hard external casing makes the grains difficult to digest, so most of them pass through the stomach and out in the faeces. For a wild teosinte plant, this is a good strategy to promote grain

dispersal via animal vectors that might eat the seeds. Before the mutation of this gene, the original teosinte grain would have been virtually inedible unless it was vigorously and repeatedly milled. If this version of teosinte was ever gathered or cultivated by ancient Mesoamericans, it was probably used for its sweet stalks, rather than the virtually indigestible grains.

In domesticated maize, the *tga1* gene has undergone a mutation so the plants no longer produce a hard grain coat and the kernels are bare on the cob. This makes maize much easier for humans to digest. The change to a naked, exposed kernel may at first sight seem disadvantageous for the maize plant because its grains will now be completely digested by people (and other animals) who eat them, rather than being disseminated via their droppings. In fact, however, this mutation has resulted in a much improved dispersal mechanism for the seeds. Rather than relying upon the digestive systems of passing animals, maize can now exploit the far more effective propagation skills of its human partners, who have spread maize globally so that it is now one of the most abundant and widely distributed plants on earth.

For example, genetic loci regulating such traits such as seed size, seed weight, short-day flowering and seed retention have been found in a range of divergent plant families, including legumes, cereals and solanaceous vegetables, including hundreds of non-crop species.

This begs the question: if most of the more than 250,000 higher plant species contain very similar domestication-related genes, why have so few crops been successfully domesticated? Part of the answer is that it is not the mere presence of such genes that is required for creation of domesticated plants but rather their chromosomal locations and regulatory mechanisms. Three interrelated genetic factors have greatly facilitated the manipulation of domestication-related traits in the major crops as follows.

REGULATION OF KEY TRAITS BY ONE OR TWO GENES One of the most important traits for early grain-crop cultivators was seed shattering. In wild plants, mature seeds are readily shed and lost before they can be harvested. In our most successful crops, this

crucial trait is regulated in a very simple manner by either one (lentil) or two (rice, wheat, barley, sorghum, oat, pearl millet) genes. For example, mutations in the *Br* gene of wheat, or the *sh4* and *Sh1* genes of rice, all of which disrupt the process of grain dehiscence, led to the retention of grains on the stalk of the ripened crop from where they could be more easily harvested. Mutations in these genes created the non-shattering cereal varieties that were the prerequisite to their successful domestication.

One of the key mutations behind the success of barley is the change from a two-row to a six-row phenotype. As shown in Fig. 7.4, six-row barley is higher yielding, relatively salt-tolerant, and responds well to cultivation under intensive irrigation systems. The six-row trait is regulated by several *Vrs* loci but the major locus is associated with the *Vrs1* gene. Maize provides more examples of important single-gene traits involved in domestication. In Fig. 7.2 and Box 7.1, we saw how the *tb1* mutation changed the architecture of teosinte from an unproductive bushy habit to the tall, erect,

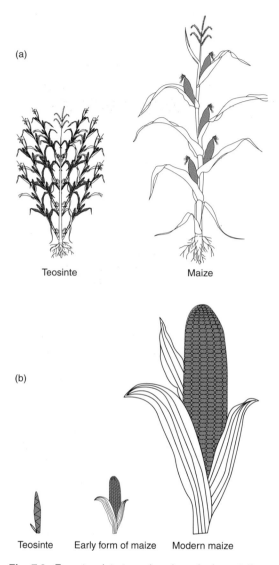

(a)

Teosinte   Maize

(b)

Teosinte   Early form of maize   Modern maize

Fig. 7.2. From teosinte to maize via a single mutation. (a) Teosinte was originally a relatively short and bushy plant with many lateral branches that bore small cobs. A single mutation in the *tb1* gene as early as 10,000 years ago resulted in the suppression of lateral branch formation and the development of a tall, erect plant that was able to bear much larger cobs on its single major stalk, namely maize. (b) Although early forms of maize had much larger cobs than teosinte, they were still significantly smaller than in modern maize. Because cob size is a genetically complex trait it took maize farmers several thousand years to select new varieties with increasingly larger cob size. About 2000 years ago, maize cobs were finally large enough to generate the high food yields required to sustain the development of advanced urbanized civilizations in Mesoamerica.

higher yielding maize plant. A second example, shown in Box 7.1, describes how the *tga1* mutation altered the grain coat of maize in a way that greatly improved its digestibility for people.

CLUSTERING OF DOMESTICATION-RELATED GENES   Chance mutations are extremely rare and two or more mutations in the same individual are even rarer. Even when a plant carries two useful mutations, if these affect genes on separate chromosomes it is unlikely that offspring arising from sexual reproduction with a wild-type individual will carry both mutations. However, if both mutations occur on genes located close together on the same chromosome, half of the progeny will carry the double mutation. And, if yet more genes in this chromosomal region undergo mutations, there is a strong likelihood that some progeny will carry all the mutations. Clearly therefore, if genes involved in domestication are clustered, the chances of early farmers being able to recognize and then select and propagate plants with multiple mutations would have been greatly increased.

An example of gene clustering can be found in maize. The tall, high-yielding cultivated version of maize and the short, small-seeded wild teosinte plant are mainly distinguished by differences in DNA sequences and expression patterns of five groups of genes. The five groups of genes regulate: (i) tendency of the ear to shatter; (ii) percentage of male structures in the primary inflorescence; (iii) internode length on the primary branch; (iv) and (v) increased numbers of kernels per cob. In the maize genome, all of these loci are tightly clustered in small genomic regions. Another example is in wheat, where the *Br* seed-shattering gene (see above) is closely linked to eight other DNA regions that regulate other domestication-related traits.

As shown in Table 7.3, similar examples of gene clustering have been found in crops such as pearl millet, sunflower, rice and common bean. Indeed, it is likely that all crop genomes exhibit some degree of clustering of domestication-related genes. Analysis of livestock genomes shows that similar sorts of domestication-related gene clustering occur in farm animals such as cattle, sheep and goats. In Fig. 7.5, examples of clustering in localized chromosomal regions are shown for emmer wheat, common bean and sunflower. Therefore, the difficulty in domesticating most wild plants and animals may be due to a lack of such genetic linkage, making it much harder to find individuals carrying multiple mutations. As we

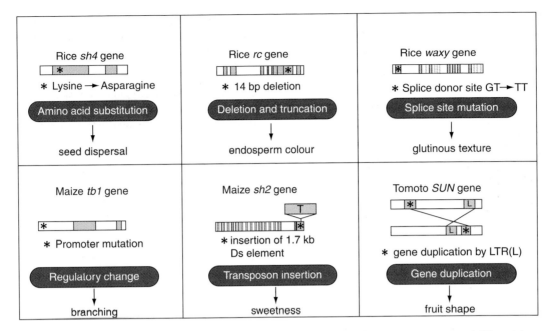

Fig. 7.3. Types of genomic changes associated with domestication-related genes. There are several different forms of genomic change and resultant phenotypes that are associated with domestication traits. Six of the most common types of genomic change, with a specific example for each, are shown. Note that two of these examples, namely transposon insertion and gene duplication, involve repetitive DNA elements. This highlights some of the useful roles that can sometimes be played by repetitive DNA. Adapted from Tang *et al.* (2010).

Table 7.1. Some major domestication-related traits in crop plants. Any given crop will not necessarily carry all of these traits and their relative importance will vary according to the crop type and farming system. Hence, grain crops invariably have much larger seeds than their wild relatives, but this may not apply to root crops like potatoes, where tuber size and absence of toxins are much more important traits than seed size.

Trait	Wild plant	Domesticated crop
Height	Tall	Short or dwarf
Growth habit	Branched and bushy	Unbranched and compact
Ripening	Asynchronous	Synchronous
Seed dormancy	Present	Absent
Seed shattering	Shattering heads	Non-shattering heads
Seed size	Small	Large
Ease of dispersal	Highly dispersible	Loss of dispersal
Threshing	Hard	Easy
Reproduction	Outbreeding	Self-fertilizing
Germination	Asynchronous	Synchronous
Hairs and/or spines	Present	Absent or reduced
Toxins	Present	Absent

will see below, methods such as comparative genomics, mutagenesis and marker-assisted selection should make it easier in the future to assemble multiple mutations, even if genes are not clustered, and thereby create new domesticated species.

'MASTER GENES' One way of avoiding the problem of manipulating dozens of separate, and not necessarily physically linked, domestication-related genes is to identify a single gene, such as one encoding a transcription factor protein, that regulates all

Table 7.2. Features of cloned domestication-related genes of rice.

Trait	Gene	Protein	Genomic change
Taste and texture of cooked rice	Wx	Granule-bound starch synthase	Mutation at first intron splice site
Seed shattering	sh4	Myb transcription factor	Mutation causing a single amino acid change
Seed shattering	qSH1	Bell-type homeobox transcription factor	Mutation causing altered expression pattern in promoter region
Seed pericarp colour	Rc	bHLH transcription factor	A 14-bp deletion causing a premature stop codon
Seed pericarp colour	Rd	Dihydroflavonol-4-reductase	Two independent mutations yielding premature stop codons
Seed width	qSW5	Unknown protein	Deletion of around 1 kb
Grain number per panicle	Gn1a	Cytokinin oxidase	A 16-bp deletion in the ORF, resulting in a frame shift mutation
Flowering time	Ghd7	CCT motif protein	Several DNA changes identified in northern rice cultivars
Plant height	sd1	GA20 oxidase	A 383-bp deletion in 'Green Revolution' cultivars, also seen in a few wild-rice accessions
Plant stature	PROG1	Zn-finger transcription factor	Mutation causing an amino acid change in an ORF
Grain filling	GIF1	Cell-wall invertase	Possible mutation in the promoter region

Data from Izawa *et al.* (2009).

of these genes. As discussed in Chapter 5, many complex processes in plant development involving large numbers of genes are regulated by 'master genes'. This also applies to several important crop traits. For example, 15 separate traits, ranging from pod number to seed yield, have been studied in the common bean. Each of these traits involves as many as several dozen genes. In each case, however, between one and four genes control the vast majority of variation in the trait in question. As more crops are studied at the molecular genetic level, we are finding more and more examples of such 'master genes'. These genes are often associated with so-called quantitative trait loci (QTL), and can be identified by molecular marker analysis. As discussed in Chapter 10, important domestication-related QTL have now been found in dozens of crops including rice, sorghum, tomato and potato.

The unfinished business of domestication

Although many important crops have been subjected to breeding for as long as 10 millennia, the process of domesticating continues to the present day. The list of domestication-related traits (Table 7.1) is relatively long and few of these traits have been optimized in every crop. This is particularly true for more 'recent' crops, such as the brassicas. For example, rapeseed is an allotetraploid hybrid of cabbage and turnip that was only formed about 2000 years ago. Unlike the vegetable brassicas, rapeseed is grown for its oil-rich seed. However, as with most wild plants, the seeds are sometimes lost due to shattering of the mature seedpods. Despite the efforts of breeders, yield losses of 10–30% can still occur due to seedpod dehiscence.

Unlike most cereals and pulses, it has been difficult to find non-shattering mutations in rapeseed where pod shattering behaves like a multigenic trait.

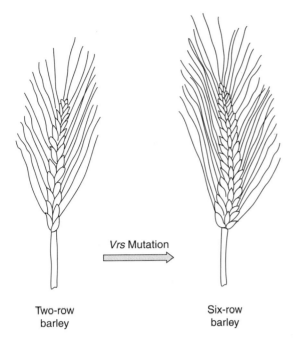

Vrs Mutation

Two-row
barley

Six-row
barley

Fig. 7.4. From two-row to six-row barley via *Vrs* gene mutations. Early forms of domesticated barley had two rows of ears. A few millennia after its first domestication, mutants with six rows of grains were discovered by Mesopotamian farmers. These higher-yielding and relatively salt-tolerant forms of barley were particularly amenable to cultivation under intensive irrigation systems and soon became a staple foodstuff. At least five independent loci control the six-rowed spike phenotype in barley, of which the most important is the *Vrs1* gene. The *Vrs1* gene encodes a transcription factor that is expressed specifically in lateral-spikelet primordia and suppresses the development of the lateral rows. Mutated versions of this gene are no longer able to stop lateral row formation, resulting in the six-row phenotype.

This behaviour may be compounded by the complex polyploid genome of rapeseed in which several alleles must be mutagenized in order to alter its pod-shattering trait. The manipulation of pod shattering in rapeseed may benefit from studies in the brassica relative *Arabidopsis*. Here, pod shattering was found to involve several genes, all of which were inhibited by a transcriptional regulator produced by the *ful* gene. Overexpression of the *Arabidopsis ful* gene in transgenic *Brassica juncea* led to inhibition of pod shattering. This form of engineered domestication has yet to be applied to commercial rapeseed but it demonstrates the potential value of such an approach.

A second example of a transgenic approach to domestication comes from rice. It is often useful to manipulate the number of tillers, or side branches, of rice plants to suit differing agronomic conditions, but tillering is difficult to alter in domesticated rice. The rice genome contains an orthologue of the maize *tb1* gene (Box 7.1), *OsTB1*, that also affects lateral branching. An active *OsTB1* gene produces a transcription factor protein that inhibits tillering. When the *OsTB1* gene was overexpressed in transgenic rice, tillering was reduced. Alternatively, a rice mutant called *fc1* was identified in which an allele of the *OsTB1* gene had suffered a DNA base deletion resulting in the creation of a premature stop codon. The new protein product was unable to bind DNA and hence *OsTB1* gene function was lost and tiller production was enhanced. These experiments suggest that manipulation of *OsTB1* gene activity, whether via mutants or transgenesis, might be a useful approach to creating different kinds of 'designer tillering' so that rice plants are better adapted to their particular growth situation.

Domesticating new crops

The 'big four' commercial cereal crops (wheat, rice, maize and barley) still make up 73% of global grain production. One of the main reasons for this is that domestication of most crops involved rare forms of genetic architecture, combined with the actions of human cultivators. People tried repeatedly to domesticate other useful plants, often persisting for many centuries, but virtually all of their attempts failed due to the recalcitrant genetics of most wild plants. However, it now appears that many, and perhaps all, major groups of higher plants have relatively conserved gene families that regulate such traits as seed size, seed weight, short-day flowering and seed retention. For example, very similar domestication-related genetic loci regulating such traits have been found in a range of very divergent plant families, including legumes, grasses and solanaceous plants, including hundreds of non-crop species.

Our current dependence on an extremely small number of major crop species has both advantages and drawbacks. On the plus side, cultivation and breeding of crops are simplified if we only need to focus on a limited number of species. This is especially true for the kinds of industrialized intensive farming that currently produces so much of our food both cheaply and efficiently. Also, because the

Table 7.3. Genomic regions showing clustering of domestication traits. The clustering of many major domestication-related traits in crops has been revealed by molecular genetic analysis following crosses between wild and domesticated plants. Such clustering would have greatly increased the likelihood of a comparatively rapid spontaneous evolution of domesticated varieties as the plants adapted to the new selection pressures created by early farmers.

Crop	Reproduction	Mapping cross (domesticated × wild forms)	Location of gene cluster*	Attributes of corresponding traits
Maize	Outcrossing, $2n=4x=20$	F2: *Zea mays* ssp. *mays* × *Z. mays* ssp. *parviglumis*	Chr 1	Shattering (ear disarticulation), growth habit, branching pattern (*tb1*), ear and spikelet architecture
			Chr 2S	Number of rows of cupules
			Chr 3L	Growth habit, ear architecture
			Chr 4S	Glume hardness (*tga1*)
			Chr 5	Ear architecture
Common bean	Self-pollinated, $2n=2x=22$	F2: *Phaseolus vulgaris* cultivated form × *P. vulgaris* wild form	LG D1	Growth habit and phenology
			LG D2	Seed dispersal (pod dehiscence) and dormancy
			LG D7	Pod length and size
Rice	Self-pollinated, $2n=2x=24$	F2: *Oryza sativa* × *O. rufipogon*	Chr 1	Growth habit (tillering and height), shattering, panicle architecture
			Chr 3	Shattering, panicle architecture, earliness
			Chr 6	Shattering, panicle architecture, earliness
			Chr 7	Panicle architecture
			Chr 8	Growth habit (height), earliness, shattering
Pearl millet	Outcrossing, $2n=2x=14$	F2: *Pennisetum glaucum* ssp. *glaucum* × *P. glaucum* ssp. *monodii*	LG6	Shattering, spikelet architecture, spike weight, growth habit
			LG7	Spikelet architecture, spike size, growth habit and phenology
Sunflower	Outcrossing, $2n=2x=34$	F3: *Helianthus annuus* var. *macrocarpus* × *H. annuus* var. *annuus*	LG17	Shattering, apical dominance, achene weight, earliness
			LG09	Achene size and weight, growth habit, head size
			LG06	Growth habit, achene size and weight, earliness, head size

*Chr, chromosome; LG, linkage group.
Adapted from Poncet *et al.* (2004).

major crops were domesticated so long ago, we have had millennia of breeding and selection to optimize their cultivation traits. On the negative side, concentration on a few major crops with narrow gene pools is ecologically undesirable as resultant monocultures are more susceptible to new pests and diseases. These ecological risks are exacerbated by the genetic impoverishment and varietal erosion that affect all our major crops. Environmental changes like salinization are also increasing in croplands for a variety

of reasons. One response to such challenges has been to try to breed stress-tolerance traits into the major crops, but another possibility is to grow new crops that are already adapted to such abiotic stresses.

As we possibly move into a period of renewed climatic instability, the option to broaden our range of domesticated crops becomes more persuasive. Also, the prospect of serious aridification episodes should make us consider domesticating new drought- and thermal-tolerant crops as an urgent priority.

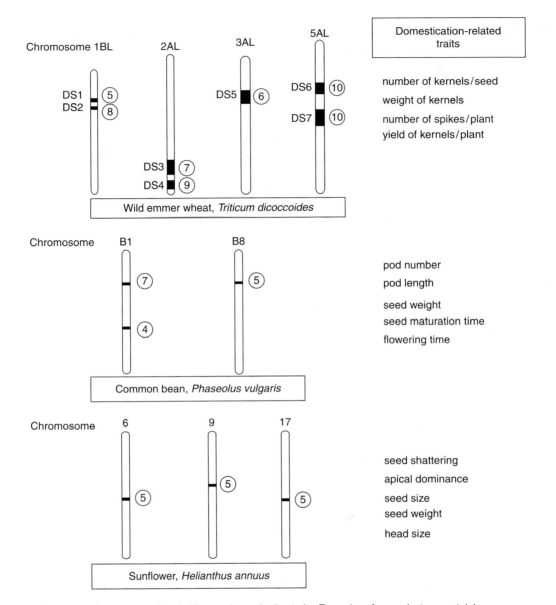

Fig. 7.5. Clustering of genes associated with crop domestication traits. Examples of gene clusters containing domestication-related traits are shown for three crops. In the wild precursor of emmer wheat, *Triticum dicoccoides*, seven gene clusters (DS1 to DS7) are shown on four separate chromosomes; in common bean, there are three gene clusters on two chromosomes; and in sunflower three gene clusters are shown on three chromosomes. In each case, the circled numbers indicate how many domestication-related traits are present in each cluster and the types of trait are listed to the right. Data from Peng *et al.* (2003).

There are tens of thousands of as-yet unused or underutilized plants that could potentially furnish us with a vast range of food and non-food products. Our new understanding of plant domestication and the availability of new breeding techniques enable us to recruit many wild plants into our portfolio of useful crops. As we continue to learn more about the genetic mechanisms regulating domestication, it will become increasingly feasible to cultivate any of the tens of thousands of potentially useful plants that have hitherto been beyond our control. The domestication strategy is also attractive in the longer term because it generates diversity, not only at the product level, but also at the primary production level. A greater number of crop species also provides an additional buffer against pests and diseases, and climatic variation, to which our current monoculture-based agriculture is relatively susceptible.

7.3 Dissemination of Agriculture

Dissemination, which literally means 'spreading seed', has been crucial to the success of agriculture. As described above, crops were only domesticated in a few locations or 'Centres of Origin'. Crops were then disseminated from these centres, first regionally but later globally. Today, staples like rice, wheat, potatoes, soybeans and maize are grown on all the inhabited continents and few people realize where such crops originated. Indeed, some of our most intensively farmed regions, including Europe, North America and Australasia, have almost no native crop species. On the other hand, some Centres of Origin, including East Africa, the Near East and the Andes, now suffer from low crop yields and regular food shortages. In this section, we will examine how a few crop species were spread around the world during the European exploration and colonization era of the 17th to 19th centuries. Some of these crops such as maize and wheat were staples, while others such as tea, sugar, coffee and cotton were cash crops that were an integral part of the colonization process itself.

Regional dissemination

The major cereal, legume and tuber/root crops were domesticated between 6000 and 11,000 years ago. Genomic evidence suggests that for some crops domestication was a fairly protracted process that might have taken hundreds or even thousands of years in each case. During the early stages of domestication it is likely that crops remained confined to their centres of origin. This would have been an important period for local human populations as they became increasingly reliant on cultivated plants and less involved with hunter-gathering. Management and processing technologies, such as ploughs, grinders and baking ovens, were invented or refined to optimize crop yields. The new farmers also took advantage of

any random variations (i.e. mutations) that might produce improved versions of their crops.

In the case of wheat, farmers selected new polyploids with larger, better quality grains. This led to the gradual move from diploid einkorn wheat to tetraploid emmer and hexaploid breadwheat genotypes in the Near East from about 10,000 to 6000 BP. In the case of barley, wild versions were gathered in the Near East and processed into flour from at least 23,000 BP and the first non-shattering, larger-seeded domesticated forms date from 11,000 BP. The original domesticated genotypes had two rows of grains on each ear and were successful in their region of origin. However, at some time after 8000 BP a new type of barley with six rows of grains on each ear was discovered (Fig. 7.4). The most important traits associated with six-row barley were its response to increased water availability due to irrigation and high tolerance to soil salinity. These qualities made six-row barley an ideal crop as people extended cereal farming away from the wetter regions of the upper Tigris and Euphrates southwards to more arid areas that could only be farmed using irrigation. Thanks to intensive irrigation-based agriculture with six-row barley as the major food staple, the first true cities and complex civilizations arose in Mesopotamia and Egypt soon after 7000 BP.

Other important staple crops were disseminated from their centres of origin in eastern Asia and Mesoamerica. For example, there are two races of rice called *indica* and *japonica*. Molecular analysis suggests that these races were genetically distinct before domestication, meaning that they must have been domesticated separately in different regions. It is likely that rice was domesticated in several parts of India, China and South-east Asia in a process that began over 10,000 years ago. One probable centre of origin of rice is the Yangtze Valley in China, where there is evidence of cultivation from at least 9000 BP. However, it took another 2000 years for rice to become a staple crop in the region as a whole. It then took a further 4000 years (i.e. until about 3000 BP) for rice cultivation to spread to adjoining regions such as Korea and Japan.

In Mesoamerica, the most important groups of ancient crop staples were maize and squash, both of which were being cultivated in parts of south-central Mexico by 8000 BP. Domesticated maize is an excellent food source but early varieties had relatively small cobs, which limited crop yield and the generation of large food surpluses. Very gradually, a series of additive mutations were selected by farmers leading to an increase in cob size from 2 cm in about 6000 BP, to 4 cm by 5000 BP and 6.5 cm by 2000 BP. This led to a huge increase in yield so that a single farmer could now feed many additional people. The domestication of common beans in Mexico around 2300 BP led to the establishment of a combined form of agriculture based on high-yielding maize, squash and beans called the *milpa* system. Thanks to the almost ideal nutritional balance of *milpa* crops, and the high yield of the new maize genotypes, Mesoamerican farmers were able to produce reliable food surpluses that directly led to the development of regional civilizations such as the Maya and Toltec. Maize then spread gradually to North and South America but remained confined to this continent until the 16th century.

The spread of agriculture into Europe from the Near East illustrates the importance of secondary additions to the original domestication traits when crops are moved into new regions. As shown in Fig. 7.6, it took well over 4000 years for farming to

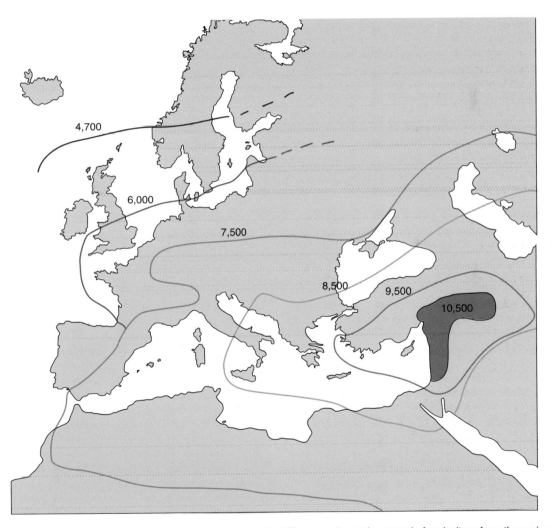

Fig. 7.6. Spread of agriculture into Europe from the Near East. The map shows the spread of agriculture from the major Near Eastern centre of origin, west and northwards into Europe. The lines represent the approximate boundaries of widespread agriculture at various dates before present. The north-westerly progression of agriculture was neither smooth nor continuous. In particular, there was a long delay before the selection of new cold-hardy cereal mutants enabled cereal farming to spread to cooler Atlantic littoral regions about 6000 years ago. Adapted and redrawn from Murphy (2007).

spread from its centre of origin in the Near East to the Atlantic coast of Europe. It is likely that crops were not introduced into Europe individually but rather as a package that included the cereals, barley and wheat, plus the pulses, peas and lentils. One factor slowing the spread of farming into cooler regions of North-west Europe was the lack of cold-adaptability of these crops, which were from comparatively warm regions in the Near East. The major cereal staples in Southern and Eastern Europe were emmer and einkorn wheat, but these warm-season crops did not grow well further north and west. However, after 7000 BP a new form of cold-tolerant barley appeared, enabling farming to spread relatively quickly to the rest of Europe. Eventually, cool-season wheat varieties were also selected, although barley and oats still perform better in harsh northern climates. As discussed in Chapter 12, current concerns about possible impacts of climate change on crops make it even more important to understand and manipulate those traits, such as warm- or cold-adaptability, that were so vital to the original dissemination of farming into Europe.

Global dissemination

By about 2000 years ago, the major staple crops had been disseminated from their centres of origin to wider regions of the various continents. However, these regions tended to develop separately and sometimes in complete isolation from each other. For example, all of the American crops including maize, potatoes, beans, tomatoes, cassava and cocoa were completely unknown to the rest of the world before the voyages of Columbus just over 500 years ago. The true globalization of agriculture only started when explorers began systematic sea-faring expeditions after the 10th century. For example, Muslim traders introduced sugarbeet from Eastern Asia to Northern Africa and then on to Spain and the Atlantic islands of Madeira and the Canaries. From the earliest days of European global exploration in the late 15th century, a key aim was to exploit some of the new crops being discovered in the Americas, South-east Asia and Africa.

Collectors and colonists

The Spanish and Portuguese were among the first Europeans to disseminate crops on a truly global basis. Soon after the conquest of the Americas, maize, tomatoes, potatoes, tobacco and common

beans were introduced into Europe. The British and others greatly expanded tea and coffee cultivation in many parts of their empires to satisfy an increasing demand for these stimulatory beverages in Europe. Cocoa was transplanted from Mexico to West Africa and the Caribbean to provide chocolate and Asian sugarcane was increasingly grown in Caribbean plantations where the resulting demand for labour led to the beginnings of the transatlantic slave trade. From the 16th to the late 18th centuries, dozens of crops were transplanted from one continent to another for a wide range of purposes.

For example, various species of palm were taken, from the jungles of Africa and Malaya, to Bengal in India to stave off famine in this new British colonial possession. Breadfruit was brought from Tahiti to the West Indies to be grown as a cheap staple to feed the recently imported African slaves. Most of the slaves were brought over to work in the new and immensely profitable Caribbean sugarcane or American cotton plantations. One of the motivations for the British conquest of Jamaica in 1655 was its agricultural potential and the colonists soon established a lucrative export industry based on sugarcane and cocoa.

The development of the global, trade-based maritime empires of the British and Dutch in the 18th century also led to a more systematic form of collection and cataloguing of potentially useful plants. Botany became more of a scientific discipline but with an increasing focus on the economic exploitation of plants. Numerous botanical gardens, often called 'economic gardens', were set up in far-flung places such as Java, Ceylon, Mauritius and Trinidad as well as more locally in London, Paris, Madrid and Vienna. These gardens, such as the Royal Botanical Gardens at Kew, established in 1759, were intended for the study and profitable exploitation of new plants. In many cases, new forms of imported plants were adapted for local conditions.

During this period, part of the mission of Kew Gardens was to study the increasing numbers of plant samples flooding into the country as the British government dispatched botanical expeditions throughout the world. Most of the great naval voyages of discovery during this period included botanists as a matter of course. On Captain James Cook's famous first circumnavigation expedition of 1768–71, there were three well-known botanists aboard who returned to Britain with 1300 new plant species. These early botanical gardens were among the first germplasm repositories, now often known as gene banks. We will return to the conservation and management of plant germplasm resources in Chapter 10.

7.4 Empirical Breeding

Although early crops were vastly superior to their wild ancestors, they had limitations that restricted their yield, quality or environmental adaptability. In order to improve their crops, farmers needed to breed them in the same way as they were breeding their newly domesticated livestock species. Breeding requires the recognition of favourable genetic variants in a population. These variants must then be selectively propagated to create improved populations carrying the favourable variation(s). Unfortunately, many variants fail to 'breed true' and lack of knowledge made it difficult to explain or manipulate many key traits. Therefore, during the pre-scientific period until the 19th century, empirical breeding was relatively limited in its scope. However, it enabled farmers to adapt many crops to local conditions under relatively non-intensive forms of management. This was important in facilitating the spread of crop cultivation into new climatic regions. Empirical breeding also helped farmers to cope to a limited extent with the arrival of new pests and diseases, although the process was slow and haphazard with crop shortages and famine as regular occurrences.

What is breeding?

Breeding is the deliberate identification and selection by a human agent of specific qualities, or traits, in an organism. For example, a field of wheat can be observed to identify the most sturdy, disease-free, highest yielding plants with the best quality grain and flour. Only seeds from these better-performing plants are then selected for planting, with a view to producing improved crops in subsequent generations. Hence, the two key prerequisites to breeding are *variation* and *selection*. Breeders seek to identify and/or create variants most suited to human use, and then select the best of them for large-scale propagation. In terms of introducing new genetic variation, empirical breeding is limited to the following mechanisms:

1. The chance appearance of favourable mutants, polyploids, or spontaneous hybrids in a population.
2. The identification of favourable clonal varieties for asexual propagation.
3. The creation of superior progeny from sexual crosses.
4. The introduction of new varieties from elsewhere.

Farmers also need to recognize and successfully propagate superior individuals within populations. However, in the heterozygous populations used to grow such traditional crops, it was often very difficult to 'fix' a new trait. After several generations, a useful trait might be lost despite the best efforts of the farmer-breeder. Despite these limitations empirical breeding had some notable successes, such as the versatile barley plant. For many thousands of years, farmers have created locally adapted barley varieties as food/feed staples or for beer brewing in such diverse regions as Iceland, Egypt, Canada and Australia. Gradually, improved genotypes would have spread through a region, giving rise to a landrace. Much of empirical agriculture was, and in some cases still is, based on selection of local landraces of crops.

Empirical breeding does not require any scientific knowledge to achieve far-reaching biological manipulations. New varieties of wheat, rice and maize were developed, and farmers bred particular landraces of crops that were adapted for specific regions, soils and climates. This resulted in some increases in crop yields, and their adaptation to new environments, as cultivation spread to new regions of the world. But there were also many setbacks for farmers, as disease, warfare and climatic changes took their toll on food production. Hence, it is likely that the varieties of major crop staples being cultivated in late-medieval Europe were in many cases only marginally superior in yield to those grown by Neolithic farmers many millennia previously. However, all this was set to change in the 18th century, as a combination of scientific enquiry and entrepreneurial activity led to the transformation of agriculture in North-west Europe.

Beginnings of a scientific approach

Modern crop breeding relies on the systematic manipulation of plant reproduction, which in turn requires some knowledge of plant biology. Beginning in the early 18th century, investigators such as Camerarius and Linnaeus established the science of botany as used by practical breeders up to the present day. Although there was little knowledge of plant genetics before the late 19th century, for two centuries prior to this time a great deal of useful information was discovered about how plants reproduced and how they could be induced to produce 'artificial' hybrids. These developments mark some of the earliest scientifically based attempts to manipulate plants.

Elucidation of plant reproduction

In about 1700, Camerarius published the proof of sexual reproduction in plants and proposed the role of pollen as the equivalent of animal sperm in plant fertilization. Camerarius also suggested that cross-breeding of different varieties, or even different species, could be used to create new types of plant. Another key achievement of this period was the creation, in 1718, of the first interspecific hybrid (see below). This greater understanding of plant reproduction was coupled with techniques such as sexual crossing and recurrent backcrossing, which enabled the 'scientific' breeders to produce dozens of new crop varieties.

Plant manipulation: early hybrids

The term hybrid is used in several different ways in biology, but it always means the progeny of genetically dissimilar parents. Two major classes of hybrid are relevant to plant breeding, namely intraspecific and interspecific. Most crop hybrids are intraspecific, i.e. the offspring of two members of the same species. Some of our most important crops and ornamental plants are hybrids. For example, since the 1930s almost all major commercial varieties of maize have been intraspecific hybrids. An interspecific hybrid is the product of mating between parents of two different species. Although such hybrids are often sterile, fertile progeny is sometimes produced, especially in plants. The earliest recorded man-made interspecific hybrid plant was called Fairchild's mule after its creator who produced the new ornamental species in 1718 by crossing the gillyflower, *Dianthus caryophyllus*, with sweet william, *Dianthus barbatus* (see Box 7.2).

Box 7.2. Fairchild's mule and modern sterile plant technologies

The earliest recorded, man-made interspecific hybrid plant was called Fairchild's mule after its creator who produced the new species in 1718 by crossing the gillyflower, *Dianthus caryophyllus*, with sweet william, *Dianthus barbatus*. As its name implies, Fairchild's mule is a sterile interspecific hybrid. It was produced just after the Dutch 'tulip mania', when new ornamental plants were in great demand and could be highly profitable for enterprising breeders such as Fairchild. The new plant was therefore created for commercial rather than scientific reasons and its distinctive and novel floral patterns greatly appealed to gardeners of the period.

Fairchild showed that plant breeders could produce novel variants by recombining existing species. Many hybrids cannot reproduce or do not breed true and must be recreated by the breeder each new generation. Even today, many annual ornamentals sold by garden centres are sterile hybrids, requiring gardeners to repurchase more seed each year, to the great profit of producers of such hybrids. Gardeners are prepared to repurchase sterile hybrid seeds because the resulting plants are superior to their fertile, but unimproved, counterparts. Nowadays, commercial breeders of ornamentals and crop plants sometimes attempt to engineer various forms of sterility into seeds in order to prevent gardeners and farmers from propagating such plants themselves.

In some cases, such as cereal hybrids, the seeds may still be fertile but are not worth propagating because they no longer maintain the hybrid vigour of the parent plants. Therefore many modern-day rice and maize farmers must buy new seed each year from the companies that produced the original hybrids. In other cases, the improved plant might not be a hybrid and its seed could potentially be saved by the grower. Seed sterility can be engineered into such plants via several strategies including the disruption of pollen or embryo development. One of these strategies, which has become highly controversial and has been labelled by opponents as 'terminator technology', involves the use of transgenes to cause seed sterility. As discussed in Chapter 11, there are concerns that such technologies might enable biotech companies to have too much control of food production in the future. However, hybrid and other sterile plant technologies have already been around for almost 300 years and seem likely to grow in popularity with growers as well as with seed companies.

One reason for the importance of hybrids in crop improvement is the poor performance of varieties that are constantly inbred in an effort to fix advantageous traits. In species that are normally accustomed to outbreeding, enforced inbreeding often leads to so-called 'inbreeding depression'. This is marked by the appearance of deleterious traits such as poor seed germination, slower growth rate and reduced disease resistance. Inbreeding causes a reduction in heterozygosity, as most alleles tend to become homozygous. As a result, many deleterious recessive genes that are normally masked become expressed, leading to the reduction in fitness and vigour associated with inbreeding depression.

Despite their poor performance, inbred crop varieties often carry useful agronomic traits that may have been selected over centuries. Therefore, breeders may wish to maintain inbred lines, but still need to resolve the problem of inbreeding depression. Crops can be rescued from inbreeding depression by crossing two different inbred lines that carry the same superior traits to produce an intraspecific hybrid. The key agronomic traits will be preserved as their alleles are present in both parents but many other alleles will be different in each parent as the progeny will be much more heterozygous. The results of such crosses are often dramatic, with the new hybrids producing as much as 25–50% higher yields than the inbred parental lines. Originally called 'hybrid vigour', this phenomenon was first described by the German botanist Kölreuter in 1761. This pioneering scientific breeder demonstrated that hybrid offspring received traits from both plant parents and were intermediate in most traits, and he also produced the first hybrid crop variety.

The hybrid crops discussed above are the result of crossing plants from the same species, but it is also possible to produce hybrids from different species (such as Fairchild's mule) and even from different genera. Interspecific hybrids are normally formed by crossing two species from the same genus and many crops will hybridize spontaneously with other species in the same genus to produce fertile progeny. An example of a spontaneous interspecific hybridization event that created a new crop is the formation of rapeseed (*Brassica napus*) about 2000 years ago. Oilseed rape is an allotetraploid formed by spontaneous hybridization between *B. rapa* and *B. oleracea*. As discussed in Chapter 3, breadwheat is another relatively recent (in evolu-

tionary terms) interspecific hybrid, dating back about 10,000 years. While spontaneously produced wide hybrids have been very useful to crop breeders, such chance events are extremely rare and in some species do not occur at all.

Inter-genus hybrids are often bizarre organisms because the genetic differences between their parents are much greater than those from the same genus. Therefore, successful inter-genus hybridization, whether spontaneous or man-made, is comparatively rare. Examples of spontaneous inter-genus hybridizations in agriculture are those between the goat grass genus, *Aegilops*, and the wheat genus, *Triticum*, which produced the polyploid species of durum wheat and breadwheat. In the 19th century, French botanist Fabre observed a similar spontaneous hybridization between wild species of *Aegilops* and *Triticum*. It was only in 1854 that Fabre's botanical colleague, Godron, conclusively demonstrated by direct experimentation that not only could *Aegilops* and *Triticum* produce viable inter-genus hybrids, but that this phenomenon could be reproducibly recreated by a careful plant breeder.

By the late 19th century, breeding was gradually moving from being an empirical art to being increasingly based on scientific principles. However, as we will see in the next chapter, the major leaps forward in applying biological technologies to plants did not occur until the rediscovery of Mendelian genetics and its application to plant breeding in the early 20th century. In the meantime, as we will now see, new methods of plant management were having significant effects on crop production.

7.5 Plant Management

Management techniques, such as burning or transplanting, and processing tools, such as sickles and mortars, are amongst the oldest methods of plant manipulation. As we saw in Chapter 1, people managed plants for many millennia, often in highly sophisticated ways, without these plants becoming domesticated. The increased dependence on a small range of domesticated crop plants stimulated the development of new methods for managing and processing crops. As early as 7000 years ago, people developed irrigation technologies that greatly increased crop yields as well as new tools like ploughs and seed drills. Prior to the 20th century, improved management and processing methods were much more important for crop production than biological improvements. This applies particularly to the first agricultural 'revolution' of the 17th to 19th centuries.

The agricultural revolution of the 18th century

The optimization of plant management was a concern to farmers and others for many centuries. Over 2000 years ago ancient Greek and Roman authors advised on the management of field and orchard crops, while Muslim writers produced detailed agricultural manuals while much of Europe was in the Dark Ages. During the medieval period, agriculture gradually recovered in parts of Europe with innovations such as crop rotation and new types of plough. In North-west Europe, the 18th century saw a more rational, evidence-based approach to agricultural improvement. This was particularly evident in Britain where demand for food was rising as the population increased during a period where harvests were regularly devastated by poor weather. The advances in crop production of the 18th century can be regarded as an 'agricultural revolution' that preceded and helped pave the way for the better-known 'industrial revolution' that reached its height in the 19th century.

Some of the earliest published systematic agronomic experiments date from the 1650s, when Childe and colleagues examined the effects of nitrates on crops and sought to domesticate new tree species. By the early 18th century, experiments were being performed across Britain to determine optimal fertilizer and crop rotation regimes. But, perhaps more importantly, the results of these studies were published and/or disseminated at agricultural fairs or other venues, which enabled the latest knowledge to reach a much wider audience. One of the most effective management innovations of the mid-18th century was the four-course turnip-barley-clover-wheat rotation promoted by Coke and Townshend. In many regions, unproductive land was converted to prime arable farmland suitable for the most profitable crops. Mechanical innovations like Tull's seed drill (1740s) and Meikle's powered threshing machine (1780) helped to increase yield and reduce manpower on farms.

The net result of these innovations was much higher arable crop yields, especially the major commercial cereal, breadwheat. Combined with the conversion of new land via drainage or expansion, these early attempts to increase crop productivity by improved management were surprisingly successful.

Between 1600 and 1800 the population of Britain almost trebled, but the agricultural workforce stayed about the same and the country remained virtually self-sufficient in food crops. Thanks to increasing agricultural efficiency, by 1800 almost a quarter of the British population was able to live in cities in contrast to less than one-tenth in the rest of Europe. The huge population increases during the 19th-century industrial revolution could not have been sustained without this earlier agricultural revolution.

Early scientific agronomy

During the 19th century, several new agronomic techniques were introduced that made key contributions to our current high levels of food output. On the scientific side, new or improved inputs were developed, including chemical fertilizers, and crop protection agents ranging from herbicides and pesticides to fungicides and antiviral formulations. On the engineering side, on-farm mechanization was accelerated thanks to steam-powered tractors and early versions of combine harvesters. Grain-processing units also became larger and more efficient, and storage conditions were greatly improved. Finally, agriculture moved from being a relatively small-scale, family or community-centred operation to the large-scale commercial ventures that are found in many of today's most productive crop-growing areas.

Fertilizers

Nitrogen availability is one of the most important limitations on crop growth, ranking in importance with such key inputs as sunlight and water. During the 19th century, increasing demands for more efficient crop production led to serious shortages of conventional sources of nitrate, such as livestock manure. Initially, horse manure and even human waste was collected from urban areas and laboriously transported to farms where it was applied in often massive quantities to the soil. For growing vegetables, as much as 1.2 t manure/ha was required. Such quantities of manure were often impractical and expensive to collect.

In a search for alternative sources of fertilizer, Europeans scoured the world for deposits of nitrate and potassium-rich guano (bird droppings). Huge quantities of guano were collected from islands and cliffs, often causing great environmental damage. The first cargo of Peruvian guano arrived at the port of Liverpool in 1835. By 1841, 1700 t were being imported, rising to 220,000 t by 1847. Between 1840 and 1880, Peru alone exported 20 Mt of guano to Britain. These unsustainable activities resulted in destruction of bird colonies and long-term impoverishment of soils in regions that supplied guano. By the mid-19th century, guano was becoming increasingly scarce and conventional animal or human manure was increasingly impractical to use on a large scale.

At this point, agriculture in industrializing countries faced a crisis in fertilizer availability. The alternative to guano was manure-based fertilizers, which were expensive and of limited availability. Moreover, the slow rate of nitrogen release from organic fertilizers imposed a significant limit on crop yields. Food production was rescued by the arrival of inorganic fertilizers. In particular, the introduction of inorganic nitrate and phosphate fertilizers greatly improved crop yields after the late 19th century and they are still mainstays of conventional farming across the world today. Inorganic fertilizers are much more rapidly taken up and used by plants, which cannot directly assimilate the organic (carbon-linked) forms of nitrogen in biologically derived fertilizers, such as manure. In contrast, manure-fed plants rely on microorganisms to gradually break down organic nitrogen to inorganic nitrates.

During the 1840s, the German chemist Liebig began to study soil fertility, and especially the effects of nitrogenous compounds on crop yields. His work led directly to the search for a way to produce nitrogenous fertilizers, which culminated in the invention, in 1908, of the Haber-Bosch process for industrial-scale fixation of nitrogen gas into ammonia. Thanks to Liebig and colleagues, chemical fertilizers are now an indispensable part of high-yield cropping systems, although in some places their misuse has caused environmental problems, such as excess nitrate runoff from fields into watercourses. Second only to nitrate as a yield-limiting mineral for crops is phosphate. Success in the search for an efficient inorganic form of phosphate fertilizer came in 1842 when Murray discovered that acid converts calcium phosphate into a soluble mixture of calcium hydrogen phosphate and calcium dihydrogen phosphate. This is the basis of superphosphate, a highly effective slow-release fertilizer. Murray and Lawes then developed an industrial process for the large-scale manufacture of an inexpensive form of superphosphate, which was soon in use around the world.

Crop protectants

After fertilizers, the other major class of inputs contributing to high yields are the crop protection agents. Every year, between one-third and a half of most crops used to be lost due to competition from weeds, to diseases caused by viruses, bacteria and fungi, and from damage caused by insects, rodents and other pests. In bad years, an entire crop could be wiped out by a disease outbreak or pest infestation. For many centuries, and with mixed success, farmers experimented with hundreds of treatments against the pests and diseases that regularly ravaged their crops. As discussed in Box 7.3, the use of chemical pesticides dates back to the ancient Greeks or earlier, but it was not until the 19th century that such agents were used in a truly systematic manner.

In several cases, newly developed crop protection chemicals prevented severe pest outbreaks that might otherwise have devastated crops on a huge scale. One example is the copper-based Bordeaux mixture that probably saved much of the French grape crop from ruin during the 1880s. Formulations based on the Bordeaux mixture are still used today, mainly by the organic farming sector. However, many of these early chemical agents are toxic to humans and may have serious effects on non-target organisms, so very few of them are still in use today. They have been replaced by a new generation of compounds that are safer, more effective and less environmentally persistent. As discussed in Chapter 9, these new crop protectants, together with greatly improved breeding technologies, helped to underpin the huge increases in crop yield that occurred during the 20th century.

7.6 Summary Notes

- Crop domestications were relatively rare occurrences that were restricted to a limited number of centres of origin.
- Major domestication-related traits include seed shattering, large seed size and a short, unbranched growth habit.
- Crop plant genomes are characterized by the regulation of key traits by one or two genes and the clustering of domestication-related genes.
- New biotechnology tools and advances in knowledge about domestication will enable new crops to be developed in the future.

Box 7.3. Development of crop protection agents

Chemicals had been used for crop protection against pests and diseases since ancient Greek times. For example, the use of sulfur as a fumigant was discussed in Homer's *Odyssey*, and other Greek writers mention the use of chalk and alum as pesticides. During the medieval period, the 12th-century Moorish writer Al-Awwam reported the use of arsenical sulfides to control pests. Although they were effective against pests, most of these chemicals were also highly toxic to other organisms, including humans.

Properly systematic studies of the effectiveness and optimal methods for the use of chemicals in the field did not begin until the scientific era. These studies led to reports, among many others, of the anti-worm properties of potash (1631), the fungicidal and herbicidal effects of copper sulfate (1761) and the antifungal effects of sulfuric acid (1799). Two of the most significant scientific pioneers of work on crop protection were Tillet in the 1750s and Prévost in the early 1800s, who laid the foundations of our understanding of fungal diseases such as cereal blasts and their control by chemical agents such as copper/lime mixtures. By 1850, several dozen crop-protection chemicals were in widespread use in European agriculture.

The modern era of chemical control in agriculture started in the late 19th century with the use of Bordeaux mixture (copper sulfate and hydrated lime) to suppress powdery mildew and weeds in French vineyards. Despite their toxicity, many of these early formulations were used on a large scale because of their effectiveness in boosting crop yields. Below is a list of some of the chemicals developed during the 19th century to control serious fungal and insect infestations of crops.

1755 Tillet established that seed-borne fungi caused bunt disease of wheat and that it could be controlled by lime or lime plus salt.

1805 Prévost described the inhibition of fungal smut spores by copper sulfate.

1814 Mixture of sulfur and lime used as a fungicide on peach trees.

1840s Rotenone derived from the roots of the derris plant used as insect control in Asia.

1850s Use of pyrethrum from the flower heads of chrysanthemum as a naturally occurring insecticide, which is still widely used today.

1867 Copper arsenate (Paris Green) used to control Colorado beetle outbreak in Mississippi.

1882 Copper sulfate and lime (Bordeaux mixture) used with great effect to treat grape vines against vine mildew.

1886 Cyanide used as a fumigant against scale insects on citrus trees. It was initially very effective but eventually there were failures in places as some insects became resistant. This is the first recorded case of the development of resistance to an insecticide by a pest organism.

1892 Lead arsenate used to control gypsy moth infestations.

- Domesticated crops were first gradually disseminated from centres of origin to wider regions, and then spread rapidly worldwide after the 16th century.
- Plant breeding and crop improvement started as empirical processes but became more scientifically based in the 18th century.
- Early scientific advances in crop production were based mainly on improved plant management, such as more efficient fertilizers and crop protectants.

Further Reading

Domestication

Doebley, J. (2006) Unfallen grains: how ancient farmers turned weeds into crops. *Science* 312, 1318–1319.

Gross, B.L. and Olsen, K.M. (2010) Genetic perspectives on crop domestication. *Trends in Plant Science* 9, 529–537.

Izawa, T., Konishi, S., Shomura, A. and Yano, M. (2009) DNA changes tell us about rice domestication. *Current Opinion in Plant Biology* 12, 185–192.

Malory, S., Shapter, F.M., Elphinstone, M., Chivers, I.H. and Henry, R.J. (2011) Characterizing homologues of crop domestication genes in poorly described wild relatives by high throughput sequencing of whole genomes. *Plant Biotechnology Journal* 10, 1–9.

Murphy, D.J. (2007) *People, Plants, and Genes: The Story of Crops and Humanity*. Oxford University Press, UK.

Peng, J., Ronin, Y., Fahima, T., Röder, M.S., Li, Y., Nevo, E. and Korol, A. (2003) Domestication quantitative trait loci in *Triticum dicoccoides*, the progenitor of wheat. *Proceedings of the National Academy of Sciences USA* 100, 2489–2494.

Pickersgill, B. (2009) Domestication of plants revisited – Darwin to the present day. *Botanical Journal of the Linnean Society* 161, 203–212.

Poncet, V., Thierry Robert, T., Sarr, A. and Gepts, P. (2004) Quantitative trait locus analyses of the domestication syndrome and domestication process. In: Thomas, B., Murphy, D.J. and Murray, B.G. (eds) *Encyclopaedia of Plant and Crop Science*. Marcel Dekker, New York, pp. 1069–1073.

Purugganan, M.D. and Fuller, D.Q. (2009) The nature of selection during plant domestication. *Nature* 457, 843–848.

Tang, H., Sezen, U. and Paterson, A.H. (2010) Domestication and plant genomes. *Current Opinion in Plant Biology* 13, 160–166.

Vaughan, D.A., Balazs, E. and Heslop-Harrison, J.P. (2007) From crop domestication to super-domestication. *Annals of Botany* 100, 893–901.

Dissemination, empirical breeding and plant management

Drayton, R. (2000) *Nature's Government: Science, Imperial Britain and the 'Improvement' of the World*. Yale University Press, New Haven.

Levetin, E. and McMahon, K. (2008) *Plants and Society*. McGraw-Hill, New York.

Smith, B.D. (2001) Low level food production. *Journal of Archaeological Research* 9, 1–43.

8 The Scientific Manipulation of Plants

8.1 Chapter Overview

In this chapter we will survey the era of scientific plant manipulation that began in the 20th century. Many biological and non-biological tools were developed for crop management and improvement with spectacular results. A range of biotechnologies based on plant genetics, physiology and biochemistry enabled breeders to manipulate plants in ways that transformed agriculture in much of the world. Modern plant breeding began with the application of Mendelian principles of heredity and was extended by tools such as quantitative genetics for the manipulation of complex multigenic traits. New technologies such as mutagenesis allowed breeders to create variation in inbred populations, to select superior genotypes and to propagate these new elite plants on a large scale in the field. Selection methods also progressed from simple visual inspection of plant phenotypes to the highly sophisticated use of DNA-based markers. In the final section, we will survey the use of transgenesis for plant breeding as first applied in commercial agriculture during the 1990s.

8.2 Plant Genetics

The scientific principles of inheritance, now known as genetics, have been used in plant manipulation for little more than a century. Indeed, the word 'genetics' was only invented in 1906 by Bateson from Cambridge, UK as a way of defining what he called the 'new science of heredity'. Since then the practical application of genetics to breeding has revolutionized our ability to manipulate plants. However, until the 1950s, the molecular basis of gene function was completely unknown to scientists. To make things more difficult, breeders discovered that many important agronomic traits were regulated by the complex interaction of numerous genes, about which they knew very little. Initially it was believed that this meant that simple

Mendelian genetics might not be applicable to most forms of practical breeding. However, despite their inability to define most traits at the molecular level, 20th-century breeders succeeded in greatly improving the yield and agronomic performance of most of the major staple crops. An outline of the principal processes involved in scientific breeding is shown in Fig. 8.1, where it can be seen that the two key aspects of successful breeding are: (i) capturing and enhancing useful genetic variation; and (ii) screening, selecting and propagating the most suitable variants.

Genetics is the science of heredity or inheritance. The inherited information that specifies the nature of an organism is called its *genotype*. The genotype includes all of the genes of an organism, plus other elements of its genome that may contribute to gene expression. During the development of an organism, its genetic information interacts with various factors in the environment to specify its eventual *phenotype*, namely its appearance, composition and behaviour. Over the past decade or so there has been much progress in genetics and genomics, which has fundamentally altered many aspects of how we view genes and their regulation. In Chapter 2, we saw that genetic information is stored as a sequence of DNA bases but that these bases could be epigenetically modified in a way that can greatly alter the pattern of gene expression.

The main principles of heredity were first outlined by Mendel in his publication on the garden pea in 1865. This described how certain plant traits are transmitted from parents to offspring and on to subsequent generations. Mendel's work was largely ignored until it was rediscovered in the early 1900s and elaborated into the science of genetics. This involves the concept of genes, which are factors that determine the nature of a trait and which can be inherited from either, or both, parent(s) in sexually reproducing species. In 1902–03 Boveri and Sutton proposed that Mendelian genes were associated with

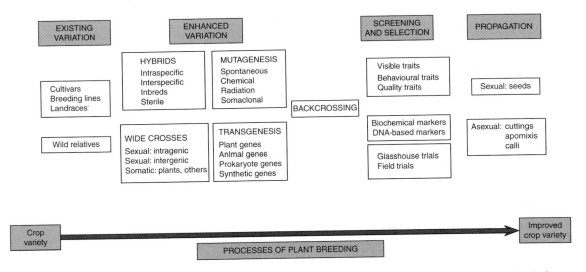

Fig. 8.1. Process of plant breeding. Plant breeding involves the selection and propagation of distinctive variants from populations. Breeders rely on access to a range of genetically diverse plants to provide them with sufficient variation in agronomic traits to select suitable candidates for crop improvement programmes. The deliberate creation of new variation began with sexual crosses between members of the same species, and gradually progressed to crosses between ever more distantly related plants from different genera. Many wide crosses beyond the species barrier can only survive if plants are subject to techniques such as chromosome doubling or embryo rescue and tissue culture. Additional variation can be introduced by induced mutagenesis using radiation, chemical and somaclonal methods. The collection or generation of a population of variants is the raw material of plant breeding. These variants, sometimes numbering tens of thousands of plants, are then screened and favourable variants are selected by various approaches ranging from straightforward visual selection to a battery of sophisticated and often automated genomic and metabolic techniques. Following selection, suitable variants are often backcrossed into existing elite breeding lines so that their favourable traits are retained, while unfavourable traits are eliminated. This produces an improved variety, which must be multiplied up and propagated for external field trials. Several years of field trials are frequently necessary to ensure that the improved variety meets the DUS (distinctness, uniformity and stability) criteria for varietal registration. Note that transgenesis (genetic engineering) only involves the initial stages of breeding, i.e. it is simply another method of creating additional genetic variation. Even if it is successful in this regard, transgenic variants must still undergo the full gamut of screening, selection, backcrossing, propagation and field trials as part of the normal breeding process that applies to all plants. Adapted from Murphy (2007b).

nuclear chromosomes and that the latter carried the genetic information in eukaryotes.

In the simplest form of Mendelian inheritance, a character or trait such as flower colour is determined by a single gene that is passed on to future generations. According to the concept of *independent assortment*, genes behave independently of each other. Mendel's work demonstrated that many traits are determined by pairs of *alleles*, one from each parent. If the alleles are different, the individual is said to be *heterozygous* and either allele might be passed on to its offspring. If the alleles are the same, the individual is said to be *homozygous* and only one version can be passed on to its progeny. When two different alleles are present in an individual, one of them tends to be dominant and the

phenotype reflects the dominant allele, as shown in Fig. 8.2. It was soon realized that most wild populations tend to be highly heterozygous and hence genetically variable. In contrast, domesticated plants and animals tend to be more homozygous due to initially low population sizes and centuries of inbreeding.

Simple inheritance

Monogenic traits behave as if they are regulated by a single gene and normally follow the classical rules of simple Mendelian inheritance. Such simple traits are the easiest to manipulate by straightforward genetic crosses. The traits selected by Mendel in his studies of inheritance, such as seed shape

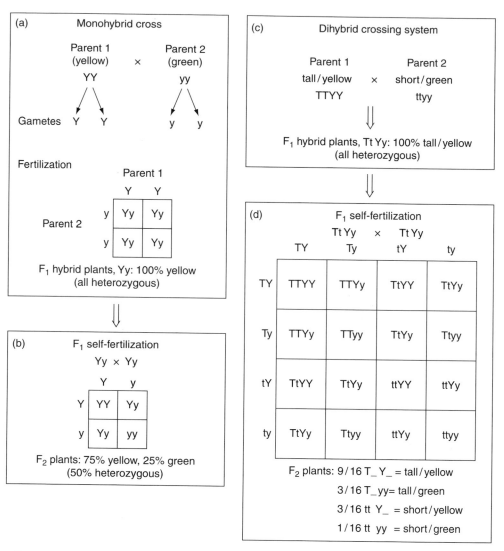

Fig. 8.2. Breeding and Mendelian inheritance. (a) The simple example of Mendelian inheritance involves a single gene from pea where the dominant Y allele produces a yellow colour and the recessive allele y produces a green colour. A cross between homozygous YY and yy individuals will produce a completely heterozygous F_1 generation, all of which will be yellow. (b) If the heterozygous F_1 generation is self-fertilized, the result is a mixture of homozygous and heterozygous progeny of which 75% will be yellow. (c) A more complex example is to consider two genes each with two alleles, T, t, and Y, y. Once again a cross between homozygous TTYY and ttyy individuals will produce a completely heterozygous F_1 generation. (d) However, in this case self-fertilization of the heterozygous F_1 generation produces a more complex set of progeny with a 9:3:3:1 ratio of phenotypes.

and flower colour, were monogenic. In Table 8.1, some monogenic traits in plants are listed, including several that are very desirable to manipulate in crops. Breeders such as Biffen and Bateson in Britain soon realized that this knowledge could be useful in breeding, although many others were

initially sceptical that simple Mendelian rules would apply to crops in the field.

One way to apply Mendelian rules practically is to develop pure homozygous lines that behave in a predictable way when crossed with other pure lines. Whereas heterozygous parents produce a mixture of

Table 8.1. Examples of plant traits regulated by a single gene (monogenic traits).

Organ	Trait	Plant species
Growth habit	Flowering photoperiod	Strawberry
	Height – tall versus short	Rice, wheat
	Annual versus biennial	Clover (*Meliotus alba*)
	Determinate versus indeterminate	Tomato
Fruit and seeds	Rachis persistence	Cereal grasses
	Winged versus wingless seeds	Tick seed
	Pod rotation	Wild lucerne
	Spiny fruit	Cucumber
	Fruit shape	Sweet pepper
	Fruit pubescence	Peach
	Fruit location	Common bean
Flower	Male sterility	Potato
	Self-incompatibility	Cabbage
	2- or 6-rowed inflorescences	Barley
	Corolla shape	Primrose
	Gender	Cucumber
	Pistil length	Californian poppy
	Petal number	Morning glory
	White versus coloured	Violet
Leaf	Angle	Innocence (*Collinsia heterophylla*)
	Cyanogenic glycosides	Bird's-foot trefoil
	Rust resistance	Wheat
	Margin	Lettuce
	Leaflets versus tendrils	Garden pea

offspring, homozygous parents that carry the same alleles always produce progeny similar to themselves. By 1903, the 'pure line' theory was developed whereby repeated self-pollination, or selfing, leads to highly homozygous lines that are true-breeding, i.e. the progeny look similar and behave similarly to the parent. This lack of phenotypic and genetic variability is a highly desirable attribute in a crop plant. A selfed, homozygous variety has superior crop performance and much greater uniformity in its appearance and quality, and as a bonus the vast majority of its progeny would inherit these desirable traits.

This process is called 'fixing' a genetic trait. Previously, fixing had occurred by a slow and uncertain trial and error process, but by applying Mendelian principles it was possible to set up crosses that would greatly enhance the speed and precision of trait fixing. It was soon realized that the practical application of Mendelian genetics had the potential to revolutionize plant breeding. In the words of Niels Hansen of the USDA in 1906: 'what was formerly a chaos of empiricism is now becoming one of the exact sciences ... No longer is heredity a jungle.' The manipulation of monogenic traits was one of the first achievements of scientific breeding in the early 1900s.

Towards the end of the 20th century, monogenic traits assumed a new importance when genetic engineering was developed. In the 1980s and 1990s, monogenic traits were the first to be modified using transgenesis technology. This was because it is technically much easier and faster to isolate, transfer and select for single genes. All of the first generation GM crops discussed in Chapter 9 carried simple monogenic transgene traits. Unfortunately, however, many useful characters in crops do not follow simple Mendelian patterns of inheritance and they normally exhibit complex behaviour in genetic crosses. As we will now see, the phenomenon of complex multigenic inheritance of traits created significant challenges for breeders, which were partially overcome by using the tools of quantitative breeding.

Complex inheritance

Although some useful traits display dominance, independent assortment and are monogenic (Table 8.1), the majority of traits that breeders wish to manipulate have more complex patterns of inheritance as shown in the following examples.

Incomplete dominance

Not all genes behave as if they are either fully dominant or recessive. There are many cases of incomplete dominance, such as flower colour in *Antirrhinum* where a red variety crossed with a white one produces some pink progeny. Red varieties are homozygous with the two copies of the *R* gene that leads to synthesis of a red floral pigment. White varieties are also homozygous, but with the two copies of the *r* gene, which is a mutated version of *R* unable to produce red floral pigment. The progeny of a red × white cross will have a ratio of 25% red (*RR*), 25% white (*rr*) and 50% pink (*Rr*). The heterozygous pink progeny have only one copy of the functional *R* gene and produce lower amounts of the red pigment, leading to their lighter colour compared with homozygous red plants. If the *R* gene were fully dominant, all *Rr* progeny would be red, so *R* is defined as an incompletely dominant gene.

Genetic linkage and gene mapping

Many genes are not inherited according to the rules of independent assortment. Instead, some genes appear to be linked with other genes so that they are inherited together. We now know that this is due to chromosomal linkage whereby the genes are physically located near to one another on the same chromosome. Genes on different chromosomes are inherited independently according to simple Mendelian principles, but linked genes will tend to be inherited as a package. This form of genetic linkage can sometimes be broken by the process of 'recombination', where homologous chromosomes exchange regions with each other. However, genes that are very closely linked on the same chromosome are much less likely to be separated by such crossing over and hence tend to retain their linkage to each other. As we saw in Chapter 7, the close physical linkage of traits involved in plant domestication has played a key role in determining our current range of crop staples.

In 1919, Haldane derived a formula to convert recombination frequencies of genes into 'map units' measured in centimorgans (cM) that represent gene locations along a chromosome. One centimorgan of genetic distance between two sites on a chromosome

is equivalent to a 1% probability of recombination between them. Genetic distances measured in cM do not necessarily correspond to physical distances measured in DNA base pairs. This is because different regions of a chromosome can have varying frequencies of recombination. Therefore, in a region of low recombination, genes might appear to be further apart as measured in cM compared with regions of high recombination. As a rough approximation, an average chromosome of length about 100 cM might contain between 1 and 100 Mb of DNA. Modern genetic mapping cannot be used to identify individual genes but is accurate down to a scale of 0.1–20 cM.

Polygenic inheritance

Breeders quickly discovered that many of the most desirable crop traits do not display monogenic inheritance. In particular, key traits such as yield, grain quality and stress tolerance behave in a complex way that indicates they are regulated by a combination of many different genes. For example, cob size in maize behaves as a polygenic trait that is regulated by over a dozen different genes. As we saw in Chapter 7, it took ancient farmer-breeders in Mesoamerica several thousand years of gradually selecting slightly larger maize varieties via many small incremental steps, each involving a single gene at a time, before large cob sizes were eventually achieved.

In a more recent example, it was originally believed that the ability to accumulate seed oils enriched in lauric acid (C12) was a monogenic trait regulated by a single *thioesterase* gene (see Chapter 4). In the 1990s, a *thioesterase* gene was transferred to transgenic rapeseed plants, but it only had limited effects on lauric acid accumulation. Eventually it was necessary to transfer several additional genes to increase the levels of lauric acid, but even these amounts were much lower than those found in plants, such as *Cuphea* spp., that naturally accumulate lauric acid. This showed not only that lauric acid accumulation was a complex polygenic trait, but also that we have yet to discover all of the genes that regulate this trait. Until these genes are known, it will not be possible to use this kind of transgenic approach to obtain crops that accumulate as much lauric acid as 'natural' accumulators.

Gene × environment effects

The phenotype of a plant is not determined by its genotype alone. As we have seen in previous chapters, plants are able to modify their developmental programmes in response to environmental factors. Genes vary considerably in their susceptibility to environmental regulation and sometimes alleles of the same gene can have differential responses to some environmental effects. This means that breeders must be aware of the range of gene × environment (G × E) effects on any trait that they wish to manipulate. In 1903, Johanssen used beans to demonstrate that environmental factors are one of the major causes of the complex behaviour of polygenic traits in plants.

Landraces and inbred lines

Many of the most important crops are grown as highly inbred cultivars that tend to be homozygous for most agronomically significant alleles. Examples include rice, wheat, barley, soybean, peas, lentils, millet, chickpea, tomato and tobacco. However, prior to the 20th century, the majority of crops were made up of a series of informally selected varieties known as landraces. Over the past century or so, these landraces have been converted by breeders into the pure lines that form the vast majority of today's highest yielding elite crop cultivars. A landrace is a locally grown population of a crop that has been selected for its adaptation to a particular set of conditions and/or end uses.

In pre-scientific farming systems, each region had its own group of landraces, often selected over many centuries. In the absence of modern breeding methods, these landraces contained mixtures of many different genotypes with high degrees of heterozygosity. The major advantage of landraces is that they are often well adapted to local environments and to a range of pest/disease threats. Moreover, their high level of heterozygosity gives them more flexibility to cope with threats such as new pathogens or periodic drought. However, this lack of genetic uniformity is also a disadvantage because agronomically superior genotypes tend to be diluted out and may have little impact on overall crop performance.

The dependence on genetically variable landraces is one reason why yields of the major staple temperate cereal grains such as wheat and barley remained close to 1 t/ha for many centuries. With the advent of scientific breeding, inbred or pure line varieties were developed by selecting the best individuals from landraces and breeding these in isolation.

Wherever possible self-fertilization was used to generate increasingly homozygous, inbred progeny. This process was assisted by the tendency of many crops to accept self-pollen in contrast to wild plants where self-incompatibility is the rule (see Chapter 5).

As shown in Fig. 8.3, a new trait can be introgressed into an existing inbred population by setting up a hybrid cross with a donor plant that displays the trait. This is followed by repeated backcrossing with the inbred parent until a plant is produced that is almost identical to that parent with the exception of a small fragment of genetic material that includes the new gene. The new improved inbred line is then propagated by self-fertilization to maintain its favourable genetic characteristics. Repeated selection of inbred lines eventually led to the disappearance of most landraces in those crops in which scientific breeding was used.

Although modern inbred varieties are vastly superior to landraces in agronomic performance and are the basis of our ability to feed a greatly increased global population, the loss of potentially valuable genetic variation is regrettable. As discussed in Chapter 10, some breeders are now attempting to save remaining landraces of wheat and barley as well as creating modern versions of landraces – so-called multilines.

Quantitative genetics

In 1906, Yule suggested that the apparently continuous variation seen in many polygenic traits could be explained by the cumulative effect of many Mendelian genes, each of which had a small effect on the trait. In 1910, Nilsson-Ehle did a classic experiment to demonstrate polygenic inheritance in wheat using a trait controlled by two genes. Polygenic traits are nowadays often called *quantitative traits*, and the genes that control them are found in chromosomal regions called *quantitative trait loci*, or *QTL*. Many QTL can contribute to the expression of a particular quantitative trait. Some of these QTL might have a very small effect, perhaps contributing less than 1% to the overall regulation of the trait. Other QTL can have larger effects and the identification and molecular characterization of these major QTL is one of the most important goals of modern breeders (see Chapter 10).

Early 20th-century breeders were able to harness their limited existing knowledge of traits and inheritance for crop breeding thanks to the development

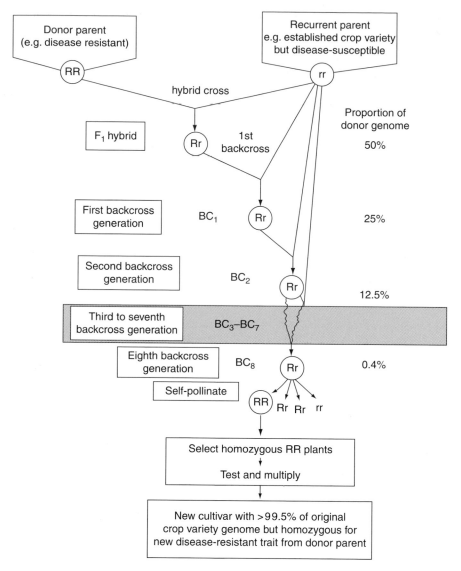

Fig. 8.3. Mechanism of backcrossing for the introduction of new genes into crop varieties. In this example, the aim is to produce a plant that resembles the recurrent parent (an established crop) as much as possible except for a tiny fragment of the genome of the donor parent that contains the locus responsible for a useful trait such as disease resistance. The donor parent might be another variety of the same crop or a completely wild species that can interbreed with the crop plant. When the donor and recurrent parent are crossed, the donor genome will make up 50% of the F_1 hybrid. If this hybrid is backcrossed to the recurrent parent, the first backcross generation will contain just 25% of the donor genome. By the eighth backcross generation only 0.4% of the donor genome will remain so the plant will be 99.6% genetically identical to the recurrent parent. At each stage of backcrossing, progeny are tested and only those that contain the desired donor trait are selected. In this way, it is possible to move useful genes from other plants and even different species into a major crop variety.

of the statistical methods of quantitative genetics by the British geneticist Fisher, in 1918. Previous work suggested that Mendelian mechanisms only gave rise to large and discrete, or quantum, changes in phenotype. Fisher showed that complex traits with continuous variation between members of a population could result from Mendelian inheritance, albeit involving many genes, plus an environmental component. Fisher's statistical framework allowed crop breeders to predict phenotypic performance from their knowledge of a given genotype. It also enabled them to design manageable and affordable field trials that had a good likelihood of detecting useful characters in a test population. From the 1930s, quantitative genetics became the basis of the scientific breeding of animals and plants.

Over the next 50 years there was only limited progress in linking quantitative traits with physical regions of chromosomes or specific sequences of DNA. One of the problems was that very large numbers of individuals in a test population were needed to detect the location of a gene able to explain >1% of observed phenotypic variation. More recently, identifying, localizing, sequencing and manipulating QTL have been greatly simplified by molecular genetics and especially the use of DNA markers, as discussed in Chapter 10. Despite its limitations in terms of molecular knowledge of genetic traits, quantitative genetics has been a formidable plant-breeding tool that is still in widespread use today. In Fig. 8.4, several examples of breeding and selection schemes are shown. The purpose of such methods is to capture valuable genes from donor plants and to introgress them into an elite variety. Such introgression/backcrossing programmes are still essential for the incorporation of novel traits into crop cultivars. For example, most newly created transgenic plants are not suitable for immediate commercial cultivation and the transgenic traits from such plants must be introgressed into local elite varieties before being assessed in field trials.

8.3 Plant Manipulation

Biotechnology is the manipulation of living systems. Although the term is often taken to mean genetic engineering, as discussed in Chapter 1, there are many types of biotechnology, most of which do not involve transgenesis. In the following sections, we will review the most significant biotechnologies applied to crop improvement during the 20th century (see Table 8.2). Some of the most powerful tools were new methods for creating genetic variation, ranging from hybrid construction to assisted wide crosses and mutagenesis. Other key biotechnologies include selection and mass propagation, plus the various tissue culture methods that underpin many aspects

of modern plant breeding. Transgenesis, which has the potential to create almost unlimited genetic variation, will be considered at the end of the chapter.

8.4 Creating New Genetic Variation

The ability of plant breeders to create new genetic variation was greatly enhanced in the mid-20th century by several technological advances. Improved knowledge of plant reproduction made it easier to set up sexual crosses with a wider range of varieties within crop species or with wild relatives from other species. The increasing use of inbred lines for commercial varieties and the loss of many landraces meant that breeders often needed to access new genetic variation from outside the race or species of the crop in question. Their efforts were assisted by the development of new forms of tissue culture as a way of manipulating and propagating plants (see Fig. 8.5), plus the increasingly refined use of synthetic growth regulators and induced mutagenesis.

These technologies enabled new traits to be added to crops via wide genetic crosses, using methods such as embryo rescue, asymmetric cell fusion, nuclear implantation and somatic embryogenesis. Previous attempts at wide crossing between distantly related species were frequently frustrated by genome incompatibility. Two important methods to overcome this are the 'rescue' of hybrid embryos that would otherwise abort in the seed, and chemically induced chromosome doubling. As well as making possible much wider genetic crosses, chromosome doubling has enabled the use of methods such as somatic hybridization and haploid breeding.

Chromosome doubling

This is a key technology for the creation of fertile interspecific hybrids. Wide-hybrid plants are often sterile, so their seeds cannot be propagated. This is due to differences between chromosome sets inherited from genetically divergent parental species that prevent stable chromosome pairing during meiosis. However, if the chromosome number is doubled, the hybrid may be able to produce functional pollen and eggs and therefore be fertile. Colchicine has been used for chromosome doubling in plants since the 1940s and has been applied to more than 50 plant species, including

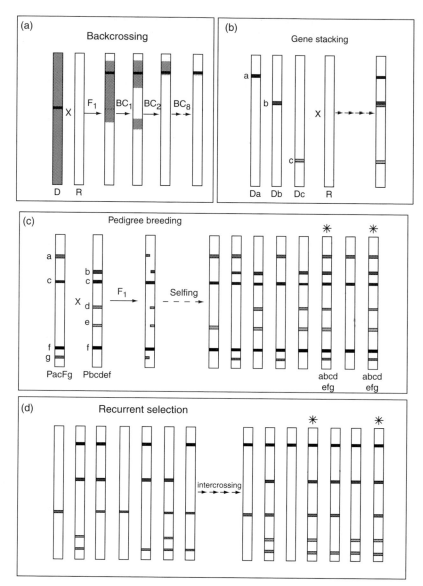

Fig. 8.4. Common breeding and selection schemes. (a) Backcrossing. A donor plant (D) featuring a specific gene of interest (dark line) is crossed to an elite variety targeted for improvement (R), with F_1 progeny repeatedly backcrossed (BC_1, BC_2 etc.) to the elite line. Each backcross cycle involves selection for the gene of interest and recovery of increased proportion of elite line genome. (b) Gene stacking. Several varieties (D_a, D_b, D_c) containing genes/QTL associated with different beneficial traits (a, b, c) are combined into the same genotype via crossing and selection. (c) Pedigree breeding. Two individuals with desirable and near-complementary phenotypes (a, c, f, g and b, c, d, e, f) are crossed and F_1 progeny are self-pollinated to fix new, improved genotype combinations. The two progeny carrying all seven desirable traits (a, b, c, d, e, f, g) are indicate by asterisks. (d) Recurrent selection. A population of individuals (seven in this example) segregate for two traits that are in turn influenced by five major favourable QTL (horizontal bars). Inter-mating among individuals and selection for desirable phenotypes/genotypes increase the frequencies of favourable alleles at each locus. For this example, no individual in the initial population had all of the favourable alleles, but after recurrent selection two out of seven of the progeny contain the desired genotype (indicated by asterisks). For hybridized crops, recurrent selection can be performed in parallel within two complementary populations to derive lines that are then crossed to form hybrids; this method is called reciprocal recurrent selection. Adapted from Moose and Mumm (2008).

Table 8.2. Deployment of crop improvement biotechnologies in the 20th century.

Technology	First used	Crops	Traits modified in field crops
Early tissue culture	Early 1900s	Many annual and perennial crops	Dozens of agronomic traits
Wide hybrids	Early 1900s	Triticale, rice, wheat, brassicas	Drought/salt tolerance, pest/disease resistance
F_1/F_2 hybrid maize	Early 1900s	Maize	Hybrid vigour (e.g. yield, disease resistance)
X-ray mutagenesis	1920s	Many annual and perennial crops	Dozens of agronomic traits
Chemical mutagenesis		Many annual and perennial crops	Dozens of agronomic traits
Seed sterility	1930s	Citrus and many other fruits	Fruit size, yield, uniformity, seedless fruit
Chromosome doubling	1940s	Many annual and fruit crops	Yield, nutritional quality, pest/disease resistance, drought/salt tolerance
γ-Ray mutagenesis	1950s	Rice, wheat, peas, lentils, soybean, grapefruit, apples	Yield, nutritional quality, pest/disease resistance, drought/salt tolerance
Shuttle breeding	1950s	Wheat, brassicas	Dwarfing, disease resistance
Somaclonal mutagenesis	1960s	Rice, wheat, sugarcane, maize, potato, tomato, poplar, banana/plantain	Nutritional value, pest/disease resistance, drought/salt tolerance, taste, yield
Somatic hybridization	1970s	Potato, rapeseed, *Citrus* spp., tomato	Pest/disease resistance, fruit size
Somatic embryogenesis	1970s	Cassava, oil palm, banana/plantain, tea	Pest/disease resistance, taste, aroma, yield
Haploids	1950s	Rice, wheat, tobacco, peppers	Pest/disease resistance, fruit size, yield
Doubled haploids	1960s	Wheat, rice, barley, brassicas, tobacco	Pest/disease resistance, taste, yield
Embryo rescue	1980s	Tomato	Pest/disease resistance
Advanced tissue culture: cell fusion, microinjection	1960s	Many annual and perennial crops	Dozens of agronomic traits
Micropropagation	1960s	Oil palm, cassava, orchard crops, banana/plantain, sweet potato	Yield, pest/disease resistance
Agrobacterium-mediated transformation	1983	Rapeseed, cotton, soybean	Herbicide tolerance, insect resistance
Hybrid rice	Mid-1980s	Rice	Hybrid vigour (e.g. yield, disease resistance)
Biolistic transformation	Late 1980s	Maize	Herbicide tolerance, insect resistance
Commercial F_1 hybrids	1980s and 1990s	Sugarbeet, rapeseed, brassica vegetables, tomato, potato, onion	Yield, pest/disease resistance, uniformity, quality

most annual crops. It has also been used to create seedless fruits and for production of wide crosses and somatic hybrids. More recently, other chromosome doubling agents, all of which are inhibitors of mitotic cell division, have been used in plant breeding. In some plant species, tissue culture techniques have also been used to induce chromosome doubling. Thanks to chromosome-doubling technology, hundreds of new crop varieties have now been produced around the world.

Somatic hybridization

This technique is used to introduce novel genes into a crop genome from a donor species with which the crop will not normally interbreed. For example, if

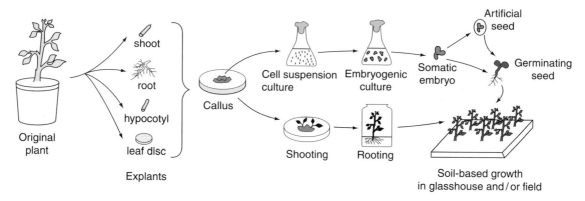

Fig. 8.5. Plant propagation via tissue culture. Several types of plant tissue can be used in the form of explants that will de-differentiate in culture to form multicellular masses of callus tissue. Individual cells or small cell clumps can be cultured in liquid media and may be induced to form somatic embryos that can be incorporated into artificial seeds, e.g. by encapsulation in a gelatin coat. Somatic embryos and seeds may then be germinated to produce adult plants. Callus tissue can be induced to form differentiated shoots and roots that can be used to regenerate adult plants.

a crop and its potential gene donor are not closely related, it may not be possible to get the pollen of one of the species to fertilize the eggs of the other. Although similar in its aims to conventional assisted hybridization, somatic hybridization involves a more radical technological approach. Somatic hybridization is another way of enhancing variation in crop species by importing genes, or even whole chromosomes, from other species that are not related closely enough for normal sexual crossing.

The development of sophisticated microinjection and cell fusion techniques in the 1960s and 1970s allowed researchers to fuse whole cells or parts of cells to create composite, or chimeric, cells from unrelated species. For example, two haploid nuclei from different species can be fused to create an artificial zygote. The resultant hybrid cells can be colchicine-treated to induce chromosome doubling so that the progeny are more likely to be fertile. Alternatively, such cells might spontaneously double their chromosome number during regeneration *in vitro*, which can stabilize the new genome. Finally, the hybrid cells are induced to divide and differentiate into new hybrid plants that might originate from unrelated parental species.

The even more radical method of asymmetric nuclear fusion involves fragmentation of the nucleus of one species and the insertion of some of these fragments into a complete nucleus of a different species. During the 1970s, this method was used in the lab to create plant/animal hybrid cells. In a few cases, chromosomes or chromosome fragments

were successfully incorporated into the genome of the recipient cell but in such extreme cases regeneration of viable hybrid organisms was not achieved. More practical forms of somatic hybridization using related plant species were introduced into breeding programmes in the early 1980s and have now been attempted with dozens of crops. The main technical hurdle has been the instability of the new genome combinations from two dissimilar species.

To a great extent, somatic hybridization was replaced in the 1990s by transgenesis, which is more precise, has fewer problems with genome instability and has a higher overall success rate. However, transgenesis is mainly useful when there is a known useful gene(s) to be transferred. Many useful traits are controlled by as yet unknown sets of genes and can only be transferred into a crop by adding an entire donor genome, or at least a substantial portion of it. In recent years, therefore, some breeders have started to explore new forms of somatic hybridization, especially in fruit crops.

The reasons for this are threefold: (i) classical transgenesis, where one or a few known genes are transferred to a plant, is not always a quick and easy option, especially when complex multigenic traits are involved; (ii) tissue culture and molecular marker techniques have improved considerably over the past decade, resulting in increasing success rates for regeneration of genetically stable progeny from somatic hybrids; and (iii) unlike transgenesis, somatic hybridization is not generally regarded by regulatory authorities as 'genetic modification' (see Chapter 11) and is not subject to the same burden of regulatory approval and testing. This has created new commercial opportunities for breeders and several new start-up companies are now developing improved forms of somatic hybridization technology.

Haploids and doubled haploids

Haploid plants can be produced using anther culture, which involves *in vitro* culture of immature anthers that contain immature pollen grains called microspores. Since pollen grains are haploid, the resulting plants will also be haploid. Doubled haploid plants were first produced in the 1960s using colchicine. Thermal shock (heat or cold), mannitol incubation or ovule culture can also be used to produce doubled haploids. Breeders value doubled haploid plants because they are 100% homozygous and any recessive genes are readily apparent. The time required after a conventional hybridization to select pure lines carrying the required recombination of characters is consequently drastically reduced. The application of this technique to plant breeding is sometimes limited by the need to test large numbers of lines, which is expensive and time consuming. Haploid methods have been used to develop new varieties of rice, wheat, tobacco and peppers and have recently been applied to oil palm.

Sterile plant varieties

Many of the crosses set up by plant breeders produce sterile varieties that cannot readily be propagated. Sometimes this is a useful trait and is deliberately engineered by breeders. For example, in water melon and citrus crops, consumers often demand sterile fruits that are seedless. Although these sterile varieties are commercially desirable, they can be difficult to reproduce, especially if sexual propagation is necessary. However, if there is sufficient demand for sterile varieties these costs can be offset by the need for a farmer to repurchase new seed or other propagules every time he replants the crop. In this way, seed sterility is analogous to F_1 or F_2 hybrids or other non-propagable plant types in its usefulness to commercial seed companies because farmers cannot use saved seed. In the case of an annual crop, this means repurchasing new stocks of seed each year.

One of the most rapid and cost-effective approaches for inducing sterility in a plant is to create polyploids,

especially triploids. This can be doubly beneficial because some triploids, such as the commercial banana, have larger fruits and higher yields as well as being genetically identical, leading to a much more uniform product. In most cases, triploid plants will grow and develop normally except for their inability to set seed. But they cannot be reproduced or propagated, except by recombining the parent lines using a method such as embryo culture. Alternatively, triploid plants can be regenerated from endosperm tissue, which is naturally triploid. The latter method has been used to create triploid varieties of numerous fruit crops including most of the citrus fruits, acacias, kiwifruit (*Actinidia chinensis*), loquat (*Eriobotrya japonica*), passionflower (*Passiflora incarnata*) and pawpaw (*Asimina triloba*).

Mutagenesis

This involves the use of mutagenic agents, such as chemicals or radiation, to modify DNA and hence create novel phenotypes. It includes somatic mutagenesis whereby tissue or cell cultures may undergo useful epigenetic modifications, provided the resultant traits are stable in future generations. First applied to barley by Stadler in 1928, induced mutagenesis has been used with great success in crop breeding. More recently, its scope and utility have been greatly extended by the new molecular-based technology of TILLING (see Chapter 10). In terms of crop improvement, the most common mutagenesis technologies involve the use of DNA-disrupting agents such as ethyl methane sulfonate, or high-energy irradiation from γ-ray sources such as cobalt-60 or caesium-137. During the 1950s, radiation mutagenesis was made available to developing countries by FAO and the International Atomic Energy Agency (IAEA). For example, portable γ-ray irradiators using cobalt-60 or caesium-137 sources were developed for use in regions that lack access to expensive large-scale mutagenesis facilities. By the year 2000, over 2000 varieties of mutation-bred crop varieties including all the major staple species had been released into cultivation in at least 59 developing countries, mostly in Asia.

Another useful way of creating mutants is somaclonal mutagenesis. Such mutations are caused by alterations in DNA induced during plant culture *in vitro*. Somaclonal mutagenesis results from tissue culture-related stresses such as: abiotic factors like cold, drought, or high salt concentrations; excess or dearth of nutrients; the effects of chemical growth regulators; and infections by pathogens. These stresses can cause: single-gene mutations; deletion or transposition of larger lengths of DNA, including chromosome segments; inappropriate methylation or de-methylation of genes; and the duplication or loss of entire chromosomes. Providing they are carefully controlled, somaclonal mutations in cultured plant cells can be a powerful new tool to generate useful genetic variation. Somaclonal mutagenesis has been used to manipulate traits such as yield, disease and insect resistance, drought and salt tolerance, and nutritional quality in crops including rice, wheat, maize, potato, sugarcane and banana.

A potential limitation of mutagenesis is that breeders can only manipulate genes that are already present in the genome. In contrast to wide crossing or transgenesis, new genes cannot be added by mutagenesis. Furthermore, most mutations result in a loss of gene function, meaning that mutagenesis often means reducing the effects of unwanted genes, rather than increasing expression of desirable genes. At first sight, this might seem like a serious limitation to the creation of useful new agronomic traits. However, recent genomic studies reveal the surprising fact that, during the 10,000-year history of agriculture, loss-of-function alleles were associated with 9 out of 19 key episodes in crop improvement and/or varietal divergence. In other words, loss of function of a specific gene can result in a much-improved phenotype, such as dwarfing in cereals or the suppression of lateral branching in maize (see Chapter 7). As we saw previously, many spontaneous, naturally occurring, mutations resulted in favourable alterations in important domestication traits that have greatly enhanced the utility of many major crops. In view of this, we should not underestimate the power of mutagenesis as a continuing tool for crop improvement.

Hybrid production

Wide hybrids

Wide crossing, or interspecific hybridization, involves crossing a crop variety with a distantly related plant from outside its normal sexually compatible gene pool. The usual purpose of wide crossing is not to produce true hybrids, i.e. progeny containing significant parts of both parental genomes, but rather to obtain a plant that is virtually identical to the original crop, except for a few genes contributed by the distant relative. In some cases, it may even be possible to use wide crossing to obtain a plant that is almost identical to an elite variety of a crop except for the presence of a single new trait or gene transferred from a different species. The strategy of obtaining useful genes from other species via wide crosses was greatly enhanced by advances in plant tissue culture.

A particular challenge was to circumvent the biological mechanisms that normally prevent interspecific and inter-genus crosses. Spontaneous rejection of hybrid embryos often occurs in higher organisms in order to ensure reproductive isolation of populations and to avoid non-viable or debilitated hybrid progeny. Therefore, many wide-hybrid seeds either do not develop to maturity, or do not contain viable embryos. To avoid spontaneous abortion, the breeder must remove embryos from the ovule at the earliest possible stage and culture them *in vitro*. Embryo mortality in wide crosses can be high, but enough will survive the stresses of removal, transfer, tissue culture and regeneration to produce sufficient adult hybrid plants for testing and further crossing.

First-generation wide-hybrid plants are rarely suitable for cultivation because only half of their genes come from the crop parent. From the other (non-crop) parent they would have received not only the small number of desirable genes wanted by the breeder, but also thousands of undesirable genes that must be removed by further manipulation. This is achieved by re-crossing the hybrid with the original crop plant, plus another round of embryo rescue, to grow up the new hybrids. As shown in Fig. 8.3, backcrossing is repeated for several generations until the breeder ends up with a plant that is >99.9% identical to the original crop parent, except that it now contains the desirable gene(s) from the donor parent. So-called 'introgression libraries' are collections of backcrossed families each carrying an introgressed segment (about 10–20 cM) from the donor parent and covering, as a collection, the entire genome. Such libraries are particularly useful both for general breeding purposes and for more research-oriented gene/QTL discovery programmes.

F_1 and F_2 hybrids

One of the most notable successes of 20th-century biotechnology was the use of intraspecific maize hybrids to increase yields per hectare by more than fivefold. Early in the century, Shull and East in the

USA began producing inbred maize lines that could be repeatedly combined to produce reproducible F_1 hybrids. The first commercial 'crossed corn' was produced by the Pioneer company in 1917. In order to produce seed on a commercial scale, it is necessary to cross two different F_1 hybrids together to create an even more vigorous F_2, or double-cross, hybrid variety. Yields of some experimental maize hybrids exceeded conventional varieties by over 30%. Following the almost universal adoption of hybrid varieties, US maize yields increased from 1.8 t/ha in the 1920s to almost 10 t/ha in the late 1990s. Although hybrid technology revolutionized maize cultivation, it is important to remember that non-biological technologies have also had major impacts on crop performance. Senior Pioneer breeder, Duvick, estimated that about 60% of the increase in maize yield was attributable to advances in breeding, with the remaining 40% resulting from improved crop management, including more effective inputs and mechanization.

By the 1990s, several other F_1 hybrid crops had been developed. The most important in terms of global food production were the 'super rice' varieties created by Chinese breeder Longping and others. These new rice varieties are intersubspecific hybrids of the normally incompatible *indica* and *japonica* races capable of yields of >10 t/ha. For comparison, Chinese rice yields in 1963 were 2.0 t/ha and the best non-hybrids of the mid-1980s yielded 4.7 t/ha. Due to the complex architecture of its floral structures, hybrid wheat has been more difficult to produce. Wheat flowers are very small and often self-pollinate before opening. This prevents controlled hybridization on a large scale and has made wheat hybrids prohibitively expensive. In the 1980s and 1990s, the use of chemical hybridizing agents to cause male sterility without affecting female fertility finally enabled commercial hybrid wheat varieties to be produced, although hybrids still only account for a tiny fraction of global wheat production. Some other crops with hybrid cultivars being used include sugarbeet, rapeseed, several vegetable brassicas, tomato, potato and onion.

8.5 Selection

Favourable genetic variants of a crop must be capable of being recognized so they can then be selected for further propagation. Historically, farmers selected suitable plants on the basis of visible characteristics such as height, branching, seed size, tuber shape, etc. However, many important attributes of a crop, such as the quality traits that determine taste and nutritional content, are invisible and can only be determined after harvest. For example, wild potato tubers tend to contain high levels of potentially toxic alkaloids. The earliest potato farmers in the South American Andes would not have known if they had a suitable low-alkaloid variety until their tubers were already cooked and ready for eating. The most important criterion at this point would have been taste. If the potatoes tasted bitter the alkaloid content was probably dangerously high and the entire harvest might have to be rejected.

In wheat, a key trait is the bread-making ability of the flour. This character depends on the presence of a particular ratio of gliadin and glutenin storage proteins in the seed. Again, the presence or absence of this kind of quality trait would not become apparent until well after harvest. One can imagine the difficulty of attempting on an empirical basis to select for any useful variation in such traits. Not only are such traits invisible in the growing crop, they are also frequently regulated by numerous unlinked genes. This made for exceedingly slow progress in the selection of many useful quality traits before the advent of scientific methods of screening and analysis.

Chemical analysis

During the 20th century, chemical analysis techniques enabled breeders to select new crop varieties that were largely free of the many toxins and other antinutritional agents that our ancestors were forced, either through ignorance or a lack of suitable breeding technology, to endure. For example, all brassica crops contain various forms of a large family of compounds called glucosinolates. When digested by animals, glucosinolates are broken down to form isothiocyanates, some of which can cause human diseases such as goitre. However, other glucosinolate derivatives can have protective effects against certain forms of cancer. Glucosinolates are especially prevalent in the seeds of oilseed rape. In the late 1980s, most industrialized countries enacted legislation setting maximum levels of glucosinolates in rapeseed.

This was only possible because of accurate and reliable technologies of chemical analysis. The ability to identify glucosinolates enabled researchers to recognize the nature of the toxicity problem in the first place. It then helped them to breed low-glucosinolate varieties of rapeseed and provided a rapid and rigorous method of mass-screening harvested seeds on a commercial scale. Finally, the same technology enabled some brassica vegetables like broccoli to be identified and marketed as sources of the beneficial anticancer glucosinolates.

Chemical analysis has also made it possible to develop new crops, more or less from scratch. One of the most impressive examples of this is the development of a new form of rapeseed in the 1960s. Prior to this time, rapeseed oil was highly enriched in a very long-chain fatty acid, called erucic acid. This oil was normally used for non-edible purposes, such as lighting, and rapeseed was very much a minor crop with a limited and not very profitable market. In the 1960s, a Canadian group used two methods to select mutant seeds with low erucic acid contents. First, the seeds were non-destructively screened by carefully removing a tiny tissue fragment for analysis, while keeping the rest of the seed to produce a new plant. Second, the fatty acid composition of each of the thousands of tiny dissected seed fragments was analysed in detail by the recently developed technique of gas-liquid chromatography.

Prior to the development of gas-liquid chromatography, it required about 200,000 whole seeds (1 kg) and about 2 weeks of lab work to perform a single fatty acid analysis. Now, it was possible to analyse a tiny fragment of less than 2 mg from a single seed in just 15 min. Thanks to this 650 million-fold improvement in analytical efficiency, breeders could accurately screen many thousands of seeds in the search for a spontaneous mutation in genes that controlled erucic acid content. By 1964 the team had developed the first zero erucic acid variety of rapeseed, which they named 'canola'. For the past 30 years, canola has been one of the mainstays of Canadian prairie agriculture and is now a major export earner for the country. Canola-standard rapeseed has also been adopted as an edible oil crop around the world, with an annual value in excess of $6 billion. This high-oleic version of rapeseed is used to make salad oil, cooking oil and margarine, as well as being a key ingredient in all manner of food products from biscuits and cakes to curries and pies.

Molecular methods of selection

These DNA-based methods were mostly developed in the 21st century and are described in Chapter 10.

8.6 Plant Propagation and Tissue Culture

Propagation

Many crops, including cassava, potato, banana, sweet potato and oil palm, are mainly vegetatively propagated and micropropagation systems are especially important for their improvement. In crops where sexual reproduction is problematic or impractical, vegetative propagation has been used for many millennia. For example, cuttings from the best yielding olive trees were being grafted on to existing rootstock to create clonal orchards by the ancient Greeks over 3000 years ago. By the late 20th century, highly sophisticated new biotechnologies had been developed for mass clonal propagation of elite, disease-free planting material using tissue-derived explants such as somatic embryos, shoots, tuber sections, or other cuttings. Because they facilitate the production of healthy planting materials at reasonable cost, these methods are especially useful in developing countries for subsistence root crops such as cassava, potato and sweet potato.

Since the 1970s, mass propagation has become increasingly useful in breeding, especially for tree crops, most of which are too long lived to be accessible to the approaches developed for annual crops. Mass clonal propagation can be a fast and cheap method for multiplying the best genetic stock in such perennial species. Tissue cuttings, typically stems or leaves from selected elite individuals, are cultivated until plantlets are regenerated. The plantlets are subcultured, often on a massive scale, until thousands or millions have been produced for transfer to the field. In this way, cuttings from a single elite tree can rapidly create an entire plantation, or even a whole series of plantations.

Tissue culture

Tissue culture and regeneration of plants is considerably easier than for animals. This is largely due to the greater developmental plasticity of plant tissues (see Chapter 5) and the phenomenon of *totipotency*. Unlike most animal cells, which become irreversibly differentiated during development, plant cells are totipotent. This means that given the right conditions they can express the entire genetic potential of their parent plant. Therefore any plant cell can, in principle, be cultured and regenerated to create a new adult plant.

In practice, the process of finding the right culture conditions to achieve totipotency for a particular plant cell or tissue can be difficult to achieve and to some extent this aspect of plant biotechnology remains somewhat empirical. Tissue culture and regeneration of adult plants are core technologies in many types of clonal propagation and in the creation of transgenic plants.

As shown in Fig. 8.6, most methods of plant transgenesis involve tissue culture. For example, tissue explants or cell cultures, rather than whole plants, are normally subjected to transformation. Even where entire plants or complete organs such as leaves are transformed, the subsequent selection

process normally involves tissue culture. In such cases, transformed cells as identified by selection must be regenerated to create adult plants. It is often found that plant cells can be readily cultured *in vitro* and transformed with new genes, but their regeneration into a healthy plant can be problematic. This sometimes limits the ability to transform useful plants. In some cases, regeneration capacity may be limited to a few varieties that may not be the best for commercial farming. In such cases, breeders must transform and regenerate one of these inferior varieties and then backcross it with an elite variety to produce a transgenic version of the elite variety.

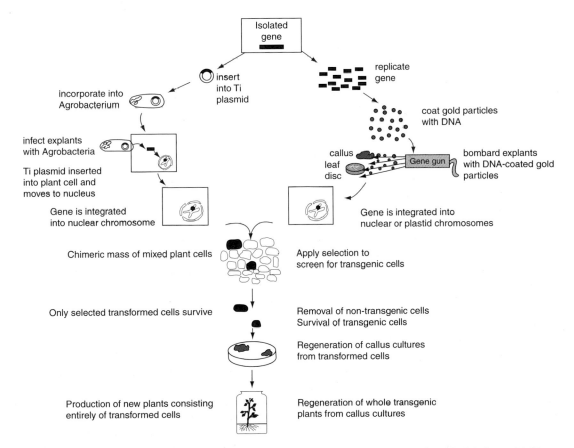

Fig. 8.6. Plant transformation using *Agrobacterium* or biolistics. Transformation begins with an isolated gene(s). For *Agrobacterium*-mediated transformation, the candidate gene(s) must be inserted into a Ti plasmid that is incorporated into *Agrobacterium* cells. When *Agrobacterium* cells infect explants, the Ti plasmid is transferred into plant cells and the gene insert becomes integrated into a nuclear chromosome(s). For biolistic transformation, many copies of the candidate gene(s) are coated on to minute gold particles, which are propelled via a gene gun into explants or whole plants. For *Agrobacterium*-mediated and biolistic transformations, the subsequent selection and regeneration steps are the same.

Culture media and hormones

Plant cells or tissues can be cultured *in vitro* by supplying them with required nutrients, by excluding potential pests and pathogens, and by providing appropriate environmental conditions such as light and temperature. Nutrients include micronutrients such as mineral ions and macronutrients such as nitrates, phosphates and sulfates. In many cases the plant cultures will be completely (heterotrophic) or partially (mixotrophic) unable to generate assimilates via photosynthesis. Therefore complex organic compounds such as amino acids, vitamins and sucrose are often added to the media.

When plant cells or tissues are cultured *in vitro*, they tend to de-differentiate and give rise to generalized cells similar to meristematic cells. In order to regenerate whole plants it is necessary for these cells to differentiate into different tissue and organ types such as leaves, roots and stems. This is achieved by treating the cultures with various cocktails of hormones and synthetic growth regulators (many of which are simply chemical analogues of hormones). The most important hormones used in tissue culture are auxins, cytokinins and abscisic acid. The effect of various hormone treatments can vary markedly between different plants and culture conditions, so it is difficult to predict the best method to apply to a particular species. However, there are some general principles that apply, such as the sequential use of auxins and cytokinins to achieve rooting and shooting as shown previously in Fig. 5.16.

Types of plant culture

Plant cultures are typically initiated by removing a piece of tissue, or explant, from a parent plant. Common explants include pollen grains, endosperm cells, or pieces of organs such as leaves or leaf discs, hypocotyls, stems and roots.

CALLUS Explants are typically grown on a solid substrate such as agar that contains an appropriate growth medium. The cells normally grow rapidly and divide to produce an unorganized tissue mass of parenchyma cells called a callus. Many callus cultures are unable to photosynthesize and require supplementation with carbon sources. Manipulation of callus cultures is a crucial aspect of plant biotechnology but, despite this, many aspects of their behaviour remain obscure. This applies especially to the process of genomic imprinting that occurs during callus differentiation and the regeneration of normal new plants. As discussed above, regeneration of callus tissue is often a bottleneck in the creation of transgenic plants. Also, as discussed in Box 5.1, epigenetic changes during tissue culture may produce serious developmental abnormalities in micropropagated crops, as in the case of oil palm in the 1980s.

CELL-SUSPENSION AND PROTOPLAST CULTURES When dense, compact solid callus cultures are grown on semi-solid media and repeatedly subcultured, they may become loose or friable, and can be cultured as single cells or small clumps. Cell-suspension cultures are maintained by agitation and repeatedly subcultured with media supplementation to keep them in a state of active cell division and growth. Cell-suspension cultures tend to be easier to study than solid cultures and are much used in research. Protoplasts are single plant cells from which the cell walls have been removed either mechanically or by enzymatic digestion. Such cells tend to be relatively fragile and are difficult to culture. Some of the earliest attempts at plant transformation used direct injection of DNA into isolated protoplasts. Protoplasts can be plated on to solid media, where they develop into callus tissue that can eventually be regenerated into whole plants.

EMBRYO CULTURE Mature or immature embryos can be removed from seeds and cultured *in vitro*. As discussed above, the removal or 'rescue' of early-stage embryos from seeds resulting from wide crosses between different plant species is sometimes the only way to ensure their survival for use in breeding programmes. Embryo culture from pollen or anther explants is also used to generate haploid or doubled haploid plants. In such cases, haploid pollen or microspore (immature pollen) cells can be induced by various physical or chemical treatments to express an embryogenic developmental programme that eventually results in production of a mature embryo. Such embryos can sometimes be dehydrated as they would be in a seed and even encapsulated in a semi-solid coating such as gelatin to create an artificial seed that can be stored for future use.

8.7 Transgenesis

Transgenesis is the transfer of genes into a recipient organism using recombinant DNA technologies. Transgenes can, in principle, be derived from any biological organism or may be completely synthetic. Prior to transgenesis, the only viable way to incorporate new genes into an organism was to set up a sexual cross with a compatible species containing the gene of interest. As shown in Figs 8.3 and 8.4, the method of wide crossing followed by repeated backcrossing can result in the insertion of one or more genes from a donor species into a plant. However, this approach is relatively time consuming and is limited to species with which the plant can be successfully crossed. Thanks to advances in tissue culture-assisted wide crossing this list is steadily growing, but it will always be restricted to relatively close relatives of the target plant. As discussed above, a wider range of potential gene donors can potentially be obtained by using somatic hybridization. But in the 1980s this technique was still quite crude and unpredictable with very low success rates. Therefore transgenesis was seen as a way to enhance variation in plants far beyond the capacity of methods then available to plant breeders.

The transfer of genes into cells that could be successfully regenerated into viable adult plants was first achieved in the early 1980s. It was made possible by advances in DNA manipulation and in the culturing and regeneration of plant cells and tissues. The core technologies of DNA manipulation are based on the use of enzymes to cleave and ligate small sections of DNA. This can create novel combinations that can be inserted into a vector for eventual transfer into a plant cell. Many of these methods were originally developed in the 1970s using microbial systems and then extended to higher plants. It was also necessary to know the identity and to have isolated copies of the genes that were to be transferred to a recipient plant. Initially, researchers only had access to a few relatively well-studied genes regulated genetically as simple traits, which limited the scope of transgenic technology in its early decades.

As shown in Fig. 8.6, gene transfer into the recipient organism can be achieved either by physical propulsion (biolistics) or via biological vectors. It is then necessary to select what is often a small number of transformed cells and to eliminate non-transformants. The next challenge is to regenerate transgenic plants in which the transgene functions in the appropriate manner. This requires a great deal of tissue culture expertise and selection to eliminate chimeric or other undesirable progeny. The rate of success in producing suitable transformants can be low and the entire process is often time consuming and labour intensive. However, in comparison with other biotechnologies such as wide crossing or mutagenesis, transgenesis can be relatively fast, although its subsequent progress may be delayed by regulatory procedures that do not apply to the other biotechnologies.

Gene identification and isolation

In the early years of transgenesis the need to be able to identify and isolate the gene(s) to be transferred was a significant constraint on the technology. As discussed in the next chapter, the first decade of commercial use of transgenesis was dominated by just two monogenic traits, herbicide tolerance and insect resistance. However, from the 1980s onwards, there was a series of efforts to extend the range of available genes by assembling gene libraries and by characterizing the functions of such genes using tools such as insertional mutagenesis.

Many of these initiatives focused on the model plant *Arabidopsis thaliana*. This small, easily grown plant has a small genome, a life cycle of about 6 weeks and can be readily transformed using several different methods. Early molecular genetic tools developed using *A. thaliana* included gene-expression libraries derived from various tissues such as developing seeds, flowers or leaves. Such libraries contained copies of most of the genes being expressed in each tissue and were useful in identifying genes involved in many agriculturally important processes such as starch, protein or oil biosynthesis. Another useful tool is the *A. thaliana* insertional mutation library. This is a collection of many thousands of plants, each of which has been mutagenized by the insertion of a piece of foreign DNA into a different place in its genome. Because the foreign DNA segment acts as a tag, the location of the mutagenized gene can be identified and its sequence determined.

If enough plants are mutagenized in this way, a saturated library can be produced in which it is likely that all of the genes have been tagged. In some cases, disruption of a gene will be fatal so the plant will not develop at all. However, in many other cases, an abnormal phenotype is produced that gives a clue about the function of the tagged gene. For example, one *A. thaliana* mutant grew normally until it flowered when it produced a curd-like inflorescence similar to that of cauliflower. These plants had a mutagenized *TFL1* transcription factor gene that caused indeterminate inflorescence. In this case, an *A. thaliana* mutant helped explain the genetic basis of the cauliflower phenotype and made it easier to manipulate the trait in crops as well as model plants.

DNA manipulation technologies

Since the 1960s, molecular geneticists have developed increasingly sophisticated methods for DNA manipulation. Key tools of recombinant DNA technology are restriction enzymes that enable DNA to be cleaved at specific sites and enzymes, such as ligases, that allow cleaved DNA fragments to be stuck back together. In principle, this enables scientists to cut and paste DNA from one place to another. At first such manipulations were limited to simple microorganisms such as bacteria, but eventually methods were developed to transfer DNA segments into complex eukaryotes such as higher plants.

In plants, DNA is normally transferred to a host cell with the aid of a delivery system, or vector, such as a plasmid. Plasmids are small double-stranded circular DNA molecules that can replicate in bacterial cells but can also sometimes be transferred from one cell to another. Using the cut and paste methods outlined above, a foreign gene can be inserted into a plasmid. This exogenous (externally derived) gene might be one that is already known to control a trait of interest in a plant, such as pest resistance. The modified plasmid is transferred to a bacterial culture where it enters the cells, which multiply to create millions of copies of the plasmid. This method can used to amplify copies of a gene of interest, although nowadays this is more usually done using PCR (see Chapter 10). Another use of plasmids in plant transformation is as vectors to transfer DNA into the genomes of intact plant cells.

Many types of plasmid vector have been developed for use in plant molecular biology. Ideally, a plasmid should be relatively small to make it easier to manipulate, it should carry a selectable phenotype so that cells carrying it can be selected from non-transformed cells and it should have several restriction sites to enable exogenous DNA to be added or removed as required. The expression of most eukaryotic genes is regulated by non-coding regions, such as the promoters and terminators, flanking the transcribed region of the gene. Promoter sequences must be present in a plasmid to enable correct functioning of the transgene. In some cases,

a non-plant promoter may be used. For example, one of the commonest promoters used in first-generation transgenic crops is the 35S promoter from the cauliflower mosaic virus. Like most viral gene promoters, 35S is constitutively expressed at high rates so the transgenes that it regulates will be active in most plant tissues most of the time. In other cases, a tissue-specific promoter might be required, e.g. to ensure that a transgene is only expressed in developing seeds. Because they are often regulated differently in different plant groups, tissue-specific promoters are normally obtained from the same or a similar species as the recipient plant. As discussed in Chapter 10, new classes of gene promoter are now being developed that will enable more precise regulation of transgene expression in plants.

Transgenes are most commonly added to plants in order to express new proteins and therefore create a new phenotype. For example, insect-resistant Bt plants contain a gene copied from bacteria that enables them to accumulate novel proteins that are toxic to insects (Chapter 9). In some cases additional copies of existing genes are added in the hope of increasing its expression. For example, extra copies of a chalcone synthase gene were added to petunia plants in order to increase anthocyanin accumulation and obtain deep-purple flower petals. In the early years of transgene technology, insertion of multiple gene copies regularly resulted in silencing of both the introduced and the endogenous genes in a phenomenon called co-suppression. In the case of petunia, chalcone synthase gene expression in some of the transgenic plants was co-suppressed and they produced white or variegated flowers instead.

Another form of gene silencing that has been useful in genetic engineering is the use of antisense DNA (see Fig. 10.6). Most genes are encoded by a sense strand of DNA, which is transcribed to sense mRNA and translated to form a protein. The complementary strand of the DNA double helix is known as the antisense strand. If a transgene made up of an antisense strand corresponding to part of an endogenous gene is inserted into a plant, an antisense mRNA is produced. The antisense mRNA forms a complex with the endogenous sense mRNA, often preventing translation of the sense mRNA and partially or completely down-regulating expression of the target gene. The earliest transgenic crop approved for commercial release, the FlavrSavr® tomato in 1992, contained an antisense copy of a polygalacturonase gene. The aim of down-regulating this cell-wall digesting enzyme was to delay fruit softening and thereby extend the shelf life of the transgenic tomatoes. Antisense-mediated gene silencing and co-suppression are now known to be part of a wider phenomenon known as post-transcriptional gene silencing, as discussed in Chapter 10.

Plant transformation methods

Transformation, or transgenesis, is the incorporation of small segments of exogenous DNA or RNA sequences into the genome of a recipient organism, such as a crop plant. The inserted genetic material is usually known as a transgene, although it may not correspond to a full-length gene. The recipient organism is most accurately termed transgenic, although the term genetically modified organism (GMO) is frequently used in the literature and in the media. Sometimes transformation involves insertion of copies of endogenously derived DNA or RNA sequences into the same species, e.g. as part of gene amplification or RNAi-based manipulation of gene expression.

For the stable, inherited transformation of plants, DNA is normally added to cells using either biolistics or biological vectors. In biolistics, DNA is attached to small particles that are mechanically propelled into plant tissues. This technique is useful because it can be applied to virtually any plant species, but is relatively inefficient and does not always result in the incorporation of the transgene(s) into the plant genome. Alternatively, DNA can be added in a more controlled fashion by means of vectors, such as *Agrobacterium tumefaciens*, which are able to insert DNA directly into the genome of a plant cell. Exogenous genes can also be delivered for transient expression using viral vectors, which is faster but less versatile than stable transformation.

Stable transformation

Agrobacterium tumefaciens is a soil-borne bacterium that causes crown gall disease in a wide range of plants, including crops such as grapes, apples and walnuts. Crown galls are tumorous growths that arise when the bacterium inserts a Ti (tumour inducing) plasmid into a plant host, followed by incorporation of the same bacterial DNA into the plant genome. Using recombinant DNA techniques, the Ti plasmid has been 'disarmed' to prevent it from causing tumour formation and re-engineered to create a suitable vector for the insertion of new genes plus their flanking regions. Once a new set of genes has been inserted into a Ti plasmid, it is reintroduced into *Agrobacterium* cells, which are allowed to multiply before being transferred to a dish containing the recipient plant material. One of the limitations of the *Agrobacterium* method is that it does not work with all plants. Monocots in particular can be difficult to transform, although new strains of *Agrobacterium* and other bacterial vectors have now been developed for a wider range of host plants, including the major cereal crops.

DNA can be added directly to plants by propelling small gold particles coated with exogenous DNA into plant tissues. This technique, called biolistics, can be used for any plant species, which makes it more broadly useful than *Agrobacterium*-mediated methods. Biolistics has therefore been used to transform monocot crops such as wheat, rice, maize and sorghum, as well as some dicots like soybean and common bean. One drawback is its low efficiency and the tendency to produce transformants with numerous copies of the transgene, which can lead to gene silencing. In addition to biolistics, several other methods of direct gene transfer have been used, such as floral dipping, electroporation, DNA uptake into protoplasts, and silicon carbide fibres. Although some success has been reported with these methods they are not widely used for commercial crop species.

In both *Agrobacterium* and biolistic transformation, exogenous DNA is inserted randomly into the recipient genome and several copies might be inserted either in the same place or into different locations in the genome. Biolistics frequently results in several transgene copies in different genomic locations, while *Agrobacterium* methods are more likely to generate multiple copies, such as head-to-head dimers plus border regions of plasmid DNA, at a single locus in the genome. Some transgenic DNA might also fragment into many smaller pieces that become inserted throughout the genome. Fragmentation of some of the introduced DNA is not necessarily deleterious for the functioning of the desired transgene. And, even if it does result in unwanted phenotypes, the latter can be screened out by subsequent selection as the transgenic plantlets are propagated and backcrossed to elite cultivars.

However, both biolistics and *Agrobacterium*-mediated methods are relatively inefficient methods for obtaining single-gene insertion into a plant genome in a way that guarantees a predictable phenotype.

Transformation technology would be greatly improved if transgenes could be targeted to a single predefined location in the genome, in order to give minimal interference with the expression of endogenous genes. As we will see in Chapter 10, new methods of transformation such as chromosome launching, and targeted transgene insertion such as zinc-finger nucleases, are now being developed.

Transient transgene expression

There are some applications of plant biotechnology where transient expression of a transgene is more useful than having a transgene that is permanently incorporated into the genome. Transient expression can be achieved if a plasmid is able to direct expression of its genes without integration into one of the plant genomes. Such plasmids can be delivered into cells using biolistics and tend to have relatively short lifetimes of a few days. Engineered plant viruses can be a more efficient and useful delivery and transient expression system. For example, in the case of biopharming (see Chapter 12), a plant RNA virus can be engineered to contain transgenic RNA encoding a useful protein such as a human antibody. The virus is sprayed or painted on to leaves, which it infects to produce millions of copies of itself. Each copy of the virus will also produce mRNA encoding the human antibody, which is therefore produced in large quantities in the infected leaves from where it can be purified. Numerous new virus expression systems using both RNA and DNA plant viruses are currently under development.

Selection of transformants

Once putative transformed cells or tissues have been produced by stable transformation, it is necessary to screen out any cells that do not carry a transgene. Selectable markers are used to distinguish between cells that carry transgenes and those that do not. In many cases after a transformation procedure has been carried out only a tiny proportion of cells in a tissue or culture might be transgenic. This requires the use of relatively powerful selection methods to eliminate the large number of non-transformed cells, while preserving the small proportion of cells that have incorporated a functional transgene(s). One of the most effective early selection techniques was based on antibiotic resistance genes that were added to the plasmid alongside

the gene(s) of interest so that they were also transferred to the recipient plant. The plant tissues or cultures were then treated with a selected antibiotic that would kill all wild-type cells: only the small number of transgenic cells that carried the antibiotic resistance gene, plus the gene(s) of interest, would survive.

Following antibiotic selection, the depleted cell or tissue culture must be regenerated to produce a mature plant. In some cases, a few non-transformed cells might survive the selection process, giving rise to a chimeric plant, parts of which are wild-type while other parts are transgenic. Such individuals must be identified and culled from populations. In other cases, as noted above, biolistic and *Agrobacterium* methods can produce multiple transgene copies that may give rise to undesirable pleiotropic effects. Again, such individuals must be identified and culled from populations. The objective is to produce as many independent transgenic plants as possible with single transgene insertions. These plants constitute the initial or T_0 generation of transformants. Transgene behaviour in the T_0 generation remains unpredictable and these plants may also have somaclonal abnormalities resulting from the tissue culture process. Therefore it is necessary to produce a first, or T_1, generation of transformants that has undergone sexual reproduction. The process of meiosis can reset methylation patterns and otherwise stabilize the genome so that the phenotype of the transformants can be assessed more reliably.

Depending on the position of the transgene in the genome, its expression can vary considerably. In addition, the transgene might have inserted into a location that affects an existing gene, possibly giving rise to additional unintended pleiotropic effects. Further selection based on the phenotype is necessary at this stage to remove abnormal individuals and to optimize the remaining population based on individuals with the best transgene expression characteristics. What might have started as a large population of transformants might be reduced to a few individuals by such repeated rounds of selection. The stability of the transgene should then be assessed by taking it through several generations of sexual reproduction, which is also an opportunity to multiply the population. Finally, in many cases involving crop species, the variety that is transformed will not be the elite variety supplied to farmers. The selected transformant must be backcrossed with the elite variety to generate a genotype that is essentially identical to the elite variety, except for the presence of the transgene.

This is termed '*substantial equivalence*'. At this stage, the new transgenic variety is ready to begin the often-lengthy field trials and other evaluations involved in gaining regulatory approval that apply uniquely to transgenic plants.

8.8 Summary Notes

- Plant breeding involves the deliberate manipulation of the genetics of specific populations for human purposes, rather than via the evolutionary processes of Darwinian natural selection.
- Scientific breeding is based on the creation and/or identification of useful variation in populations and the selection of such variants for propagation.
- Simple Mendelian inheritance involving unlinked monogenic traits is relatively rare for the most important agronomic characters in crops.
- Complex polygenic traits for many key agronomic characters can be manipulated, even without knowing the identity of the genes involved, by applying the principles of quantitative genetics.
- New genetic variation can be created via technologies such as mutagenesis, hybrid production and transgenesis, all of which rely on enabling technologies such as tissue culture and chromosome doubling.
- Newly created variants are often unsuitable as crops and must be repeatedly backcrossed with elite lines to generate new improved elite varieties.
- Transgenesis involves the insertion of genes, potentially from any source, via methods such as biolistics and *Agrobacterium*.

Further Reading

Scientific plant breeding

Acquaah, G. (2007) *Principles of Plant Genetics and Breeding*. Blackwell, Oxford, UK.
Brown, J. and Caligari, P. (2008) *Plant Breeding*. Blackwell, Oxford, UK.

Duvick, D.N. and Cassman, K.G. (1999) Post-green revolution trends in yield potential of temperate maize in the North-Central United States. *Crop Science* 39, 1622–1630.
Kang, M.S. and Priyadarshan, P.M. (2007) *Breeding Major Food Staples*. Blackwell, Oxford, UK.
Kingsbury, N. (2009) *Hybrid. The History and Science of Plant Breeding*. Chicago University Press, Chicago.
Murphy, D.J. (2007a) *People, Plants, and Genes: The Story of Crops and Humanity*. Oxford University Press, UK.
Murphy, D.J. (2007b) *Plant Breeding and Biotechnology: Societal Context and the Future of Agriculture*. Cambridge University Press, UK.
Narain, P. (2010) Quantitative genetics: past and present. *Molecular Breeding* 26, 135–143.

Biological technologies

Chilton, M.D. (2001) Agrobacterium. A memoir. *Plant Physiology* 125, 9–14.
Dunwell, J. (2010) Haploids in flowering plants: origins and exploitation. *Plant Biotechnology Journal* 8, 377–424.
Forster, B.P. and Thomas, W.T.B. (2005) Doubled haploids in genetics and plant breeding. *Plant Breeding Reviews* 25, 57–88.
Halford, N. (ed.) (2006) *Plant Biotechnology: Current and Future Applications of GM Crops*. Wiley, New York, USA.
Kikkert, J.R., Vidal, J.R. and Reisch, B.I. (2005) Stable transformation of plant cells by particle bombardment/biolistics. *Methods in Molecular Biology* 286, 61–78.
Moose, S.P. and Mumm, R.H. (2008) Molecular plant breeding as the foundation for 21st century crop improvement. *Plant Physiology* 147, 969–977.
Primrose, S.B., Twyman, R.M. and Old, R.W. (2001) *Principles of Gene Manipulation*. Blackwell, Oxford, UK.
Slater, A., Scott, N.W. and Fowler, M.R. (2008) *Plant Biotechnology: Genetic Manipulation of Plants*, 2nd edn. Oxford University Press, UK.
Stewart, C.N. (2008) *Plant Biotechnology and Genetics, Principles, Techniques and Applications*. Wiley, New York, USA.

Crop Improvement
in the 20th Century

9.1 Chapter Overview

This chapter is concerned with the use of biological and non-biological technologies to improve crop production in the 20th century. During the first half of the century, increases in food production were mainly due to use of additional land and chemical inputs such as fertilizers. However, after 1950, improvements made possible by the new biotechnologies had a more decisive impact on global food production. The earliest biological breakthrough was the development of hybrid maize, which transformed yields in North America from the 1920s. In terms of increasing global crop production, by far the most important breeding achievements came from the Green Revolution of the 1960s and 1970s. This principally involved the breeding of shorter wheat and rice varieties with enhanced disease/pest resistance that were responsive to fertilizers and to mechanical technologies such as harvesters. After the 1970s, the pace of crop improvement accelerated as technologies such as wide crossing and mutagenesis were deployed. Another development was the cultivation of transgenic crops. Although first-generation GM crops had only modest impacts on yield or quality they were significant in opening up new possibilities to create genetic variation for breeders, and in the way they changed commercial agriculture and its perception by the general public.

9.2 Population, Land and Food Supply

When considering 20th-century agriculture it is important to be aware of the broader context of population growth and food supply during this period. As discussed in Chapter 7, the 2.5-fold increase in global population from 1700 to 1900 was made possible by the more efficient dissemination of crops around the world, the cultivation of more land, and agronomic improvements such as more efficient crop rotations. As shown in Fig. 9.1

these trends continued in the early part of the 20th century as population growth and land cultivation increased more or less in parallel with each other. By 1960, the global area of cultivated land had reached 1.26 billion ha, up from 0.9 billion in 1920 and 0.47 billion in 1860. Over the same period the population had increased from 1.7 billion in 1900 to 2 billion by 1925 and 3 billion by 1960. Hence, during the first half of the 20th century, the increasing world population was largely being fed thanks to the cultivation of more land. This was especially true for developing countries where most of the world population lived and where most of the population growth and increased demand for food were occurring.

This situation changed between 1960 and 1997 with a mere 9.5% increase in cropland cultivation while the global population increased by 100%, from 3 to almost 6 billion, and food production almost trebled. How was it possible for crop output to treble while the arable land area increased by less than 10%? The answer is that there was a massive increase in crop yield in much of the world. This was the major achievement of 20th-century agriculture, and it enabled an ever-increasing population to be fed to a better extent than at any other time in the history of agriculture. The increasing efficiency of crop production in the second half of the 20th century was largely thanks to two related groups of technologies. These were mechanical/chemical (machinery, fertilizers, biocides) and biological (scientific breeding) technologies.

Although crop biotechnologies became more significant during and after the Green Revolution of the 1960s and 1970s they remained highly connected with mechanical/chemical technologies and both are indispensable to high-yield crop production today. The period after 1970 also saw an important shift in the application of crop technologies from developed countries and more towards developing countries, especially in Asia. As shown

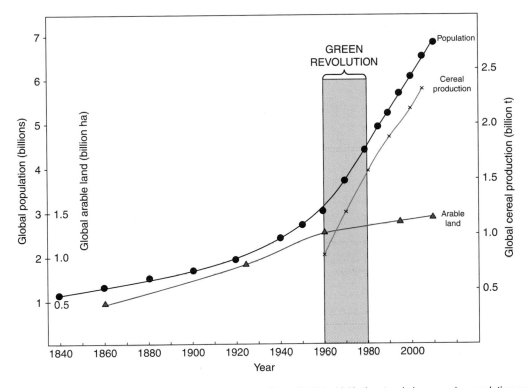

Fig. 9.1. Population, land and cereal production, 1840–2010. From 1840 to 1940, the steady increase in population was matched by an equivalent increase in the amount of arable land. However, during the second half of the 20th century, a much more rapid population increase was sustained by the biological improvements of the Green Revolution, which made possible huge increases in cereal yields without corresponding increases in land use. Data from UN-FAO.

in Fig. 9.2, there was a modest increase in *per capita* food production in developed countries between 1970 and 2000, but much greater increases in developing countries. During the same period, average yields of the three major cereal crops, wheat, rice and maize, rose dramatically (see Fig. 9.3). Therefore we can regard the later decades of the 20th century as the period when agricultural technologies, both biological and non-biological, became truly global in their impact.

9.3 Mechanical and Chemical Technologies

The 20th century witnessed the first large scale use of mechanical and chemical technologies in crop production. Mechanization was made possible by the internal combustion engine and the availability of relatively cheap fuel after the early 1900s. By the 1920s, the first powered machines specifically designed for crop management were

being mass-produced in developed countries, a process that went global after the 1950s. This period also saw a huge increase in the use of chemically based fertilizers and crop control agents, which largely replaced earlier practices such as manuring and hand weeding.

To some extent, mechanical and chemical technologies must fit in with the biological attributes and limitations of each crop. Hence, different types of mechanical sowers and harvesters are often needed for the various kinds of pulses or cereals that often grow in very different ways. However, since the advent of scientific breeding, the reverse has also been true and some crops have been bred to adapt them to existing mechanical and chemical technologies. For example, as we will see below, the Green Revolution crops of the 1960s and 1970s were selected specifically to be responsive to fertilizers and other inputs, while other crops have been redesigned to fit in with harvesting machinery. We will also see that first-generation transgenic crops,

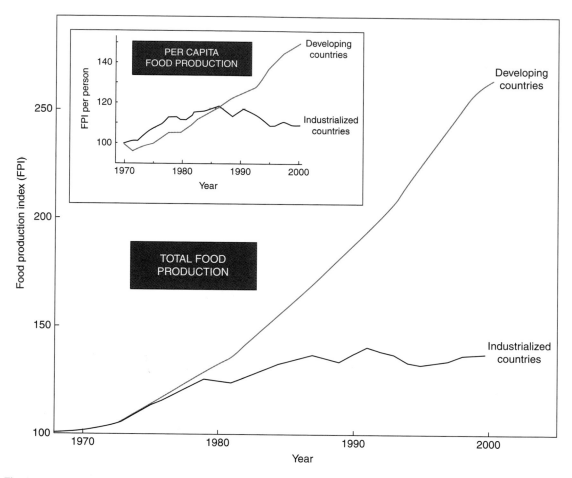

Fig. 9.2. Global food production, 1970–2000. Since the 1970s, both total and *per capita* food production in developing countries has increased much more rapidly than in industrialized countries. Data from UN-FAO.

as developed in the 1980s and 1990s, were designed specifically to complement existing chemical herbicide regimes.

Mechanization

Hunter-gatherers, and then farmers, have always relied on non-biological technologies to facilitate the management, harvesting and processing of the plants they exploited. As we saw in Chapter 1, hunter-gatherers developed tools for collecting and processing wild plants many thousands of years before crop domestication. In Chapter 7, we saw how agriculture began to benefit from new mechanical innovations during the 18th and 19th centuries. These innovations helped to increase

productivity at a time of considerable population increase, especially in Western Europe. These changes accelerated considerably during the 20th century and their effects spread across much of the world. Today, the vast majority of global food production involves the use of mechanical rather than human or animal power for most of its operations.

The process of adapting crops to suit harvesting machinery dates back to the 1920s, when maize breeder and founder of the seed company Pioneer Hi-Bred, Henry Wallace, proposed to a manufacturer of harvesting equipment that he would develop a 'stiff-stalked, strong-rooted hybrid' variety of maize specifically to suit the new machinery. This process accelerated in the 1940s and 1950s as most aspects of crop management were mechanized in developed

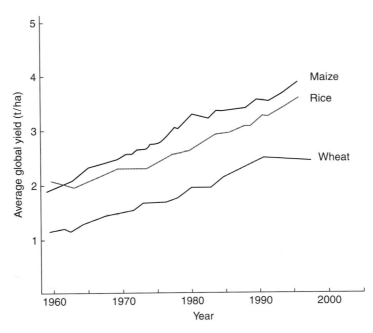

Fig. 9.3. Global yields of the three major cereal crops, 1960–2000. Thanks to a combination of biological and non-biological technologies, global yields of maize, rice and wheat approximately doubled between 1960 and 2000, although in some regions there were far greater yield increases. Data from UN-FAO.

countries, including ploughing, sowing, fertilizer and biocide applications and harvesting. Some of these processes made new demands on crop architecture that were met by breeding new varieties.

For example, traditional varieties of sugarbeet had a fruit structure with mutltigerm seeds. These plants grew in a way that required careful thinning of the crop by hand during the growing season. Increasing labour costs led farmers to look for mechanized alternatives. Suitable machinery was available but new varieties were needed that could cope with a fully mechanized process. In the 1950s, breeders developed new monogerm varieties that avoided the requirement for hand thinning and facilitated automated harvesting. This enabled farmers to mechanize the entire production process and helped sugarbeet to become an important source of home-produced table sugar in much of Europe.

A second example is the tomato, which was originally grown as a hand-picked salad fruit. Traditional tomatoes had relatively thin skins and were sweet and tasty when ripe. But because they were picked close to ripening, these fruits had a short shelf life. New technologies were therefore developed for storing unripe fruits and using

ethylene to ripen them before sale. In the mid-1960s, thick-walled tomatoes were bred in California to adapt the fruits for mechanical harvesting and prolonged transportation and storage of this hitherto soft, hand-picked crop. As a result, modern commercial tomatoes in the USA are thick-walled, robust and relatively tasteless. More recently the technology-created problem of lack of taste led to a search for a biotechnological solution. This happened in the 1990s when the agbiotech company Calgene developed transgenic FlavrSavr™ tomatoes, designed to retain the thick-walled trait for easy harvest while also allowing other aspects of ripening to improve fruit flavour (see below).

The process of mechanization was almost complete in industrialized countries by the 1970s, but was still being applied in developing countries well after this. By 1998, there were 26 million tractors in use around the world, with 11 million in Europe, 6 million in North America, 6.5 million in Asia but only about 0.5 million in Africa. The sustained increase in food production in Asia between 1970 and 2000 was assisted by greater mechanization, especially in the 1990s. On the other hand, the lack of mechanization in much of sub-Saharan Africa is one of several factors responsible for the relatively poor productivity of agriculture in this region.

Chemicals for crop management

The centuries-long history of using chemical agents to manage crop performance was discussed in Chapter 7. By about 1900, naturally occurring and synthetic compounds were being widely used for crop management. Throughout the 20th century, these fertilizers, pesticides, fungicides, herbicides etc. transformed the productivity of crops in developed counties. After the 1960s, their increasing use in developing countries also led to huge yield gains that repeatedly averted potential famines (see below). In many respects, the application of crop breeding biotechnologies has been shaped by the use of these chemical technologies and the two approaches continue to coevolve today.

There is a tendency to regard chemical inputs in agriculture as inherently undesirable, but without these agents our modern agro-urban civilization probably would not be possible. In the case of biocides (e.g. herbicides, pesticides, fungicides, bactericides, molluscicides and nematicides), there have been environmental and safety concerns especially about some of the earlier agents. Many of these concerns, which also applied to other forms of technology (e.g. lead in petrol or asbestos in buildings), were appropriate and have helped stimulate development of safer alternatives. Since the 1970s, much progress has been made in producing more specific biocides that are less toxic to non-target species and do not persist in the environment and very few of the early generation agents are still used. Some of the progress in increasing the safety of crop control agents is shown in Table 9.1.

By 2000, global consumption of biocides was split as follows: 47% herbicides, 29% insecticides, 19% fungicides and 5% growth regulators and other agents (see Fig. 9.4). Although herbicides were the dominant group worldwide, pesticides were more important in developing countries. In many cases, insecticides had a particularly large impact on developing-country crops. For example, cocoa production in the world's leading exporter, Ghana, almost trebled after the capsid bug was controlled, and sugar yields in Pakistan increased by 30% after the introduction of insecticides.

Table 9.1. The decreasing toxicity of biocides to non-target mammals. Continuing research has led to the discovery of new pest control products that offer improved environmental profiles and higher safety standards for consumers and users. These products are much more specific, targeting single pests or pest families and having greatly reduced application rates compared with earlier types of biocide.

Insecticides	Oral LD$_{50}$ (mg AI/kg)* (Mammalian)	Year of discovery or registration
Nicotine	50–60	1690
Rotenone	132–1,500	1840s
Paris Green	22	1880s
Lead arsenate	150	1890s
DDT	113	1930s
Carbaryl (Sevin)	246–283	1950s
Chlorpyrifos (Dursban)	96–270	1970s
Cypermethrin (Cymbush)	250	1970s
Imadacloprid (Admire/Merit)	450	1990s
Indoxacarb (Avaunt)	687–1,867	2000
Fungicides		
Lime sulfur	400–500	1800
Copper sulfate	472	1880s
Mercuric chloride	37–100	1860s
Pentachlorophenol	50–500	1930s
Captan	9,000	1940s
Benomyl (Benlate)	>10,000	1960s
Mancozeb (Dithane)	11,200	1960s
Chlorothalonil (Bravo)	>10,000	1970s
Vinclozolin (Ronilan)	>16,000	1990s
Herbicides		
Arsenic acid	48–100	1900/1920s
Copper sulfate	472	1890s
2,4-D amine	1,492	1940s
Atrazine	1,600	1950s
Glyphosate (Roundup)	>5,000	1970s
Fenoxaprop-ethyl (Excel)	2,565	1980s
Imazethepyr (Pursuit)	>5,000	1980s
Nicosulfuron (Accent)	>5,000	1990s

*The relative toxicity of a pesticide can be measured by its LD$_{50}$. This is the amount of the active ingredient of the chemical in milligrams used per kilogram of test animal (usually rats) that kills 50% of the test animals with a single high dose. Data from Croplife Canada (2003).

Fertilizers

As we saw in Chapter 8, the 19th century witnessed a crisis in fertilizer availability that was especially marked in industrialized countries. Experiments with inorganic forms of nitrogen and phosphorus demonstrated huge effects on crop yields in many soil types.

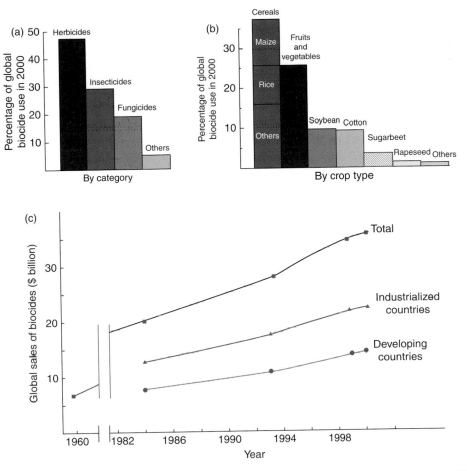

Fig. 9.4. Global use of biocides, 1960–2000. (a) Major categories of biocide in 2000. (b) Biocide use for different crops in 2000. (c) Global sales of biocides. Data from UN-FAO and AGROW (1999).

These chemical inputs were much faster acting than organic fertilizers such as manure, guano or wood ash. In the early 20th century, new industrial technologies such as the Haber-Bosch process for nitrogen fixation into nitrates provided relatively cheap fertilizers. This revolutionized agriculture as yields increased and farmers were able to grow crops on poorer soils that were enriched with chemical fertilizers. In some cases, the need for frequent rotations between legumes and cereals was reduced or completely removed, which further increased yields of the more productive cereals.

The major three nutrients in crop fertilizers are nitrate, phosphate and potassium, commonly abbreviated as NPK. While nitrate availability is often a major constraint on plant growth, phosphate and potassium are also vital nutrients for many crops. Hence a typical maize crop yielding 6–9 t/ha of grain requires 30–50 kg/ha of phosphate fertilizer. Also, while legumes can generate much of their own nitrate thanks to their rhizobial symbionts, a legume crop such as soybean still requires 20–25 kg/ha of phosphate fertilizer for optimal yield. One of the problems about agriculture in Africa is that soils throughout the continent tend to be moderately (annual deficit of 30–60 kg/ha NPK) to severely (>60 kg/ha NPK deficit) depleted in NPK nutrients. These data are from the mid-1990s when most farmers in Africa were unable to afford to buy imported NPK fertilizer.

Between 1960 and 2000, production of nitrate increased more than tenfold to almost 100 Mt/year while use of phosphate increased from 9 Mt/year to almost 40 Mt/year. Overall, annual NPK usage rose from just over 30 Mt in 1960 to almost 150 Mt in the 1990s, with most of the increase in developing countries. As shown in Fig. 9.5, fertilizer use levelled off in the 1990s as more efficient methods were developed to apply and manage what was becoming an increasingly expensive chemical input. Greater efficiency was also encouraged by concerns over excess fertilizer runoff and pollution of watercourses. As we will see later in this chapter, the inability of many African farmers to use fertilizers was one of the principal reasons behind the limited success of the Green Revolution in sub-Saharan Africa. Despite these problems in Africa, the increased use of chemical fertilizers in the rest of the world helped to transform crop production during the 20th century and was essential in maintaining food production in the face of accelerating population increase.

Herbicides

Herbicides, also known as weedkillers, are chemicals that are toxic to plants and can be either selective (affecting some plants) or non-selective (affecting many or all plants). Crop plants in the field often suffer from competition with other plants for resources such as light, nutrients and water. Any plant that competes with a crop is classified as a weed. Globally, weeds account for about 25% of crop losses due to biotic factors, and in some regions losses are much higher. Before the development of herbicides,

weed could only be removed by laborious, time-consuming and expensive manual methods. Herbicides have had an enormous impact on food production, both by increasing crop yield and by reducing the need for farm labour. By 2000, herbicides made up almost 50% of all crop protection chemicals. It has been estimated that the additional yield from herbicides is at least three times more than their cost to the farmer. A further benefit from herbicides is the reduction in the need for mechanical cultivation, which saves fuel and reduces the risk of soil erosion.

The first widely used herbicide was 2,4-dichlorophenoxyacetic acid, or 2,4 D, which is an auxin analogue introduced in the 1940s. It is relatively cheap and kills many broadleaved plants while leaving grasses, such as cereal crops, largely unaffected. During the 1950s, the bipyridium herbicides, diquat and paraquat, were introduced as non-selective, broad-spectrum agents. These photosystem I inhibitors are especially useful for removing foliage during land clearance or prior to sowing. Although it is still in widespread use, paraquat is acutely toxic to people and is regularly involved in poisoning episodes. The 1950s also saw the introduction of triazine herbicides including atrazine, which is a low cost, broad-spectrum photosystem II inhibitor. Despite its environmental persistence and potential for groundwater contamination, atrazine is still used in some regions because of its effectiveness in controlling broadleaf and grassy weeds in crops such as maize and sorghum.

During the 1970s, a new generation of much less persistent and less toxic (to animals) herbicides was

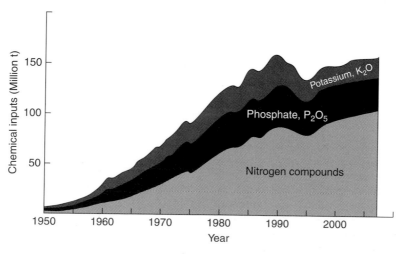

Fig. 9.5. Global production of synthetic fertilizers, 1950–2010. Fertilizer production increased more than tenfold from 1950 to 1990 and then stabilized. Data from UN-FAO.

introduced. Probably the most notable of these was glyphosate, often sold as Roundup®, which was first produced in 1974 as a non-selective weed control agent that is still in widespread use today. As we will see in more detail in the section on transgenic crops, glyphosate tolerance was the first commercially successful transgenic trait transferred to a plant. Glyphosate is an inhibitor of aromatic amino acid biosynthesis and is therefore toxic to all plants. However, unlike earlier broad-spectrum agents such as paraquat and atrazine, glyphosate decomposes rapidly in the soil and has a very low toxicity to animals. These qualities have helped to make glyphosate one of the most successful herbicides ever produced.

A similar group of herbicides is the ALS (acetolactate synthase) inhibitors, which disrupt the formation of branch-chain amino acids and, like glyphosate, are toxic to all plants. ALS inhibitors include sulfonylureas, imidazolinones, triazolopyrimidines, pyrimidinyl oxybenzoates and sulfonylamino carbonyl triazolinones. Because the ALS pathway is only found in plants, these herbicides are among the safest in terms of their effects on animals. Another plant-specific herbicide is glufosinate, which inhibits photosynthesis and therefore kills all vegetation. As we will see below, glufosinate tolerance was the second major transgenic trait used in commercial crop production in the 1990s. The importance of herbicides in modern agriculture is shown by their widespread use and huge impact on crop yields. Like other classes of biocide (see below and in Table 9.1), many earlier herbicides had undesirable side effects. Nowadays, however, new herbicides are formulated to minimize their environmental persistence and toxicity to non-target organisms.

Pesticides

By far the most destructive group of crop pests are the insects and the term 'pesticide' is used here interchangeably with 'insecticide'. The term 'fumigant' is sometimes used for a pesticide that is applied as a vapour. Insecticidal fumigants, such as burning sulfur, have been used since antiquity. A more recent fumigant is nicotine, which is a highly toxic alkaloid that blocks synapses associated with motor neurones in insects. Nicotine was used in solution form as an insecticide as early as 1690, but it is also used today in the form of a dust or volatile agent where it is especially effective for pest control in confined spaces such as glasshouses. As with many pesticides, nicotine affects verte-brates as well as insects and must therefore be used with caution. As noted in Chapter 7, farmers and others have used many forms of often toxic and non-specific pesticides for millennia. During the 19th century, several new, but still fairly toxic, pesticides such as rotenone and lead arsenate were produced for the first time on a commercial basis (see Table 9.1). However, the development and widespread systematic use of pesticides in agriculture only dates from the mid-20th century.

During the 1930s, several new groups of plant-derived and synthetic pesticides were developed by organic chemists. While some of this research was done in public sector labs, an increasing proportion of the R&D was being carried out in the rapidly expanding commercial agrochemical sector. Agrochemical companies continued to expand after 1945 as chemical inputs became cheaper and more effective and demand from farmers increased. During the 1990s, many of these companies purchased or merged with seed companies to form chemical/biotechnological conglomerates such as Bayer, Dow, Monsanto and Dupont. By 1999, the top ten agrochemical companies accounted for almost 90% of global sales of non-fertilizer biocides (see Table 9.2).

Table 9.2. Global non-fertilizer biocide sales by company, 1999.

Major biocide manufacturers	Country	Value of sales ($ billion)	Percentage of global market
Monsanto	USA	4.0	12.9
DuPont		3.2	10.3
Cyanamid		2.2	7.1
Dow Agro		2.1	6.8
	USA subtotal:	11.5	37.1
AgrEvo	Germany	2.4	7.7
Bayer		2.3	7.4
BASF		1.9	6.1
	German subtotal:	6.6	21.3
Novartis	Switzerland	4.2	13.5
Zeneca	United Kingdom	2.9	9.4
Rhône-Poulenc	France	2.3	7.4
	Top ten subtotal:	27.5	88.7
	World total:	31.0	100

Data from OECD, Development Cooperation, Report 1998; AGROW *World Crop Protection News*, 1999.

Some of the early pesticides dating from the 1930s included insecticides such as alkythiocyanate, thiodiphenylamine, phenothiazine and naphthalene, as well as fumigants such as ethylene dibromide, ethylene oxide and carbon disulfide. A landmark discovery made by Müller in 1939 was the well-known and exceptionally potent insecticide dichlorodiphenyltrichloroethane (DDT). DDT was one of the first of the new generation of pesticides to be used on a truly global scale with immense effectiveness, especially against crop pests in developing countries and to eradicate malarial mosquitoes from many regions of the world.

From 1960 to 2000, the value of traded pesticides increased from less than $100 million to $12 billion. By the late 20th century, the four main classes of insecticide in order of market value were organophosphates (40%), carbamates (30%), pyrethroids (15%) and organochlorines (15%). Organophosphates are synthetic pesticides that inhibit synaptic transmission of nerve impulses by blocking the active site of acetylcholinesterase. While they are highly toxic to insects, these pesticides can also affect animals and, as with most other agrochemicals, much of the R&D after the 1950s was concerned with synthesizing new classes of organophosphate that were more specific in their insecticidal effects with fewer impacts on non-target species, including humans. Examples include malathion and parathion. Carbamates, such as carbaryl and carbofuran, have a similar mode of action to organophosphates but can be more selective in their toxicity and tend to be less persistent in the environment.

The huge family of pyrethroids, which are still some of the most effective agents against flying insects, were originally obtained from plants such as chrysanthemum. Pyrethroids bind to sodium channels in insect neurons, causing hyperexcitation leading to convulsions and death. Most naturally occurring pyrethroids are fast acting but also break down rapidly, which can limit their use in agricultural settings. From the 1930s, synthetic derivatives with greater stability were developed against a range of crop pests. The best known member of the organochlorine group is DDT, which is very effective against domestic insects and mosquitoes but less effective with mites and much slower acting against flying insects than pyrethrum or thiocyanates. DDT acts in a similar way to pyrethroids but is much more persistent and can gradually accumulate in the food chain, sometimes reaching damaging levels in predators such as insectivorous birds and raptors. The widespread use, and sometimes misuse, of DDT led to the development of resistance in many target insects as early as the 1950s. The use of DDT then declined sharply and it was largely replaced by alternative organochlorines, such as lindane and the cyclodienes, although in the 2000s DDT was reintroduced for more targeted mosquito control in several developing countries.

Many other types of pesticide are used in agriculture but the four main classes described above made up the vast majority of globally traded agrochemical pesticides in 2000. A relatively minor insecticide group that became more prominent in the 1990s was the crystalline endotoxin proteins produced by the soil bacterium *Bacillus thuringiensis* (Bt). Sporulating Bt bacteria produce large quantities of these toxins, some of which are especially effective against lepidopteran pests (see Fig. 9.14). Live Bt was originally applied on to crops as a powder or spray and was especially popular on 'organic' farms where synthetic chemicals were not permitted. As discussed later in the chapter, several Bt toxins were successfully expressed in transgenic crops in the 1990s and this became the second most widespread form of transgenic technology after herbicide tolerance.

Fungicides (plus oomycetes)

Fungi and oomycetes are the most serious crop pathogens and were still causing catastrophic infestations, such as the Irish potato blight that led to the death of 1 million people, in the mid-19th century. Until the 1940s, control of fungal diseases relied largely on cocktails of relatively toxic inorganic chemicals. From 1940 to 1970, dithiocarbamates and later the phthalimides were introduced into widespread use as more active, less phytotoxic, and more user-friendly agents than the older inorganic fungicides. From the 1970s, a much wider range of more effective and specific fungicides was introduced, including the triazoles, strobilurins, anilides, pyrimidines and benzimidazoles.

One of the problems with many of these agents was the build-up of resistance in fungal populations, especially if the fungicides were misused. For example, poor spray coverage could result in the application of sub-lethal doses over a wide area, which favours the development of resistant fungal genotypes. The first case of resistance to benzimidazoles occurred in powdery mildew in greenhouses in 1969, only 1 year after their introduction.

By 1984, resistance had been reported in many pathogens against which benzimidazoles had previously been active. The reason for the rapid development of resistance was that these fungicides were single-site inhibitors of microtubule assembly during mitosis. Therefore, a single mutation that disrupted benzimidazole binding to tubulin would confer resistance to the fungicide, which would rapidly spread through the adjacent population.

Fungicide resistance has been found in many pathogens and can appear in the field within a few years of a particular agent being used on a wide scale. The issue of unwanted resistance in target populations is common to all biocides and, as with antibiotic resistance in medicine, requires a combination of improved application and management measures, plus constant surveillance to detect outbreaks of resistance. The increasing occurrence of multiple resistance to several products from different manufacturers led to the establishment of an industry-wide body, the Fungicide Resistance Action Committee, to tackle this growing threat in conjunction with international bodies such as FAO. By 2000, the value of global fungicide sales was over $7 billion and the major crop groups were the cereals (22%), grapes (10%), rice (8%), soybean (8%), potato (7%) and pomme fruits (6%).

Impact of biocides on crop production

The historical yields of UK wheat from 1200 to 2000, shown in Fig. 9.6a, reveal the dramatic impact of 20th-century technologies such as biocides on this key staple, which is the major UK and European food crop. In Fig. 9.6b, we can see the close correlation between biocide sales and wheat yields in the UK. Between 1945 and 1975, biocide use was relatively low and yield gains were mainly due to improved breeding and higher fertilizer use. The rapid increase in wheat yields after 1975 was the result of a combination of a near-tenfold rise in biocide use and the introduction of the 'Green Revolution' semi-dwarf varieties as discussed in the next section.

9.4 Breeding Achievements in the 20th Century

During the latter half of the 20th century, much of the increase in crop productivity was due to improved inputs such as fertilizers and biocides.

However, biological improvements via breeding also played a vital role in the improvement of crop quality and yield. The use of modern breeding methods resulted in the creation of high-yielding cultivars of all the major commercial crops. Such cultivars were specifically adapted to optimal growing conditions that were made possible by the intensive use of inputs such as fertilizers, biocides and in some cases irrigation. Although the replacement of traditional landraces by inbred elite cultivars led to loss of genetic diversity and inbreeding depression, many of these problems were overcome using biotechnologies such as hybridization, wide crossing and mutagenesis.

Early in the century, the power of hybrid technology was demonstrated in commercial maize crops in the USA (see Fig. 9.7). But it was not until the 1990s that it was adopted more widely with spectacular results, as in the new hybrid rice in China. For much of the period from 1910 to the 1950s, there was steady progress in developed countries in identifying useful traits and incorporating these into elite crops, often using the methods of quantitative genetics. In association with the widespread use of chemical inputs and mechanization, these biological improvements led to increasing yields and higher quality of the major commercial crops such as wheat and barley. However, in global terms, biological improvements were not applied to most crops in developing countries until the 1960s. Before this time, increases in food production were mainly due to expansion of the arable land area.

The most important breeding achievement of the 20th century was the extension of biological improvement methods to developing countries and their use alongside chemical and management methods in the Green Revolution of the 1960s and 1970s. As shown in Fig. 9.1, the period after 1960 witnessed an unprecedented rise in global crop and food production that was due in large part to biological improvements. During the remainder of the century the older biotechnologies continued to be refined while new DNA-based methods, such as marker-assisted selection, were developed. One of the newer technologies that emerged in the 1980s was transgenesis. Although not yet as significant as earlier biotechnologies in terms of its impact on crop yield or quality, transgenesis is important in its capacity to increase genetic variation in crops.

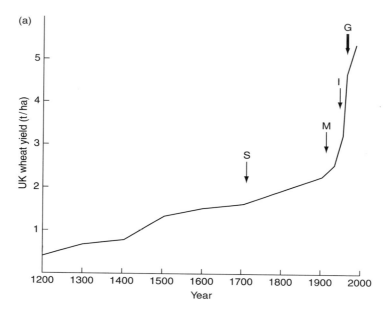

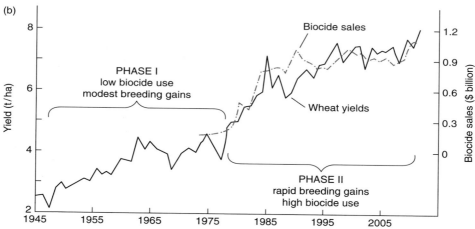

Fig. 9.6. Historical wheat yields in the UK and the recent correlation with biocide sales. (a) Average UK wheat yield increased very slowly from 1200 to 1700. Yields grew faster during the first agricultural revolution from 1700 (S) to 1900. Yield gains accelerated dramatically in the 20th century as the benefits of modern scientific breeding (M) were combined with greater use of chemical inputs (I) and the Green Revolution (G). Data from UN-FAO. (b) From 1945 to 1975, biocide use was relatively low and the modest yield gains resulted from a combination of breeding success and greater use of fertilizer. After 1975, yields increased dramatically thanks to Green Revolution traits and other breeding successes, plus greater use of chemical inputs and improved management methods. Data from Defra, UK.

Hybrid maize

Before the advent of hybrid maize, yields had remained essentially static since it was first grown commercially in the 1700s. Because maize is normally an outcrossing open-pollinated plant, the constant forcing of self-fertilization to produce elite lines with desirable homozygotic traits can lead to inbreeding depression and a reduction in crop performance (see Chapter 8). However, when two inbred lines from outcrossing species like maize are hybridized together, they often exhibit *heterosis* as characterized by dramatic increases in yield and vigour. Even in the absence of appreciable inbreeding depression, hybrids

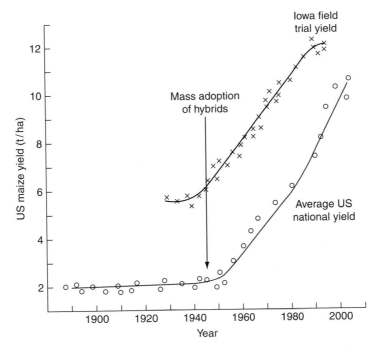

Fig. 9.7. Historical yields of US maize. Yield of hybrid maize varieties in USA according to their year of release. Note the dramatic increase in yields shortly after mass adoption of hybrid varieties in the 1950s. Iowa values are based on field trials at three locations in the central maize-growing belt of the state. Data from Duvick and Cassman (1999).

between two inbred lines can exhibit substantial heterosis. As early as 1878, the superior yield of intervarietal hybrids of maize was demonstrated by Beale. However, because each F_1 hybrid had to be recreated every year, breeders were faced with the challenge of constantly generating massive quantities of hybrid seed to supply to farmers. By the 1920s, the problem of producing hybrid seed had been solved by Shull and East and developed on to a commercial scale by Wallace. In many cases two different F_1 hybrids were crossed to generate a highly vigorous F_2, or double-cross, variety.

However, despite the 20–30% yield advantages of hybrid maize, very few farmers adopted these varieties for the next two decades. As late as 1934, hybrids accounted for a mere 0.4% of US maize cultivation. There were several reasons for this reluctance: (i) the existing conventional varieties gave sufficient yield for farmers to obtain a profit from their crop; (ii) because hybrid varieties did nor breed true, farmers were unable to use saved hybrid seed and had to repurchase expensive seed each year from seed companies; and (iii) hybrid crops were still a new technology and an unknown quantity

for most farmers. Since they did not absolutely need the extra yield (and potential risk) to make a profit, most farmers kept using conventional inbred lines. This situation changed with the droughts, dust bowl and depression of the mid-1930s. The hybrids considerably outperformed other varieties, especially in drought conditions, and a fall in prices meant that higher yields were now essential if farmers were to make a profit from maize. By 1944, 59% of US maize was hybrid and by the 1960s hybrids accounted for virtually all North American production. The resulting four- to fivefold increases in crop yield between 1950 and 2000 are shown in Fig. 9.7. Following the almost universal adoption of hybrid varieties, US maize yields had increased from 1.8 t/ha in the 1920s to 7.8 t/ha in the 1990s.

A drawback of hybrid technology in maize is that it became possible for industry to generate new hybrids by continually reworking the few inbred lines developed in public sector labs early in the 20th century. This meant that commercial maize had a very narrow genetic base, which almost led to disaster in the early 1970s, when much of the US crop was struck by the Southern corn leaf blight, *Helminthosporium maydis*. The highly inbred maize varieties had little or no genetic disease resistance to this virulent fungus and crops across the country were devastated. The unusually severe nature of this disease episode was exacerbated by the fact that 85% of the US maize crop contained Texas male-sterile cytoplasm, which made the plants uniquely susceptible to the leaf blight fungus. Luckily, a few resistant lines were found and were crossed into the hybrid varieties. However, the basic issue of poor genetic diversity was not resolved and in the late 1990s a project called GEM, or genetic enhancement of maize, was launched. The GEM project is a public/private sector collaboration between breeders to survey germplasm from throughout the Americas for traits that could be used in commercial hybrids. Selected germplasm can then be produced in a form that can be supplied to industry for introgression into their hybrid breeding programmes.

Hybrid maize technology gradually spread to other maize-growing areas of the world and by the late 1990s hybrids made up two-thirds of the maize area in the main cultivation regions of Asia. Maize hybrids were also a commercial success in parts of Latin America and Africa. However, the costs of repurchasing seed every year, coupled with the chemical inputs required to obtain maximum yield, have led to a much lower rate of adoption of the technology by poorer farmers in these regions. A potentially useful development for such farmers is transgenic hybrids with Bt insect resistance traits that can reduce the requirement for chemical pesticides (see below).

The Green Revolution

The Green Revolution probably saved many tens, and possibly hundreds, of millions of people from severe hunger or starvation during the explosive population growth in developing countries following World War II. The Green Revolution was mainly concerned with the development of shorter varieties of wheat and rice. A major factor limiting the yield of some cereals is their tendency to grow tall, especially in response to fertilizer. As we saw above, the increasing use of fertilizers in the first half of the 20th century had a huge impact on crop yields. However, the tendency of fertilized cereals to grow tall meant that these long-stemmed plants are more likely to 'lodge' (fall over) in response to wind or rain. Tall stems also meant that less carbon was available for grain production than in shorter plants.

Wheat

The key research that led to the Green Revolution was carried out in the 1940s and 1950s in Mexico by a team of breeders led by Borlaug. At the time, one of the major constraints on wheat yields in many parts of the world was the stem rust fungus, *Puccinia graminis*. Due to stem rust and other agronomic shortcomings the average yield in Mexico was only 0.75 t/ha, even though much of the land was irrigated and well suited to wheat production. Borlaug's team collected many different wheat varieties and eventually found one that carried genes for resistance to the stem rust fungus. These plants were hybridized with a new short-stemmed Japanese variety called 'Norin 10' and crossed into local Mexican wheat varieties to produce short, locally adapted plants that also carried rust resistance. Thanks to their shorter stems these wheat varieties were also much higher yielding than the old tall varieties.

The process of making the genetic crosses needed to produce the new wheat varieties was made more difficult because wheat is a self-pollinating species and it was necessary to prevent or screen out progeny from self-fertilization. It took more than 9 years and over 6000 crosses before a series of high-yielding, resistant hybrids could be tested in large-scale cultivation. To speed up the breeding, researchers worked out a system called 'shuttle breeding'. This involved using climatically different upland and lowland regions to grow two field crops a year, halving the time needed for plant production. In addition to breeding, agronomists in the team established a high-input system of fertilizers, irrigation and mechanization to ensure that yields were able to approach their biological potential. This programme was so successful that Mexico went from importing half of its annual wheat requirement in the early 1950s to complete self-sufficiency by 1956. By 1964, Mexico was exporting an annual 500,000 t of wheat and average yields are now 6.5 t/ha, a ninefold increase since the mid-1940s.

The next stage was to apply these technologies to the wheat-growing regions of Southern Asia, where India and Pakistan had suffered a succession of poor harvests in the early 1960s, at a time of rapid population increase. There was much concern in India about the inability of the country to feed itself from its own produce. As well as the very real threat of famine, the necessity to import food from overseas imposed a financial burden that most developing nations could not afford. The process of improving the Indian wheat crop started in 1966 when the government imported 18,000 t of the new semi-dwarf, disease-resistant seed from Mexico. In a single year, from 1967 to 1968, the Indian wheat harvest increased from 11.3 to 16.5 Mt. Between 1966 and 1971 Pakistan and India each doubled their wheat production.

Yields then increased further as breeders crossed the Mexican wheat with their own locally adapted wheat varieties and production in India alone rose to almost 70 Mt in 2000 (see Fig. 9.8). The predicted famine in the Indian subcontinent had been averted and the whole region was well on its way to regaining self-sufficiency in food production. Today, despite a population of over 1.2 billion, India is self-sufficient in the major food grains and earns valuable foreign exchange by exporting some of its surplus crops, including wheat. Its wheat production has tripled since 1968, which contributed to the ninefold growth in the Indian economy over the same period.

Rice

During the 1960s and 1970s, Beachell and colleagues in the Philippines pursued a similar breeding strategy to the wheat breeders in Mexico, with the aim of improving rice yields in eastern Asia. After screening thousands of plants and crossing promising varieties for several years, the researchers produced a new variety that would become known as 'miracle rice'. This short, thick-stemmed plant, called IR8, had much higher yields than conventional tall rice and was derived from a cross between a popular rice variety called 'Peta' and a new Chinese variety called 'Dee-geo-woo-gen' (*Dee-geo* means 'short leg' in Chinese). The Chinese variety carried a spontaneous mutation in the sd_1 gene, causing a dwarfing phenotype (see Fig. 9.9).

The shorter, semi-dwarf rice or wheat plants were much sturdier than their tall cousins and therefore less likely to fall due to wind or rain. They were also more responsive to added fertilizers, making it worthwhile to apply higher doses to increase grain yield. Following the release of the new 'miracle rice', yields across Asia more than doubled from the mid-1960s to 1990. By 2000, almost 90% of the semi-dwarf rice varieties grown in tropical Asia were derived from the *Dee-geo-woo-gen* mutation. Between 1966 and 2000, worldwide rice production more than doubled, from 257 to 600 Mt. Over 90% of this increase occurred in Asia and was

directly attributable to improved breeding and related achievements of the Green Revolution.

Impact of the Green Revolution

The success of the Green Revolution was largely due to genetically improved plant varieties but it was also helped by three additional factors, namely greater use of irrigation, better use of inorganic fertilizers, and reforms in government regulation and management that facilitated market growth. As shown in Fig. 9.10, the Green Revolution had a huge impact on global food prices, which more than halved during the 1980s. This price reduction was especially significant for the billions of poorer people in developing countries who use most of their income to buy food. The result was a considerable reduction in hunger despite a doubling of the global population between 1960 and 2000. Despite its immense achievements in most of the world, the Green Revolution was much less successful in many parts of Africa (see Chapter 12).

9.5 First-generation Transgenic Crops

Hybrid maize and semi-dwarf wheat and rice varieties were responsible for some of the most impressive achievements of 20th-century breeding technologies. In both cases, the biological improvements to the crop itself were closely coupled with the more effective use of mechanical and chemical technologies. By using this mixed approach, breeders and agronomists created multi-technology packages that resulted in a quantum increase in crop productivity during the final third of the century. In this section we will see how, during the 1980s, another potentially groundbreaking technology, namely transgenesis or genetic engineering, was developed.

The principal techniques involved in transgenic plant production were described in Chapter 8. Production of the first experimental transgenic plants was reported in 1983 and over the next two decades many hundreds of transgenes were inserted into dozens of plant species. From the earliest days it was realized that this might be a revolutionary new technology able to create almost unlimited genetic variation in plants, including crops. Many new agbiotech start-up companies were established during the 1980s and 1990s to exploit the new technology. Despite this exciting potential,

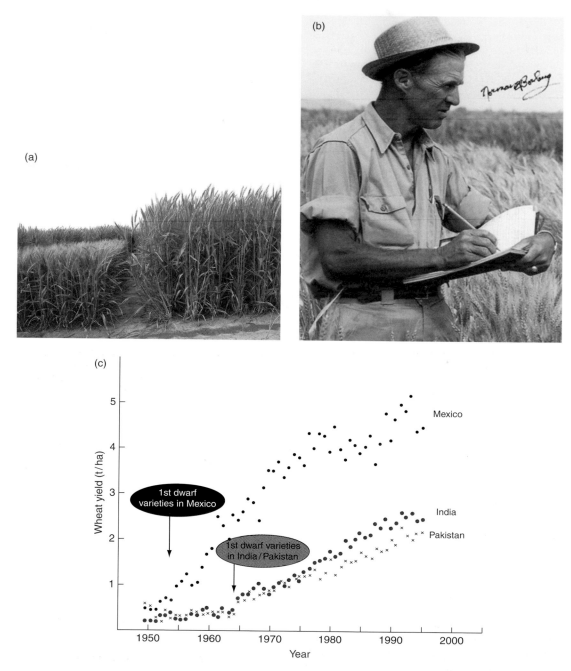

Fig. 9.8. Dwarf cereal crops and the Green Revolution of the 1960s and 1970s. (a) Modern dwarf wheat (left) is less than half the height of older varieties (right) and has a far higher grain yield. (b) Norman Borlaug was one of the pioneers of the disease-resistant, semi-dwarf wheat that led to the Green Revolution (photo courtesy of Texas Agricultural Experiment Station). (c) Over the last few decades, access to the new wheat short varieties has transformed wheat farming in developing countries across the world, with yield increases from two- to fivefold in India, Pakistan and Mexico. Data from UN-FAO.

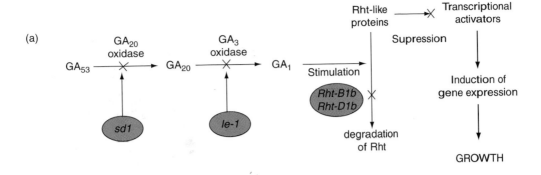

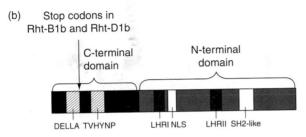

Fig. 9.9. Molecular genetics of the Green Revolution. (a) Effect of Green Revolution mutations on the gibberellin signalling pathway. In wild-type plants, GA_{53} is converted to GA_{20} and then GA_1, which causes degradation of Rht, thereby preventing it from suppressing the gibberellin-induced genes that lead to normal growth. In wheat, the *Rht-B1b* and *Rht-D1b* mutations are not susceptible to GA_1-induced degradation while in rice the *sd1* mutation prevents the conversion of GA_{53} to GA_{20} due to insertion of a premature stop codon into the GA_{20} oxidase gene. In peas, the *le-1* mutation that blocks conversion of GA_{20} to GA_1 was responsible for the dwarf phenotype originally observed by Mendel. (b) Structure of *GAI/Rht* and related DELLA-domain proteins implicated in cereal dwarfing traits. The dwarfing alleles of wheat, rice and maize are determined by mutations in *GAI/Rht* genes that result in a reduced response to the hormone gibberellin. The *GAI/Rht* genes encode proteins with highly conserved nuclear localization signals, and repressor and gibberellin-signalling domains. The arrow shows the site of the two stop-codon mutations in 'Norin 10', the semi-dwarf variety used to develop Green Revolution wheat. Adapted from Hedden (2003).

however, the commercially successful first-generation transgenic varieties were nearly all created by long-established agrochemical companies and involved relatively mundane input traits that could often, in principle, be created by alternative, non-transgenic breeding methods.

The earliest commercial transgenic crop, the FlavrSavr® tomato first cultivated in 1992, was engineered to have delayed fruit senescence so that the fruits would have a longer shelf life. However, FlavrSavr® was a commercial failure and it was not until 1996 that two new transgenic traits, namely herbicide tolerance and insect resistance, were introduced into commercial agriculture. By 2000, four commodity crops expressing these two traits were being grown on more than 50 Mha in several countries, mainly in the Americas. In addition to

these two major commercial transgenic traits, crops expressing several other traits were released on a much smaller scale. Among these minor traits was virus-resistant papaya, which was introduced into Hawaii in 1998. This was a modest commercial success and was crucially important in rescuing part of the Hawaiian agricultural industry.

The major commercial traits

The first commercially successful transgenic crops were developed by agrochemical companies such as Monsanto and Bayer (then trading as Hoechst). For their initial targets, the companies selected simple monogenic traits for which the relevant genes were readily accessible. By the early 1990s, several genes had been identified that altered key

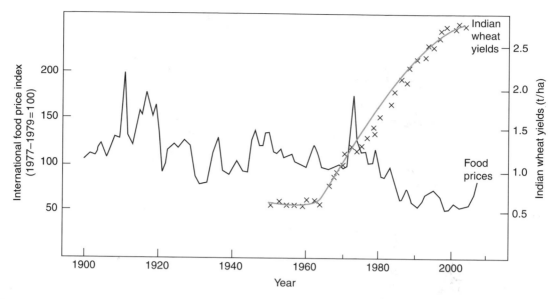

Fig. 9.10. Economics of the Green Revolution. The inflation-adjusted international food price index was relatively stable from 1900 until the 1970s when it fell due to greatly increased global cereal production as exemplified in this case by Indian wheat. Despite a brief spike in the mid-1970s due to a global oil crisis, food prices halved between the late 1970s and 2005. More recently, food prices have risen again but are still at historically low levels. Data from Anderson (2010).

input traits in crops. An input trait can be defined as a genetic character that affects how the crop is grown without changing the nature of the harvested product. Examples include herbicide tolerance and insect resistance, which are useful traits for the farmer because they simplify the process of crop cultivation and can reduce yield losses due to weeds or insect pests. Because a modified input trait does not affect the product of the crop, e.g. its seed composition, the new varieties could be harvested and processed in the same way as conventional varieties. In other words there were no additional costs associated with downstream processing of the crop.

The input trait strategy was commercially attractive for several reasons: (i) companies could sell a package consisting of their proprietary herbicide + their patented transgenic seed to farmers, who were then obliged to use only this particular herbicide; (ii) the utility patents that applied to the transgenic seeds gave the company much stronger ownership rights than would be the case for non-transgenic seeds carrying a similar trait (see Chapter 11); and (iii) because the transgenic seed belonged to the company, farmers could not save seed from their crop and were required to repurchase new

seed and herbicide from the company each year. From the early years of transgenic crop cultivation, herbicide tolerance traits dominated the market (see Fig. 9.11) and this domination continued into the 2010s.

Herbicide tolerance

As outlined above, traditional weed control was largely replaced by chemical agents during the 20th century and hundreds of herbicides are now available to farmers. Some of the most effective herbicides are non-selective (also called broad-spectrum) agents, which kill most or all weeds. The drawback of broad-spectrum herbicides is that they can also affect growing crops, especially at critical stages of development such as establishment or grain filling. This led to the development of crop varieties that are tolerant to one or more broad-spectrum herbicides. Such varieties can be treated with the herbicide in question at any time during the growing season to control weeds without adversely affecting development of the crop itself.

For several decades, herbicide-tolerant crop varieties were developed by breeders using existing

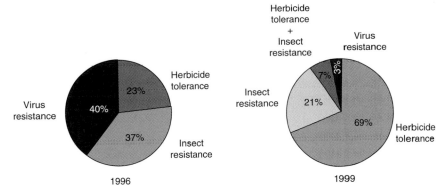

Fig. 9.11. Categories of transgenic traits 1996–1999. The first-generation transgenic crops were dominated initially by three and later by only two major traits, namely herbicide tolerance and insect resistance. Data from ISAAA.

genetic variants or via mutagenesis. For example, Clearfield® is a non-transgenic maize variety sold in North America from the early 1990s that is tolerant to the imidazolinone herbicide Patriot®. Somaclonal variation was used to create another non-transgenic variety of maize that is tolerant to glyphosate. However, it was often difficult to find such varieties using non-transgenic approaches. Also, unlike conventional varieties, transgenic varieties could be patented, which meant that farmers could be legally prevented from saving seed and were obliged to buy new seed and herbicide from the company each year. After the mid-1980s, the availability of transgenic technologies enabled agrochemical companies to engineer tolerance to broad-spectrum herbicides that they already produced. Agbiotech companies then developed a portfolio of different crops, all with tolerance to the herbicide in question.

The biotechnological part of this strategy was relatively straightforward because herbicide tolerance often behaved as a simple monogenic trait. Providing the gene in question was available to the company and the crop could be readily transformed, transgenic herbicide-tolerant varieties could be produced in a few years. By 2000, about 40 Mha of transgenic herbicide-tolerant crop varieties were under cultivation around the world. Almost all of these varieties expressed either of two forms of transgenic trait, namely glyphosate or glufosinate tolerance. Some of the potential risks and benefits of herbicide-tolerant crops are explored in Box 9.1.

GLYPHOSATE Since 1996, by far the most common transgenic herbicide tolerance trait in commercial agriculture has involved the broad-spectrum herbicide

glyphosate, also known as Roundup™ (Fig. 9.12). Glyphosate (N-[phosphomethyl] glycine) is a derivative of the simple amino acid glycine. Although it is toxic to plants, glyphosate has no effect on non-photosynthetic organisms, including animals. Indeed, in animal feeding trials glyphosate was less toxic than table salt. Unlike some earlier herbicides, glyphosate is rapidly broken down in the soil and has low environmental persistence. Glyphosate was originally developed by Monsanto and became one of the most successful herbicides in the 1980s and is still widely used today. Examples include uses for crop protection by farmers, for domestic weed removal by gardeners and to clear unwanted vegetation in urban areas such as public footpaths and parks.

When sprayed on to plants glyphosate is readily taken up into most tissues, where it inhibits the activity of enolpyruvyl shikimate phosphate (EPSP) synthase, an enzyme required for the synthesis of phenolic amino acids. A reduction in phenolic amino acid biosynthesis leads to reduced rates of protein synthesis. Glyphosate also strongly inhibits the induction of nitrite reductase, which disrupts amino acid formation. Within a few days of glyphosate application, most plants turn brown and die. It is especially effective against grasses, including volunteer cereals. Due to its effectiveness against weeds and favourable environmental credentials, glyphosate was an attractive target for the creation of herbicide resistance in crops. Moreover, in the 1980s the manufacturer of glyphosate, Monsanto, was also one of the most advanced developers of transgenesis in plants.

The most common method for creating glyphosate tolerance in commercial crops involves the

Box 9.1. Benefits and risks of herbicide-tolerant crops

Benefits

Transgenic herbicide-tolerant varieties of four crops, maize, soybean, cotton and rapeseed, were rapidly adopted by farmers after their introduction in 1996. By 2010, crops with such traits were being grown on almost 150 Mha in 29 countries. The success of this technology indicates that it has been beneficial to farmers as well as to the companies involved. It is estimated that US farmers benefited due to reduced overall herbicide applications and higher yields per hectare reportedly worth $15–28/ha in some cases. The agbiotech companies benefited from a greatly increased market share and from selling a profitable package of seeds + herbicide to growers. On the other hand, there is little evidence of significant consumer benefit in terms of lower prices or improved product quality.

A potential environmental benefit of herbicide-tolerant crops is that farmers do not need to plough their fields before sowing, which is normally necessary to limit weed emergence. Instead the crop can be sprayed with herbicide to remove weeds at any time. The reduction or cessation of ploughing, as in no-till farming, can improve soil texture, increase rhizosphere biodiversity and reduce the likelihood of soil erosion.

Risks

There are concerns that in some crops like rapeseed or sugarbeet, which have closely related weed species, herbicide-resistance traits may spread into the weed population via cross-pollination. In principle, this could lead to the emergence of so-called 'superweeds', resistant to herbicides. However, research has shown that the direct transfer of herbicide-resistance transgenes into weedy crop relatives is very unlikely to result in emergence of more virulent weedy hybrids.

Herbicide tolerance is a perennial problem that has grown with the increasing use of herbicides. By 2009, over 300 weed species had acquired resistance to at least one commercial herbicide. The most frequent forms of resistance were to photosystem I and II inhibitors (120 species), AHAS inhibitors (100 species), ACCase inhibitors (35 species), synthetic auxins (25 species) and glyphosate (20 species). Due to the diversity of available herbicides, however, the development of resistance of one weed to one herbicide need not necessarily affect overall crop production.

One of the risks in focusing so much crop production on just two herbicides, namely glyphosate and glufosinate, is that farmers become increasingly reliant on these two agents. In transgenic crops, glyphosate has replaced most other herbicides or alternative means of weed control. From an evolutionary viewpoint, this leads to an intense selection for glyphosate resistance in weed species. The most common mechanism for resistance in such weeds involves mutations in the EPSP synthase gene that enable the plants to produce a glyphosate-tolerant version of the enzyme. For example, new forms of *Eleusine indica* and several *Lolium* spp. that have serine, threonine or alanine substitutions at Pro-106 in a highly conserved region of EPSP synthase have been found in different parts of the world. In the longer term, therefore, it would be useful to develop new crop varieties (transgenic or non-transgenic) that are tolerant to a much wider range of herbicides than just glyphosate and glufosinate.

---

insertion of a bacterial EPSP synthase gene. Although all known plant versions of EPSP synthase are highly sensitive to inhibition by glyphosate, some bacteria have a slightly different form of the enzyme that is insensitive to the herbicide Therefore, if a copy of the bacterial gene is inserted into a crop plant, the resulting transgenic variety should be resistant to applications of glyphosate while all non-transgenic plants in the same area will be killed. Monsanto transgenic crop varieties carrying glyphosate resistance are known as RoundupReady®.

Field trials demonstrated the effectiveness of the glyphosate tolerance trait in several crops and in 1996 Monsanto released the first RoundupReady® varieties of soybean for commercial cultivation in the USA. Despite the higher price of RoundupReady® seed, US farmers rapidly adopted the new transgenic soybean varieties and within 2 years they made up almost 40% of the crop. By 2001, almost two-thirds of US soybean carried RoundupReady® traits. This rapid rate of technology uptake contrasts with the much slower adoption of hybrid maize earlier in the 20th century, as discussed above.

GLUFOSINATE The second commonest transgenic herbicide tolerance trait is associated with glufosinate or phosphinothricin, which is a low-persistence,

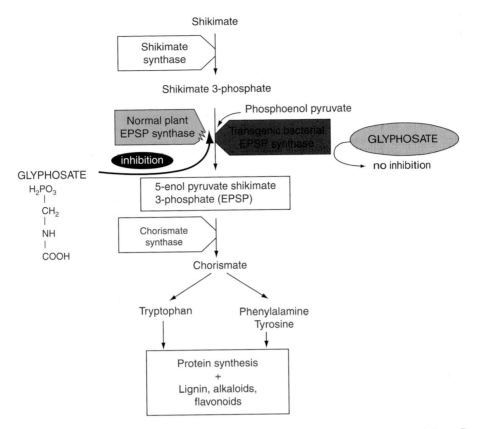

Fig. 9.12. Herbicide tolerance I: Glyphosate. The systemic herbicide glyphosate, known commercially as Roundup®, is a non-selective inhibitor of the plant enzyme EPSP (5-enolpyruvyl shikimate 3-phosphate) synthase. Inactivation of this shikimate pathway enzyme prevents synthesis of aromatic amino acids leading to the fatal cessation of protein synthesis in glyphosate-affected parts of the plant. Transgenic plants expressing a bacterial EPSP synthase gene are tolerant to glyphosate because the bacterial enzyme is unaffected by the herbicide.

non-selective agent most effective against broadleaf weeds. There are several formulations of glufosinate including Basta® and Liberty®, both currently owned by Bayer. Because glufosinate is a non-selective herbicide its use is normally restricted to total eradication of vegetation or to controlling weeds before or shortly after crop emergence. Glufosinate is an inhibitor of glutamine synthase in plants, disrupting photosynthesis and causing death of the entire plant within a few days (see Fig. 9.13).

In transgenic plants, resistance to glufosinate is due to the presence of a gene from the soil-dwelling bacterium *Streptomyces viridochromogenes*, which encodes the enzyme phosphinothricin acetyltransferase. This enzyme converts glufosinate into an inactive acetylated derivative. Plants expressing this transgene are able to grow normally, even after the application of relatively large doses of glufosinate. Some crop varieties carrying this transgenic trait are known as LibertyLink®. The first LibertyLink® rapeseed was grown commercially in Canada in 1995. Soybean and maize were approved in 1997, shortly followed by other crops such as sugarbeet. Although not as widespread as RoundupReady® varieties, LibertyLink® crops were commercially successful and are still widely grown.

Insect resistance

After herbicide tolerance, the second most common modification in transgenic plants in 2000 was insect resistance, which alone accounted for 8 Mha of crops plus a further 3 Mha where the trait was expressed in combination with herbicide tolerance. Insect resistance in transgenic maize, cotton and potatoes was

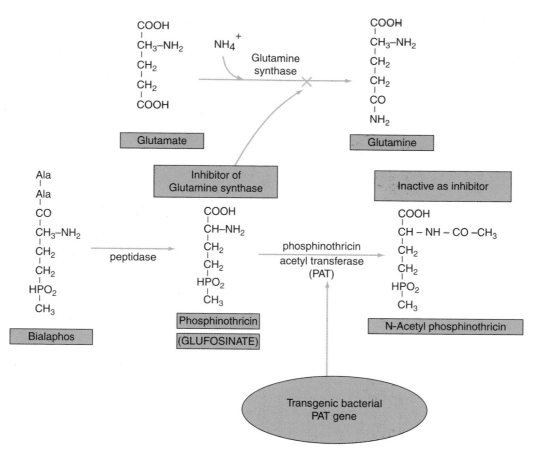

Fig. 9.13. Herbicide tolerance II: Glufosinate. Glutamine synthase is a key enzyme of amino acid biosynthesis that is targeted by several herbicides including bialaphos and its derivative phosphinothricin (glufosinate). Transgenic plants expressing a bacterial phosphinothricin acetyl transferase gene are able to convert the herbicide into an inactive acylated derivative and are therefore tolerant to this normally toxic herbicide.

conferred by insertion of a gene encoding a protein toxin from the Gram-positive soil bacterium *Bacillus thuringensis* (Bt). As we saw above, insect-control sprays containing live toxin-producing *Bacillus thuringensis* suspensions were used for over 30 years in organic farming but the widespread use of Bt toxins in transgenic crops is much more recent. Bt toxins are a family of so-called crystalline (cry) proteins that are converted into their active form during digestion in the gut of a range of insect larvae, causing a lethal disruption of potassium ion transport (see Fig. 9.14). Mammals do not convert the toxins into their active forms and so are unaffected by them.

As shown in Table 9.3, many Bt toxins are available from various bacterial strains. Some Bt toxins

affect specific groups of insects or nematodes, which makes Bt technology an attractive proposition for pest control. As we have seen previously, one of the major challenges for all classes of biocide is the development of resistance by the target population. As discussed in Box 9.2, the establishment of refugia and the addition of several different Bt toxin genes (gene stacking) can markedly reduce the likelihood of resistance in the population.

In 1996, the first transgenic Bt crops were introduced into commercial agriculture in the USA by several agbiotech companies (with Monsanto taking the lead) and were soon widely adopted by farmers. The three major cotton insect pests in the USA were tobacco budworm, cotton bollworm and

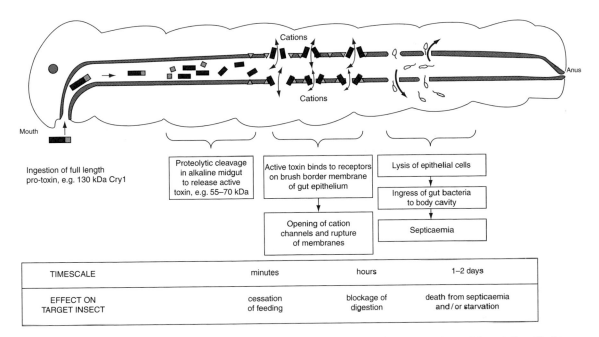

Fig. 9.14. Mode of action of Bt toxins in invertebrates. The Bt group of toxins includes many bacterial crystalline (Cry) proteins. Transgenic plants normally produce a high molecular weight pro-toxin form of the Cry protein. The pro-toxin is ingested by an insect pest and converted into an active toxin in the midgut. The activity of the toxin results in blockage of digestion and rupture of the gut, resulting in rapid cessation of feeding and eventual death of the insect.

Table 9.3. Some insecticidal and nematicidal crystalline (cry) Bt proteins and their target pests.

Cry proteins	Target insects or nematodes
Cry1A – 1L	Lepidoptera
Cry2A	Lepidoptera and Diptera
Cry3A – 3C	Coleoptera
Cry4A – 4B	Diptera
Cry5Aa, 5Ab	Nematodes
Cry5Ac, 5Ba	Hymenoptera
Cry6Aa, 6Ab	Nematodes
Cry7A	Coleoptera
Cry8A – 8C	Coleoptera
Cry9A – 9E	Lepidoptera
Cry10, 11	Diptera
Cry12 – 51	Various

pink bollworm and these were effectively tackled by Bt constructs expressing various forms of *Cry1A* gene. The major maize pest was the European corn borer, which was combated by toxins encoded by *Cry1A*, *Cry 9C* and *Cry1F* transgenes. At this time, Bt technology was credited with saving the cotton industry in Alabama. Developing countries also took up Bt technology, with China growing its first Bt crops in 1997, South Africa in 1998, followed a few years later by India. The two most successful Bt crops in these countries were cotton and maize, where local insect pests were effectively combated by Bt toxins.

Transgenic Bt maize and cotton varieties were, and still are, highly successful and are now grown around the world with benefits that include a reduction in the requirement for costly and sometimes toxic chemical pesticides. However, in some crops, Bt varieties were not successful, although this was not always for scientific reasons. For example, in 1995, transgenic Bt potatoes carrying a *Cry3A* gene for resistance against the Colorado potato beetle were released by Monsanto for commercial cultivation in the USA. Some of these varieties also contained transgenic virus coat protein for resistance to the potato leaf roll virus (see below for more on virus resistance). The transgenic potatoes were initially well received and were approved for export to major markets in Europe, Japan and Australia. However, they were abruptly withdrawn after a few years when several major potato processors decided to avoid using transgenic products following several instances of adverse publicity for GM crops in 1999 and 2000 (see Chapter 11). As a consequence,

Benefits

Transgenic Bt varieties are effective in controlling target insects and average crop yields can increase by as much as 7%. They also reduce the need for spraying with more toxic and less desirable pesticides that often affect beneficial organisms such as insectivorous birds. This can save growers about $40/ha. In developed countries major benefits include reduction in input and management costs and greater resilience in the face of sporadic pest outbreaks. In developing countries there are added health benefits as farm workers face lower risks from applying hazardous chemicals, often with inadequate protective clothing. Since 2002, Bt cotton has been widely adopted by millions of small farmers in India, leading to >65% higher yields and >50% more cotton production.

Risks

The danger in relying on a single class of toxins, such as Bt, is the strong selection pressure favouring the survival of insects able to sequester the toxin or otherwise render it harmless. As long ago as 1995, at least two insect species had already become resistant to Bt toxins in the field with at least another ten species showing the potential for resistance in laboratory studies. One strategy to prevent or delay the acquisition of resistance by insects is the inclusion of several unrelated toxin genes in a transgenic crop. This form of 'gene stacking' is expensive to develop but is an effective long-term option that could greatly reduce the incidence of pest resistance.

Another practical way to combat the development of resistance to Bt toxins in the field is to set aside refugia, which are areas adjacent to the main transgenic crop that are sown with non-transgenic seeds of the same crop. In the refugia, non-tolerant insect populations can continue to thrive and hopefully will outcompete conspecifics that develop Bt tolerance. This strategy relies upon the cooperation of growers and can fail if it is not implemented on all farms. However, refugia appear to have been effective where they are used and to date there have not been any widespread outbreaks of Bt resistance in major pest species.

Monsanto discontinued what had been a productive potato-breeding programme.

Other input traits

Although insects and weeds can exert significant biotic stress on crops, the most important biotic threats come from viruses, bacteria, fungi and nematodes. These are major crop pathogens and there has been much research aimed at producing resistant varieties using transgenic approaches. To date, these efforts have involved relatively small crop areas and have not yet led to commercial transgenic modification of any of the major crops, especially the cereals. In the medium term, however, the production of virus resistance is an area where transgenesis may be the best option for combating many serious crop diseases.

Virus resistance

As discussed in Chapter 6, viruses are serious plant pathogens that cause considerable crop losses each year. Breeders normally rely on developing endogenous resistance within the plant itself but this is not always possible. Because there are no chemical agents available for the control of viral pathogens the use of transgenic methods is particularly attractive in this case. This has stimulated efforts to engineer viral resistance into crops. One of the major strategies is to express complete or partial proteins from the viral pathogen, typically part of the viral coat protein complex. This causes plants to become sensitized to subsequent infections with the same virus. When such plants are attacked by the viral pathogen, they mount a successful defence response. Although the exact mechanism of viral immunity in plants is not yet known, it is analogous to immunization in animals by injection with an attenuated virus.

Research by Beachy *et al.* in the 1980s demonstrated the feasibility of this approach in lab studies that were extended to large-scale field trials with crops. Since then, the technique has been used successfully in many plants against dozens of viruses, with some examples listed in Table 9.4. The strategy of expressing virus coat protein in plants confers resistance against the virus in question, and often against related viruses as well. One of the earliest examples of the successful use of a transgenic virus-resistant crop was Hawaiian papaya in the 1990s, as outlined in Box 9.3. Although transgenic virus-resistant papaya was successful and the technology was adapted for

Table 9.4. Transgenic virus-resistant crop projects.

Crop	Virus	Inserted gene	Location
Papaya	Papaya ringspot virus	Coat protein and replicase	USA, India, Malaysia, Brazil, Mexico, China
Squashes and pumpkins	Watermelon mosaic virus Cucumber mosaic virus Zucchini yellow mosaic virus	Coat protein	USA, Egypt
Citrus spp.	Citrus tristeza virus	Coat protein	USA
Banana	Banana bunchy top virus Banana bract mosaic virus	Coat protein	Australia, Belgium, Egypt
Rice	Rice tungro spherical virus	Coat protein	USA
Cassava	Cassava mottle virus	Coat protein	USA/Africa
African rice	Rice yellow mottle virus	Replicase	USA/West Africa
Faba bean	Faba bean necrotic yellow virus	Coat protein	Egypt
Potato	Potato leaf roll virus	Coat protein	USA, Mexico
Sweet potato	Feathery mottle virus	Coat protein and replicase	USA/Kenya
Wheat	Barley yellow dwarf virus	Coat protein	USA
Raspberry	Raspberry bushy dwarf virus	Coat protein	USA
Maize	Maize streak virus	Coat protein	USA
Cabbage	Turnip mosaic virus	Coat protein	China
Tobacco	Cauliflower mosaic virus	Coat protein and satellite RNA	China
Sugarcane	Sugarcane mosaic virus	Coat protein	USA
Stone fruits: plum, peach, apricot	Plum pox virus	Coat protein	USA

Box 9.3. Transgenic virus-resistant papaya in Hawaii

Papaya, *Carica papaya*, is a tree crop grown throughout the tropics. For decades, plantations in the Hawaiian islands suffered serious infestation from the papaya ringspot virus that was spread by aphids from other plants. Infection by the virus leads to misshapen, tasteless fruits and eventual death of the plant. Control measures had a limited effect and on some islands such as Oahu papaya cultivation was abandoned. Between 1992 and 1995, the industry faced disaster as the virus became widespread in the major remaining area of cultivation in Puna.

Research on transgenic methods of virus resistance started in public sector labs in Hawaii in the 1980s. The strategy was to express complete or partial proteins from the viral pathogen (typically part of the viral coat protein complex), which causes the plants to become sensitized to subsequent infections with the same virus. In the case of papaya, the coat protein from a mild mutant strain of the virus was transferred to plants using biolistic transformation of embryogenic cultures. Due to the mildness of the viral strain, the plants did not become diseased but still became sensitized to the coat protein of the more virulent wild-type virus. Attack by this viral pathogen resulted in a successful defence response.

By 1992, the first transgenic virus-resistant papaya plants were field-tested and by 1998 they had completed all regulatory stages and were ready for commercial release. This project involved use of proprietary technologies owned by major biotechnology companies, but licensing agreements ensured that the transgenic seeds were made available free of charge to farmers. The impact of the new varieties was impressive. Between 1992 and 1998, annual papaya output had declined from 25,000t to under 12,000t, but by 2001 it was back to 20,000t, most of which was from transgenic varieties. Research is now focused on expressing additional forms of the ringspot virus coat protein to combat the potential threat of the virus overcoming resistance in transgenic papaya plants. Meanwhile the technology developed in Hawaii was adapted for use in more than a dozen other countries in the tropics with several field trials being conducted in the late 1990s.

use elsewhere, it was only used on a relatively small scale in comparison with the other major input traits described above. During the same period, transgenic varieties of virus-resistant squash were also released on a small scale (under 2000 ha) in the southern USA, where they were effective at combating several mosaic viruses.

Fungal, bacterial and nematode resistance

Development of transgenic methods to engineer endogenous resistance to bacterial, fungal and nematode pathogens has been much more difficult than manipulating viral or insect resistance, although several promising approaches have been demonstrated, at least in principle. Research approaches include the induction of so-called 'suicide genes' in plant cells infected with a nematode, and the expression of protease inhibitors that inhibit nematode growth. To date these studies are still some way from commercial application, but they remain promising options for the future. Another possibility is to insert resistance genes, such as the *Xa21* bacterial blight resistance gene that was transferred to several Chinese rice varieties. Antifungal agents like phytoalexins or chitinases have also been expressed in plants and a gene encoding an anti-fungal protein from lucerne was transferred to potatoes. The transgenic potatoes became resistant to the soil-borne fungus *Verticillium dahliae*, which causes some $70–140 million in damage to US potato crops each year. However, the resistance was quite specific and the potatoes were still susceptible to the late-blight fungus *Phytophthora infestans*.

This highlights a major problem in engineering fungal resistance into crops, namely the difficulty in producing broad-spectrum resistance, without transferring numerous resistance genes. Also, fungal resistance often evolves spontaneously in the field and can sometimes be found in other varieties of a crop, or in sexually compatible wild relatives, from which it can be transferred to an elite crop cultivar by conventional breeding. It is possible that, in the longer term, more useful forms of transgenic fungal resistant traits may be developed but, at present, the non-transgenic approach seems to be the more realistic option for most crop breeding programmes.

Output traits

In addition to the traits discussed above, dozens of output traits were targeted during the 1990s by researchers seeking to create new commercial transgenic varieties. In almost all cases, these traits proved to be considerably more complex and difficult to manipulate than herbicide tolerance and insect resistance. Output traits normally involve many genes and give rise to a new crop product that requires segregation from existing forms of the same crop, which involves extra expense for farmers and processors.

The major output traits targeted were genes that regulated the accumulation of the principal classes of crop product, namely carbohydrates, oils and proteins (see Chapter 4). Other important output traits involved minor but useful products such as vitamins and minerals that may be deficient in some crops. Finally, a more radical step was to use plants as production systems to generate large amounts of high value compounds such as industrial enzymes or pharmaceuticals for human use. While none of these traits were successfully commercialized as part of first-generation transgenic crop technology, many of them were subsequently refined and may eventually form part of the portfolio of 21st-century biotechnology, as discussed in the next chapter.

9.6 Summary Notes

- During the first half of the 20th century most of the increasing crop production required to feed a rapidly expanding population came from cultivating more land and greater use of fertilizers.
- After the 1950s, increases in crop production were mainly due to more effective use of biocides and advanced breeding technologies, coupled with mechanization.
- Biocide use increased markedly from the 1980s, especially in developing countries.
- Major crop yield gains came from maize hybrid technology in the USA and the Green Revolution varieties of semi-dwarf, disease-resistant wheat and rice that were mainly grown in Asia.
- The vast majority of first-generation transgenic crops were developed by a few large agbiotech companies with agrochemical and breeding operations.
- Over 99% of first-generation transgenic crops consisted of either or both of two trait modifications: herbicide tolerance and/or disease resistance.

Further Reading

Population, land and food supply

Anderson, K. (2010) Globalization's effects on world agricultural trade, 1960–2050. *Philosophical Transactions of the Royal Society B* 365, 2835–2851.

FAO (2009) available online at: www.go.nature.com/DdNYvk.

Mechanical and chemical technologies

AGROW (1999) *World Crop Protection News.* Burrill & Co.

Cooper, J. and Dobson, H. (2007) The benefits of pesticides to mankind and the environment. *Crop Protection* 26, 1337–1348.

Cremlyn, R.J. (1991) *Agrochemicals: Preparation and Mode of Action.* Wiley, New York.

Croplife Canada (2003) A History of Crop Protection and Pest Control in our Society. Available at: www.croplife.ca/english/pdf/Analyzing2003/T1History.pdf.

OECD (1998) Development Cooperation Report, New York, USA.

World Bank (2007) *Agricultural Mechanization: Issues and Options.* World Bank, Washington, DC.

Breeding achievements

Duvick, D.N. and Cassman, K.G. (1999) Post-green revolution trends in yield potential of temperate maize in the north-central United States. *Crop Science* 39, 1622–1630.

Hedden, P. (2003) The genes of the Green Revolution. *Trends in Genetics* 19, 5–9.

Kang, M.S. and Priyadarshan, P.M. (2007) *Breeding Major Food Staples.* Blackwell, Oxford, UK.

First-generation transgenic crops

Gonsalves, D., Gonsalves, C., Ferreira, S., Pitz, K., Fitch, M., Manshardt, R. and Slightom, J. (2004) Transgenic virus resistant papaya: from hope to reality for controlling papaya ringspot virus in Hawaii. *APSnet* July 2004.

Kaniewski, W.K. and Thomas, P.E. (2004) The potato story. *AgBioForum* 7, 41–46.

Powell-Abel, P., Nelson, R.S., De, B., Hoffmann, N., Rogers, S.G., Fraley, R.T. and Beachy, R.N. (1986) Delay in disease development in transgenic plants that express the tobacco mosaic virus coat protein gene. *Science* 232, 738–743.

Powles, S.B. and Yu, Q. (2010) Evolution in action: plants resistant to herbicides. *Annual Review of Plant Biology* 61, 317–347.

Ülker, B., Li, Y., Rosso, M.G., Logemann, E., Somssich, I.E. and Weisshaar, B. (2008) T-DNA–mediated transfer of *Agrobacterium tumefaciens* chromosomal DNA into plants. *Nature Biotechnology* 26, 1015–1017.

10 Plant Biotechnologies in the 21st Century

10.1 Chapter Overview

In this chapter we will examine recent developments in plant biotechnology and their applications in basic research and crop improvement. We will begin by surveying underpinning technologies, such as improved analytical devices supported by developments in computing, robotics and informatics, before moving on to look at how new biotechnologies are being used in practice. Many of these technologies involve the creation of novel genetic variation or 'genome enhancement', which might involve either transgenic or non-transgenic methods. We will then examine some of the DNA-based selection technologies used in agriculture, many of which are substantially reducing the timescale and costs of breeding new crop varieties. New methods for culturing and manipulating plant cells and tissues *in vitro* are also assisting areas such as micropropagation and regeneration of adult plants free from somaclonal abnormalities. Finally, we will briefly survey the collection and maintenance of plant germplasm resources both for biodiversity conservation and for practical use in crop breeding.

10.2 Technology Categories

In addition to its core biological methods, modern biotechnology relies on a host of non-biological technologies. Overall, we can divide these technologies into hardware, software and wetware. The term hardware is often used to describe useful non-living systems or devices. Software includes the information systems used in the function and management of complex hardware and wetware. Wetware applies to biologically based living systems used and manipulated by people, and includes whole organisms, cultured cells/tissues and cell-free *in vitro* systems. In the context of this chapter, hardware provides our instruments and tools, software enables us to run complex devices and analyse vast quantities of data, and the wetware is the biological system itself, i.e. the plant or its biological derivative that we are seeking to exploit.

Hardware

Modern biotechnologies are built on a foundation of improved hardware that enables researchers and breeders to carry out tasks that would have been impossible just a few decades ago. Such hardware includes new instrumentation, such as ultra-rapid DNA sequencing, small and powerful analytical systems such as desktop GC-MS for metabolite profiling, *in silico* technologies such as DNA arrays for genome profiling, and robotic devices to automate many research and breeding tasks. It also includes computing and communication systems, such as hugely powerful personal computers and centralized servers, and the rapid and easy connectivity of such devices, e.g. via the Internet. Robotic systems allow the automated and around-the-clock performance of many repetitive functions involved in plant analysis, phenotyping and selection.

Sequencing crop genomes

Probably the most dramatic example of hardware improvement in the 21st century has been in DNA sequencing technologies, where the cost per base has decreased 100,000-fold since 2000, as described in Box 10.1 (see also Fig. 10.1). The first plant genome to be sequenced was the model species *Arabidopsis thaliana*, as published in 2001. The first crop genome to be sequenced was rice, where a high quality sequence was published in 2005. However, this was exceptional because rice has an unusually small genome – just 400 Mb in contrast to 2300 Mb for maize and 16,000 Mb for wheat. The sequencing of the maize genome required a massive effort by company and public labs and the results were published in a series of papers in 2009.

Box 10.1. Advances in DNA sequencing technologies

By 2011, more than ten plant genomes had been sequenced, including the crops rice, maize, potato and soybean. After 2005 there were considerable advances in what is often referred to as 'next-generation' DNA sequencing technology. The 'traditional' Sanger method was refined using high-throughput shotgun genomic sequencing methods developed by Ventor. This delivered genomic sequences at the remarkably low cost of $500/Mb. However, since most plant genomes are of the order of several Gb in size, it still cost $1–50 million per genome depending on its size and the extent of 'oversampling' (required to link overlapping DNA fragments).

The next important advance was so-called '454 pyrosequencing', based on emulsion PCR methods able to rapidly read short DNA strands of 250 bp and costing about $60–80/Mb. By 2010, even cheaper methods had been developed, such as Solexa and SOLiD technologies, with read lengths of only 36 bp and a cost of $2–6/Mb. Meanwhile, newer technologies such as Polonator and HeliScope promise even faster sequencing speeds costing less than $1/Mb. In addition to their speed and cheapness, these 'next-generation' technologies have much lower require-ments for robotics or bioinformatics infrastructure and will therefore be more affordable for smaller labs.

These new technologies will soon enable plant genomes to be sequenced for under $1000 each. This will make it feasible to sequence numerous individuals from the same crop species to discover key polymorphisms and mutations. It will facilitate 'metagenomic' sequencing, which is the discovery of cryptic infectious and commensal flora or fauna within a plant. It will greatly speed up the sequencing of the 'transcriptome', which is the portion of a genome that is expressed at a particular time. Transcriptome profiling can identify genes involved in key processes such as flowering, crop architecture, or pathogen resistance. The identification and profil-ing of small RNA sequences will be greatly facili-tated, as will the determination of methylation patterns in genomic DNA as part of research into epigenomics. In the future, the major challenge is likely to shift from simply generating and assembling vast amounts of raw DNA sequence data to extract-ing more meaningful information whether for basic research or crop improvement.

Other large-scale projects are under way for crops such as potato, sorghum, foxtail millet and oil palm and sequence data are being released at an increasingly rapid pace.

Advances in next-generation sequencing tech-nologies are now enabling the genomes of even comparatively minor crops to be characterized. In some cases a single method has been used but, more commonly, several sequencing technologies are used in combination for best results. For exam-ple, Roche 454 technology was used to sequence the 430 Mb genome of cocoa, *Theobroma cacao*, and the 1700 Mb genome of oil palm. In contrast, a combination of Sanger and Roche 454 sequenc-ing was used for the apple and grape (500 Mb) genomes. A combination of Illumina Solexa and Roche 454 sequencing was used for the genomes of polyploid cotton. Roche 454 sequencing has been used for *Miscanthus*, while Sanger, Illumina Solexa and Roche 454 sequencing are being used for banana. Illumina GAII sequencing has been used for the *Brassica rapa* genome, while Sanger and Illumina Solexa technologies were used for the cucumber genome.

These powerful combined approaches are now making it feasible to tackle very large cereal genomes such as barley (5500 Mb) and bread-wheat (17,000 Mb), where their massive size had previously ruled out full-genome sequencing. The cheapness and speed of genome sequencing are also making it possible to sequence not just single reference genomes, but many individual genomes in a population. This approach will be used to uncover genome-wide variations that underlie some of the more complex developmental and agronomic traits of interest to researchers and breeders.

'Omic technologies

Moving beyond gene composition to gene expres-sion and its ultimate manifestation as a phenotype in an organism, it is often necessary to analyse structural and functional molecules, such as pro-teins, membrane lipids and structural carbohy-drates in particular plant cells or tissues. At a more detailed level, there are many thousands of smaller metabolites whose composition differs greatly

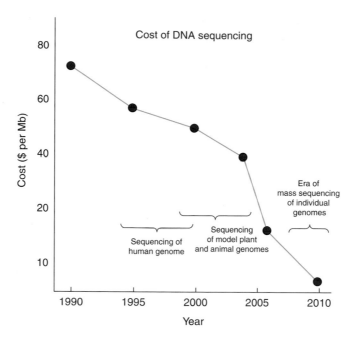

Fig. 10.1. The declining cost of DNA sequencing. There was a steady reduction in DNA sequencing cost during the first human genome sequencing project. After 2005, costs fell even more sharply as 'next-generation' sequencing technologies were increasingly used. In the 2010s, mass sequencing of the genomes of individual organisms or even cell types is becoming more routine. Adapted from Shendure and Ji (2008).

according to tissue, developmental stage and in response to environmental conditions. The ability to simultaneously analyse large numbers of often-complex molecules is the basis of the so-called 'omic technologies. Hence, transcriptomics is the analysis of transcribed genes in the form of mRNAs; proteomics is the analysis of protein composition; lipidomics is the analysis of lipid composition; metabolomics is the analysis of small metabolites, and so on. Several automated analytical techniques have been developed to separate and identify each of these classes of biomolecules.

The *metabolome* is the complete list of metabolites found in a particular organelle, cell or tissue under a specific set of conditions. The identification of important plant metabolites, such as sterols, flavones and acyl derivatives, used to be a very slow process relying on bulky, expensive equipment that could only be operated by a few skilled specialists. However, new lightweight devices, supplemented by robotic and informatics approaches, now make it possible to automate the process and even to assign accurate identities to complex mixtures of such molecules.

Metabolome analysis can help uncover how plants are reacting at the molecular level to specific stimuli. For example, by studying the metabolome in stressed and unstressed plants, we might gain important information about how some plants can tolerate certain stresses while others cannot. In other cases, metabolome analysis can give useful information about molecular changes caused by the addition of transgenes to plants. This kind of analysis is used as part of the process of regulation of transgenic crops where it may be necessary to test whether a transgenic variety is 'substantially equivalent' to non-transformed varieties of the same crop.

The *proteome* is defined as the expressed protein complement of an organism, tissue, cell or subcellular region (such as an organelle) at a specified stage of development and/or under a particular set of environmental conditions. Perhaps the most important molecules in cells are the proteins, some of which are structural while many others act as enzymes responsible for the biosynthesis of most of the other molecules in a cell. Proteins are the direct products of gene expression and the timing and spatial distribution of their accumulation and function results in the phenotype of a particular organism. However, patterns of gene expression as measured by transcription, i.e. the formation of mRNA, are not always reflected by patterns of accumulation or activity of the corresponding proteins.

In some cases, the mRNA may not be efficiently translated to protein. In other cases the protein might be synthesized but then either is broken down or remains inactive. A protein might be present in a cell but is inactive due to incomplete post-translational processing, e.g. failure to bind a ligand, or due to inhibition, e.g. by phosphorylation. Ideally, the information in the proteome should include any post-translational processing undergone by each protein analysed. First-generation proteomics was mainly concerned with identifying the gross protein composition of samples, but new-generation technologies are

beginning to focus on questions such as post-translational processing and the biological activities of such proteins.

It is only by addressing these latter questions that we can verify not only that a particular protein has been synthesized and is in the right location, but also that it has the appropriate biological function. Therefore, we can learn a lot about the actual function of a genome in a specific cell or tissue by examining its proteome. Like the metabolome, the proteome in a plant sample can vary greatly according to genotype, tissue location, developmental stage and environmental conditions. The full proteome will comprise thousands of proteins, some of which may be present in abundance while others are at very low levels. The analysis of low abundance proteins poses considerable difficulties for proteomics that have yet to be resolved. Some of the major proteomic technologies are described in Box 10.2.

Box 10.2. Proteomic technologies

The analysis of complex mixtures of proteins has been made possible by more accurate and reproducible methods of polypeptide separation, fragmentation and identification, plus associated software and automation developments. Hundreds of individual polypeptides can be separated at great resolution by highly reproducible two-dimensional gel electrophoresis. This can be automated so that many samples can be run and analysed simultaneously. Each spot can then be excised from the gel for further analysis, such as its molecular mass, amino acid composition and post-translational modifications (e.g. phosphorylation). Excised spots can be split into smaller fragments by proteases such as trypsin and the fragments separated by further electrophoresis or chromatography and identified by mass-spectrometry.

A common method of peptide identification is mass fingerprinting. Here, the peptide fragments are separated according to size using mass-spectrometry with an electron spray ionization or matrix-assisted laser desorption ionization source coupled with a time-of-flight analyser (so-called MALDI-TOF). This enables the mass of each fragment to be determined with great accuracy. The masses of the fragments are then used to search

databases to find amino acid sequences that could have generated the observed patterns. These steps can be automated so that the proteome of a specific sample can be analysed relatively quickly by technicians.

One of the limitations of two-dimensional electrophoresis, and other separation methods such as multidimensional liquid chromatography, is that they can only resolve a maximum of several thousand of the more abundant proteins in a particular proteome. This is a problem because a complete proteome might include more than 10,000 proteins. Moreover, some of the key proteins that often have major phenotypic effects, such as transcription factors, tend to be present in tiny quantities – possibly as low as a few dozen molecules.

Another form of proteomic technology is the protein microarray. This is similar in principle to nucleic acid microarrays but the probes are proteins or molecules that bind proteins. For example, an array of many different proteins can be used to screen various mixtures for ligands, enzyme substrates, or interacting proteins. Alternatively, an array might be made up of many different antibodies that can be used to screen a protein mixture for candidates that bind to the various antibodies.

Software

Many devices used in biotechnology labs rely on software to carry out their functions. Software is also crucial for the data analysis and manipulation systems that enable researchers and breeders to deal with the huge volumes of raw data generated by the new hardware systems described above. Bioinformatics is a relatively new discipline that enables biologists to deal with the avalanche of data from genome sequencing and profiling programmes, and from other 'omic technologies. The sheer volume of data generated by the new generation of hardware often makes it virtually impossible to analyse raw results manually. For example, a next-generation DNA sequencer can generate thousands of sequence fragments making up millions of base pair readouts per day. These fragments need to be analysed for overlaps and then assembled into 'contigs', or continuous sequences of many fragments that will eventually be collated to make up an entire chromosome. This process is now done automatically using algorithms, or repetitive step-by-step mathematical procedures.

Other algorithms are used in genome annotation. This involves the identification of putative genes, including their promoters, regulatory elements, introns, exons and mRNA/protein products. Other software can detect possible small RNA-encoding regions and specific repetitive elements in DNA sequences. Software is also used to drive the robotic and other automated systems used in tasks such as mass-profiling of large populations. Advances that enable non-specialists to use sophisticated software have been facilitated by improved computing technology and more powerful linked networks. This has been especially crucial in enabling massive amounts of data, often measured in many terabytes (10^{12} bytes), to be generated by some of the new technologies. For example, a single 2 h run on an Illumina GAII DNA sequencer can generate 10 terabytes of data.

One potential problem here is that the vast amounts of raw data generated by DNA sequencers are beyond the ability of many labs, or even companies, to archive. Therefore, the raw data are normally processed by proprietorial software developed by instrument manufacturers and only the much-reduced processed data are saved. Even with the most advanced computing technology, the costs of storing the original raw data are greater than the cost of repeating the entire sequencing run. Another challenge for future software development is to improve the assembly of processed sequence data for the increasingly diverse applications required by researchers. To address this, new forms of open-source bioinformatics software, such as SOLiD, are being developed where members of the community can adapt and improve software tools to fit their own applications.

Wetware

Derived from hardware and software, the term wetware describes living systems used for a specific purpose. In some cases, the use of wetware is the only way of making products, such as most foods and many organic materials such as cotton, wood or linen. The term encompasses the use and manipulation of biological systems, whether free-living or *in vitro*. For plants, it includes the systematic exploitation of crops via scientific agriculture as well as emerging biotechnologies such as bio-pharming. Wetware development is achieved via the core biotechnologies of tissue culture, recombinant DNA manipulation, plant culture and other aspects of crop breeding.

As we have seen above, advances in wetware development (i.e. improved crops and/or a better understanding of plant biology) are underpinned by developments in hardware and software and the three forms of technology are therefore closely linked. The remainder of this chapter will focus on how modern biotechnologies are enabling us to improve our exploitation of plant wetware. Although *in vitro* or high-containment plant production systems are increasingly being developed for the biosynthesis of high-value products, these are largely in the research phase. Therefore the overwhelming majority of plant wetware is still in the form of mass-produced crops grown in conventional agricultural settings.

10.3 Research and Biotechnology

Biotechnologies are sets of methods developed from basic research discoveries that are then applied for practical purposes. To begin with, newly discovered techniques of recombinant DNA manipulation in the early 1970s were mostly used to address basic research questions. Similarly, the revolution in genomics and other 'omic technologies that occurred in the 2000s has so far been mainly applied for research purposes. As we saw above, one of the limitations of the genomics revolution has been the generation of massive amounts of DNA sequence data for dozens of plant species, without a corresponding increase in our knowledge of gene function, regulation, or expression as a phenotype.

For this reason, there is now a greater focus on going beyond the genome to characterize and manipulate those aspects of the phenotype of plants that make them useful or interesting to people. These so-called 'post-genomic' technologies are highly interconnected, often incorporating genetics, biochemistry, DNA analysis, physiology, ecology and derived methods such as the various 'omics. In this section, we will look at some examples of research approaches that are assisting practical crop manipulation as well as addressing fundamental biological questions.

Genomics and phenomics

In 2010, almost a decade after the first plant genome had been sequenced, as many as 15–40% of the genes in each genome still remained to be identified. In model species such as *Arabidopsis*, many hundreds of these 'unknown' genes have been knocked out in various ways in an attempt to discover their function. However, a surprisingly high proportion of these gene knockouts have been reported to result in 'no visible phenotype'. These failures to identify sequenced genes and/or their corresponding phenotypes have led to efforts to link basic genomics much more strongly with gene expression, particularly in terms of observable phenotypes.

This approach is termed 'phenomics'. In many cases phenotypes might only be present under certain circumstances, such as drought stress or infection by specific pathogens. Also, only a small proportion of phenotypes are readily visible to the human observer. Most of them involve more subtle changes in plant composition that can only be detected by large-scale screening methods using the tools of informatics, robotics and metabolite analysis, in additional to genetic methods such as QTL analysis and association genetics. The 'phenome' of a plant can be defined as the expression pattern of the genome and its manifestation as phenotypic traits within a particular environment. The phenome is therefore similar to the traditional concept of gene × environment (G×E) interactions in plant breeding, although it involves a much broader and more detailed description of such G×E interactions.

Plant phenomics is a relatively new approach aimed at studying plant development, composition and performance using the latest high resolution, high throughput 'omic technologies throughout the life cycle of a plant. The phenome can then be related to the genome to allow genetic information to be better exploited for basic knowledge of plant biology and for practical applications in agriculture. The development of phenomics has benefited from technical advances ranging from new algorithms to novel non-invasive imaging methods. For example, new standardized data sets and retrieval systems have been created to facilitate the storage and accessing of a huge amount of phenotypic data. New forms of non-invasive imaging technology can be used for high throughput screening of plant populations to extract useful phenotypic data. For example, pulse modulated chlorophyll fluorescence imaging has been used to examine photosynthetic responses to drought stress in intact *Arabidopsis* plants. Some of these non-invasive imaging technologies can also be used in the field and will eventually allow the real-time phenotyping of crop plants growing under agricultural conditions, rather than in experimental plots or glasshouses.

QTL analysis

Quantitative trait loci (QTL) are chromosome regions containing genes that regulate complex traits (see Chapter 8). In several cases, genetically complex traits of agricultural interest are mostly regulated by one or a few major QTL. For example, it was known that the grain-shattering trait in rice is regulated by numerous genes and several potentially important QTL were found. However, more detailed analysis showed that just one of these QTL is responsible for 69% of the genetic variation in grain shattering. This locus coincides with the *sh4* gene, which encodes a transcription factor regulating the expression of several other genes involved in grain shattering. By isolating the *sh4* gene, researchers were able to learn a great deal about the process of grain shattering in rice as well as possible new ways to manipulate this key trait in rice and other crops. QTL analysis has now been applied to many other crop species and their wild relatives.

In order to carry out QTL analysis, two parents with widely different genotypes are crossed to create a segregating population. The parent plants might be different varieties of the same species, or a crop plant crossed with a wild relative. This method is known as biparental crossing. Using such populations, a series of genetic markers can be assembled at intervals along each chromosome. Major QTL involved in traits of interest can then be pinpointed with respect to these markers. The initial resolution of such mapping is relatively low as it can only localize QTL to chromosome regions of 10–30 cM, which might correspond to several

hundred genes (see Fig. 10.2). In order to localize a gene of interest, it is necessary to perform finer resolution genetic analysis, e.g. by crossing near-isogenic lines that only vary in the QTL region. This might narrow the region to just a few genes that can then be tested for their effects on the original trait by knockouts or overexpression studies. Eventually, a single gene might be characterized that could enable a relatively complex trait to be manipulated in order to improve crop performance.

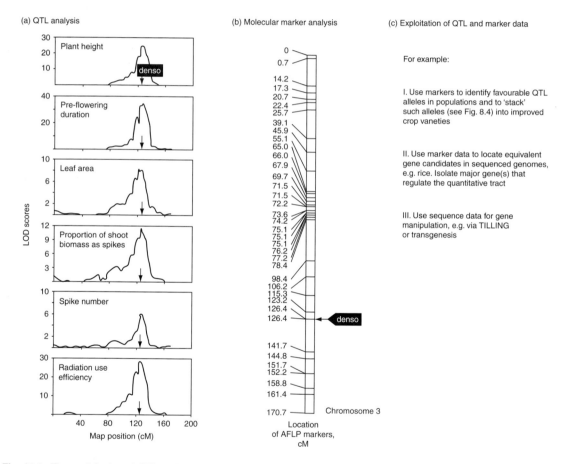

Fig. 10.2. The exploitation of QTL and molecular marker data. (a) The location of a quantitative trait as a map position on a linkage group, or chromosome, can be plotted against its LOD (least odds of divergence) score, which is a measure of the likelihood that a particular QTL phenotype is linked to physical markers mapped as in panel (b). The LOD score is an indication that a given marker may be associated with a QTL. In this example from barley, six different complex traits had LOD scores with peaks around 126.4 cM on chromosome 3. (b) Molecular marker map of barley chromosome 3. The marker at 126.4 cM corresponds to the *denso* gene, which is therefore a strong candidate as a major regulatory locus for the six traits analysed in (a). *Denso* is a major dwarfing gene that has many additional pleiotropic effects and may therefore be a good candidate for further analysis and manipulation. (c) QTL and molecular marker data can also be used to locate and isolate unknown genes that regulate important traits. Data from Yin *et al.* (2003).

Association genetics

Association genetics, or association mapping, was initially developed as a research tool for mapping and characterizing genes of interest, especially those regulating complex traits. It is therefore related to methods such as QTL mapping. Previously, genetic mapping in plants was normally done by selecting two dissimilar parents to create a biparental segregating population. In contrast, association genetics uses collections of individuals from diverse sources, such as wild populations, germplasm collections, or specific breeding lines. It uses new methods to accelerate crop genetic profiling, including large-scale single nucleotide polymorphisms (SNP) discovery and high throughput sequencing, and the availability of increasing availability of plant genomic sequences.

Association genetics involves searching for statistically significant associations between changes in a DNA sequence and changes in the phenotype of a trait in a large panel of unrelated genetic lines of a species. So far it has been used as a research tool to study the genetic basis of complex traits in human and animal systems, and more recently in plants. It was initially focused on single gene traits in plants, but is increasingly used to analyse quantitative traits. Unlike traditional linkage mapping studies of populations created by crossing two parents, association genetics can explore all the recombination events and mutations in large and diverse populations. It can also achieve higher mapping resolutions, which facilitates identification of major genes regulating complex traits.

More recently, association genetics has been used for more practical applications such as commercial crop breeding. So far, it has mostly been applied to maize for the analysis of such traits as starch composition, anthocyanin biosynthesis, oleic acid content and carotenoid content. It has been used to study flowering time in pearl millet and barley. In the future, as genome characterization becomes more detailed and informatics tools become more sophisticated, association genetics will be applied to more agronomic traits and to more crop species. The technique is especially useful for characterizing desirable alleles that are moderately abundant in populations. For the identification of rare alleles, such as some disease resistance genes or genes introgressed from exotic germplasm (such as wild relatives of a crop), conventional segregating populations formed from biparental crossing will still be required for genetic mapping.

Apomixis

Apomixis technology is still at the research phase but it has huge potential for application in crops and is therefore important to understand. The mechanism of apomixis, or asexual propagation via seeds, was described in Chapter 5. At present, apomixis is confined to a restricted range of plants and no major crops are apomictic. However, there are numerous possible applications of apomixis if the trait could be transferred to a wider range of useful species, including crops and ornamental plants. Some practical examples include the following:

- Huge numbers of hybrid cultivars could be created from newly apomictic crops, enabling breeders to exploit hybrid vigour in a much wider range of species.
- Heterozygous genotypes, including the results of wide crosses, could be maintained: this would speed up breeding and enable it to be more responsive to micro-environmental conditions, including specific pest or pathogen problems.
- By replacing vegetative with apomictic propagation in crops such as cassava, potato and yams, the incidence of propagule-borne disease (especially viral diseases) would be dramatically reduced.
- Farmers could propagate uniform, true-breeding plants from elite apomictic hybrids, enabling them to use their own saved seed rather than buying new seed each year.

The final example is of particular interest to breeders of major crops where hybrids have enormous potential for yield improvement via heterosis. Apomixis would save commercial breeders some of the huge costs involved in recreating hybrid crosses for each generation. However, apomixis also would enable farmers to save seeds from hybrids in the knowledge that they would breed true and maintain their advantageous heterotic traits. The fact that farmers could save their own apomictic seed is a disincentive for commercial biotechnology companies to develop apomixis, as they would lose control over seed varieties in which they might have invested large amounts of R&D. In contrast, companies marketing transgenic seeds own the rights to these seeds and can prevent farmers from saving and replanting their own seed. As discussed in Chapter 11, this relatively strong form of technology ownership has been implicated in the slow spread of transgenic technologies to non-commercial crops in developing countries and for the lack of public R&D in such technologies compared with the private sector.

In the early days of apomixis development, many scientists hoped that access to the technology would be available to all, especially in developing countries, and not subject to restrictive patenting arrangements. This aspiration was announced by leading researchers in 1998 at the Bellagio Apomixis Declaration. One way to ensure that both public and private sectors can continue to develop, and eventually profit from, apomixis technology is for companies to market apomictic varieties that are also engineered with genetic use restriction technologies (GURTs) for sale in large-scale commercial agricultural systems, such as in much of the Americas and Europe. Other apomictic varieties of developing country crops could be developed in the public sector without GURTs. Developing country farmers could then reuse seed from such non-GURT apomictic varieties without restriction.

10.4   Genome Enhancement I: Non-transgenic Technologies

As discussed in Chapter 9, several 20th-century biotechnologies were developed to enhance genetic variation in crops. In many cases this was necessary because elite varieties of crops such as maize had become dangerously inbred and potentially susceptible to pathogens. The gene pool of such elite cultivars needs to be regularly 'refreshed' by genetic crosses with other varieties, followed by repeated backcrossing and selection, to create an enhanced version of the elite cultivar. In other cases, it has been useful to bring in new genes from related species using assisted crosses followed by backcrossing. Non-directed (i.e. random) variation can also be created by various forms of induced mutagenesis, including chemical and radiation exposure and somaclonal effects originating from tissue culture.

In all of these examples of non-transgenic technologies, many forms of useful genetic variation have been and still are being created by manipulating the existing genomes of the crops and their near relatives. In the 21st century, these technologies remain the main source of genome enhancement for most crops and are likely to remain so for the majority of major yield- and quality-related agronomic traits for the immediate future. As we will see in the next section, transgenesis is likely to become increasingly useful in crop breeding as more trait-regulating genes are identified in the coming years, but it will tend to complement rather than replace non-transgenic technologies. Meanwhile, many new methods are being developed to enhance genetic variation without transgenesis. Two of the most powerful of these technologies are new forms of wide hybrid production and TILLING/mutagenesis.

Wide hybrids

As we saw in Chapter 9 and elsewhere, plants form inter- and intraspecific hybrids much more easily than do animals. Hybrid crop varieties were behind many of the most significant achievements of 20th-century agriculture, including the Green Revolution and inbred-hybrid maize. We have also seen that many of the most widely used and highest yielding crop varieties have relatively restricted levels of genetic variation. This means that the production of hybrids will remain one of the most important methods to enhance genetic variation and thereby create new genomic combinations that will serve as the raw material for crop improvement. The creation of wide hybrids between crops and other species, including wild plants, has always been problematic for breeders. However, this process is now benefiting from new tissue culture and selection technologies as described in the following two examples.

NERICA – a new hybrid rice for Africa

NERICA (New Rice for Africa) is a series of new interspecific hybrids produced by crossing two different rice species, namely African rice *Oryza glaberrima* and Asian rice *O. sativa*. African rice is a partially domesticated crop with relatively low yields that is prone to seed-shattering and lodging. It produces relatively heavy grains that are less palatable than those of Asian rice. However, this traditional African crop is well adapted to the local climate and to the pests and diseases of tropical West Africa where it is mostly grown. In contrast, while Asian rice is potentially much higher yielding and of better quality, it is much more susceptible to the biotic and abiotic stresses found in West Africa. The solution was to create a new hybrid between the two rice species that had the agronomic advantages of Asian rice plus the stress tolerance of the more robust African rice.

Because these two species do not naturally interbreed, it was necessary to use a range of advanced tissue culture technologies to enable the hybrid plants to survive. In particular, embryo rescue and anther culture methods ensured that crosses survived to produce plantlets able to grow on to full maturity. As with many other hybrids of two relatively inbred lines, NERICA varieties display very good heterosis. For example, they grow faster, yield more and/or resist stresses better than either of their parents. Other useful features of NERICA varieties include: increased grain-head size from 75 to 100 grains to 400 grains per head; yield gains from 1 t/ha to 2.5 t/ha and up to 6–7 t/ha with fertilizer application; 2% more protein than their African or Asian parents; better pest and weed resistance; and more tolerance of drought and infertile soils than Asian rice.

In total, about 3000 NERICA lines were developed, many of which were released, and by 2009 they were being grown by millions of farmers in 30 African countries. Their unique adaptations to growing conditions in West Africa have helped increase yield and have led to record rice harvests in the region. Many NERICA varieties are particularly suitable for use in rain-fed upland areas where smallholders lack the means to irrigate or to apply chemical fertilizers or pesticides. The NERICA example shows how breeders can use modern biotechnologies to create new hybrid plant species capable of improving the lives of many millions of people.

New hybrid sorghum varieties in Africa

Sorghum is one of the most important crops in Africa. Two of the main challenges it faces are periodic drought and competition from the often-devastating plant parasite Striga or witchweed (see Chapter 6). Striga affects 40% of the arable savannah land in Africa, while recurring drought and resultant crop failure still lead to hunger and even famine in regions of Eastern Africa. There was evidence that Striga resistance might be a simpler genetic trait than drought tolerance so Sudan-based breeders initially focused on Striga. They used a broad-based approach involving molecular genetics, biochemistry and agronomy to screen a wide range of sorghum accessions in order to identify genes for Striga resistance. Genetic crosses were

then used to introgress these genes into locally adapted traditional varieties and into more modern commercial sorghum varieties. This meant that many new hybrid sorghum lines were produced that were broadly adapted to a wide range of African farming systems.

In parallel with these breeding approaches, the researchers developed an integrated Striga management system. This management system has further increased sorghum productivity through improved weed resistance, soil-fertility enhancement and water conservation. By 2002–03, the new hybrids were being grown across Eastern Africa, from Sudan to Zimbabwe. Thanks to heterosis, some of the hybrids produced 50–100% higher yields than traditional varieties. This helped to lay the foundations of a commercial seed industry in Sudan. More recently, newer drought-tolerant hybrid varieties in Niger have yielded 4–5 times the national average. Some of the new sorghum hybrids have been so successful that, in a reversal of the normal route for technology transfer, African breeders used germplasm produced in Niger and Sudan to develop elite inbred lines of sorghum to generate commercial sorghum hybrids for the US and international markets. Once again, the imaginative use of a mixture of biotechnologies enabled breeders to improve a staple developing-country crop in a way that benefited both small farmers and the wider agribusiness community.

TILLING: an updated version of mutagenesis

We have already seen that during the 20th century induced mutagenesis was of great value to breeders and resulted in the creation of thousands of new crop varieties. However, induced mutagenesis is a relatively crude, slow and labour-intensive way of creating new genetic variation. Mutagenesis breeding in the 21st century is taking advantage of new biotechnologies, such as TILLING, which is non-transgenic, and zinc-finger nucleases (see below), which involves transgenesis. Targeting Induced Local Lesions IN Genomes, or TILLING, is a relatively new molecular biological method of analysing mutagenized populations. An outline of the method is shown in Fig. 10.3. As well as screening mutagenized populations, TILLING can be used to screen variation in natural populations in what has been termed EcoTILLING. By using TILLING, large pools of genetic variation can be produced in a short time for introduction into breeding programmes.

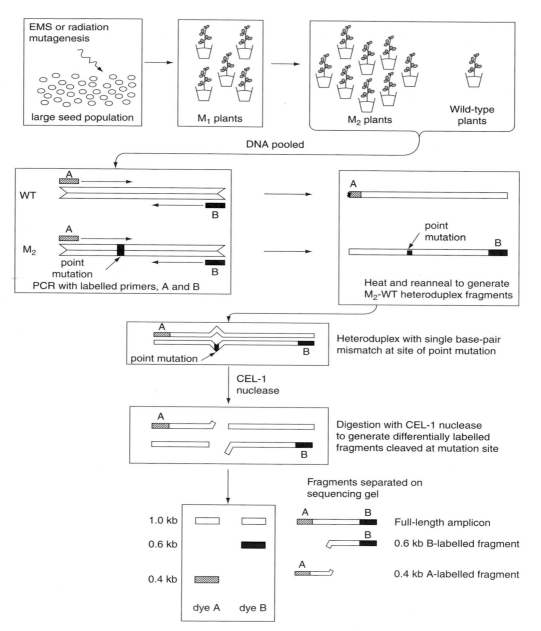

Fig. 10.3. Targeting Induced Local Lesions IN Genomes (TILLING). In TILLING, mutagenic agents, such as EMS or radiation, are used to create a large population of mutagenized plants (M_1). Next, the second (M_2) generation of these mutants is screened by a semi-automated, high-throughput, DNA-based method in order to detect mutations in genes of interest. Screening involves polymerase chain reaction (PCR) amplification of gene fragments of interest, plus rapid identification of mutation-induced lesions by looking for mismatches in heteroduplexes with non-mutagenized DNA sequences obtained from wild-type plants. The third step is to evaluate the phenotypes of a limited number of selected mutant plants. TILLING is also amenable to automation, including high-throughput robotic screening systems, making it especially suitable for the large polyploid genomes found in several major crops. Adapted from Slater *et al.* (2008).

The potential of the technology was demonstrated in 2005 when TILLING was used to identify variants in the *Waxy* gene, which plays an important role in determining flour and bread quality. In this study, the authors used TILLING to identify a series of variants of the granule-bound starch synthase I (GBSSI) gene in hexaploid and tetraploid wheat. GBSSI or *Waxy* plays a critical role in the synthesis of amylose, which, together with amylopectin, makes up the starch fraction of the seed. Reduction or loss of GBSSI function results in starch with reduced or absent amylose. This is a desirable trait because it leads to improved freeze-thaw stability and resistance to staling compared with conventional starch. Therefore the new variants identified by TILLING can be used for breeding of wheat varieties with improved starch contents.

TILLING is now being used in soybean and other major crops. One of the drawbacks of TILLING and many other high-tech approaches to crop improvement is their high cost and technical complexity. However, as with other technologies, TILLING will eventually become cheaper and more accessible, especially to developing countries. In the future, TILLING will both speed up and broaden the scope of crop improvement programmes. TILLING is a non-transgenic form of molecular mutation breeding. In Fig. 10.4, it is contrasted with transgenic methods that are also increasingly being used in combination with DNA marker and automated selection technologies to create useful mutagenized crop varieties.

10.5 Genome Enhancement II: Transgenic Technologies

Transgenic technologies give breeders additional tools to manipulate genomes using recombinant DNA methods that are continually being improved and refined. Breeders never employ transgenesis on its own; instead they use it in combination with other technologies such as tissue culture/regeneration, hybrid creation, mutagenesis, backcrossing and marker-assisted selection. This means that it can be misleading to speak of a new crop variety as 'transgenic' or 'GM' as if it had only been created using transgenic technologies. In 2010, almost 150 Mha comprising about 10% of the global arable land area was reported as being planted with transgenic/GM crops (see Fig. 10.5). However, each of these crops has also benefited from one or more of the non-transgenic technologies listed above. For example, well over three-quarters of all crops

grown, including most transgenic varieties, have resulted from some form of hybridization and backcrossing.

Despite the fact that transgenesis is simply one of several alternative strategies for variation enhancement in breeding programmes, the resultant plants are treated very differently from almost identical non-transgenic varieties. As we will explore in the next chapter, transgenic varieties have a different legal status and are subject to much more complex regulatory systems in various regions of the world. They are also perceived very differently by the general public and are even banned or heavily restricted in some countries. For this reason, we need to look at the development of transgene technology in a different way from other technologies. As we will see below, some developments such as so-called 'clean gene' technologies are aimed more at satisfying generalized public concerns rather than addressing proven safety issues or wider aspects of crop improvement *per se*.

With these points in mind, there are several ways in which transgene technology can be improved to make it technically easier, more efficient, wider in its scope, and better able to address concerns expressed by certain sections of the public. Some technical issues and areas of public concern are listed below:

- In the future, it will be desirable to generate transgenic crops that do not contain selection markers, such as genes for antibiotic or herbicide tolerance.
- First-generation transgenic plants were created using random insertion of transgenes, which can lead to variations in transgene behaviour and other unpredictable pleiotropic effects.
- In order to achieve stable and predictable transgene expression under a variety of field conditions, transgene introduction technologies need improvement.
- The spread of transgenes into wild populations via cross-pollination can be prevented using genetic use restriction technologies (GURTs).
- Biocontainment strategies should be incorporated into biopharmed plants to prevent risk of contamination of human or animal food chains.

In addition to tackling these and other areas of public concern, it will be important to increase the number of traits that transgene technologies can be used to manipulate. As shown in Table 10.1, almost two decades after transgenesis was introduced

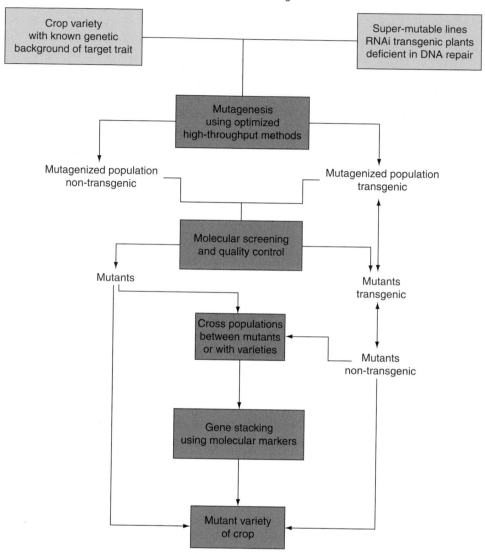

Fig. 10.4. Molecular mutation breeding. The first stage in mutation breeding, the induction of mutations, can be slow and rate-limiting, but knowledge of DNA damage/repair systems will enable specific mutagenic strategies to be devised, such as combinatorial chemical/radiation induction, that greatly increase mutation frequency. Super-mutable lines might include plants with enhanced mutagenic frequencies due to knockouts of DNA-repair genes. Having created useful mutations from super-mutable lines, the mutated genes could be separated from transgenes by self-crossing to generate non-transgenic mutagenized progeny. DNA-marker assisted backcrossing with established varieties can then place the mutagenized gene(s) into an elite cultivar for further propagation. Adapted from Shu (2009).

into commercial agriculture, over 99% of GM crops carry either or both of just two groups of transgenic traits, namely herbicide tolerance and/or Bt insect resistance. It is likely that during the 2010s, dozens of new output and stress tolerance traits will be released. Newer technologies such as RNAi, trait stacking, chromosome engineering and pathway engineering will play important roles in expanding the scope of these second-generation transgenic crops.

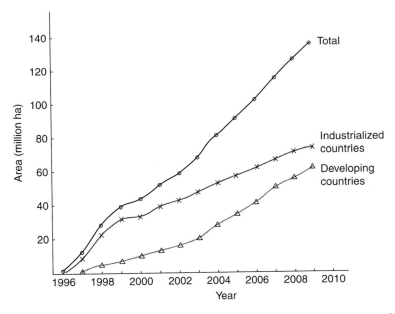

Fig. 10.5. Uptake of first-generation transgenic crop varieties, 1996–2010. Cultivation of transgenic crop varieties increased steadily to reach more than 10% of global arable land by 2010. However, >99% of transgenic crops were from the four species soybean, maize, cotton or rapeseed and >99% expressed either of two traits: herbicide tolerance or insect resistance. Data from James (2011).

Table 10.1. Major global transgenic crops and traits in 2010.

	Area (Mha)	Percentage
Crop		
Soybean	73.3	50
Maize	46.8	31
Cotton	21.0	14
Rapeseed/canola	7.0	5
Sugarbeet	0.5	<1
Lucerne	0.1	<1
Others	<0.1	<1
Total (approx.)	148	100
Trait		
Herbicide tolerance	89.3	60
Insect resistance (Bt)	26.3	18
Herbicide tolerance + Insect resistance	32.3	22
Others	<0.1	<1
Total (approx.)	148	100

Data from James (2011).

In many cases, improved transgene technologies will continue to complement rather than replace alternative non-transgenic approaches. But there will also be cases where the genes responsible for a desirable trait are neither present in the crop in question, nor in any plant relative that could serve as a gene donor via wide crossing. In such cases, transgenesis, where copies of the required gene(s) are brought in from an exotic location, is the only option. An example of this is the modification of oil crops to produce nutritionally desirable very-long-chain ω-3 polyunsaturated fatty acids (VLCPUFAs). Because higher plants do not contain genes necessary for VLCPUFA production, the only way to incorporate this trait into crops is to add a series of genes from sources such as mosses, fungi, algae and fish. In the remainder of this section, we will survey some of the improved transgenic tools that are being developed or are already in use for research and crop improvement.

Removal of unnecessary DNA

Transgenes are introduced into plants as parts of larger DNA regions that contain additional genes or regulatory regions that are required for the transformation process, but are not necessary for expression of the transgene itself. In most first-generation transgenic crops, the additional DNA included selectable markers such as antibiotic resistance or herbicide tolerance genes. As discussed in Chapter 11, public concern about antibiotic resistance genes has led to attempts to avoid the use of such markers or to remove them from transgenic plants. Several so-called 'clean gene' technologies have been developed to achieve this aim. An ideal approach might be to insert the transgene without a selectable marker and use DNA markers to screen putative transformants for presence of the transgene. With the help of robotized screening methods, this approach might become practical, but it will always be difficult to select out chimeric plants in which some cell lineages are transformed but others are not.

A more practical method would be to add the gene of interest and the selectable marker on separate T-DNA molecules so the two T-DNA molecules become inserted into different genomic locations. Some of the progeny of the transgenic plants will inherit only the gene of interest, which can be identified by using DNA markers. These progeny, which contain no DNA from the selectable marker, can be kept and the remainder discarded. An alternative approach is to excise the selectable marker from the genome using specific DNA-cleaving enzymes. This can be done by engineering flanking regions around the marker that are recognized and cleaved by site-specific recombinase systems. These enzymes will cut the DNA at each flanking region, remove the unwanted marker DNA, and rejoin or ligate the cut ends to restore the segment of genomic DNA. Several recombinase systems are available including phage P1 Cre-*lox*, yeast FLP-*frt* and the Meganuclease system. Such systems may become standard in the future, as shown in 2006 when the agbiotech company BASF Plant Science announced the licensing from Cellectis of its Meganuclease Recombination technology for removal of marker genes from transgenic crops.

Gene silencing

There are several mechanisms of transgene-induced gene silencing in plants (see Fig. 10.6). Gene silencing may be desirable if we wish to knock out an unwanted gene. However, it can also be an unwelcome but avoidable side effect of the transformation process itself (see Box 10.3).

Avoiding transgene silencing

Unwanted transgene silencing can be avoided by improving the design of inserted sequences and

their function in the transgenic plant. A major cause of gene silencing is the presence of repeated DNA regions. Therefore T-DNAs should not contain inverted repeat structures and only a single intact copy of the transgene should be present. In many cases, a high proportion of primary transformants can contain multiple transgene copies and these plants must be detected and eliminated from further selection. In general, any superfluous sequences, e.g. from the plasmid, or any fragments of transgenic DNA in the genome should be avoided. The transgene should have a similar base composition to the host plant and its locus should be protected from position effects. Finally, very high rates of transgene transcription can be counterproductive.

Despite these measures, both biolistic and *Agrobacterium*-mediated transgene insertion can result in partial fragmentation of the inserted DNA and integration of the resulting fragments into multiple genomic locations. Also, significant deletions of genomic DNA (between 25 and 1980 bp) have been reported at *Agrobacterium* T-DNA insertion sites in plants such as aspen and *Arabidopsis*. Sometimes superfluous DNA is inserted, such as extra whole or partial copies of the transgene, or vector backbone DNA. Similar findings have been reported for transgenic plants generated by biolistics. For example, in one of the commercialized lines of RoundupReady® herbicide-tolerant soybeans, at least two additional fragments of the EPSP synthase transgene were present, plus several other misplaced DNA sequences. These results show that transgene fragments can be formed when current transformation methods are used, with the attendant risks of unwanted side effects, including gene silencing.

Additional procedures such as tissue culture and regeneration can also result in somaclonal mutations that must be screened out of the resulting transformants. One of the problems in seeking to eliminate unwanted transgenic DNA from a genome is the difficulty in identifying smaller fragments without going through extensive time-consuming PCR-based screening procedures. Any potential risk is balanced to some extent by the decreased likelihood that very small fragments will have serious phenotypic consequences compared with more easily screenable gross artefacts. Nevertheless, these examples demonstrate that first-generation transgenesis is far from being a precise technology. Apart from possible

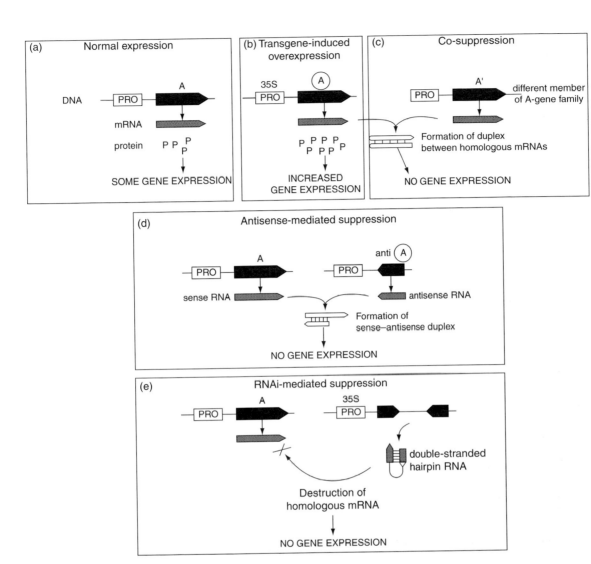

Fig. 10.6. Transgene-induced gene overexpression or gene-silencing strategies. Comparison of normal gene expression (a) with transgene-induced overexpression (b), co-suppression (c), or antisense-mediated suppression of gene expression (d). (e) RNAi, or RNA interference, is a more effective form of gene silencing that is an updated version of antisense technology. Antisense RNA, which is a linear molecule often several dozen bases in length, is far less effective than a smaller double-stranded hairpin RNA. This RNA is created by engineering a DNA sequence made up of a short inverted repeat that is complementary to part of the target gene sequence. Expression of this DNA in a transgenic plant is driven by a strong promoter, such as 35S for dicots or ubiquitin-1 for monocots. When the DNA is transcribed, the RNA forms a double-stranded hairpin that is extremely effective in silencing any complementary gene(s) in the transgenic plant.

pleiotropic effects that may take several generations to identify and screen out, the fragmentation of transgenic DNA can potentially affect regulatory procedures and the wider public perception of the technology, as discussed in Chapter 11.

Deliberate silencing/down-regulation of unwanted endogenous genes

While unwanted gene silencing can sometimes be a problem in transgenesis, there are also many cases where breeders actually want to knock out a particular gene in order to create a favourable phenotype. As we have seen above, in the many cases where the identity of the target gene is unknown, mutagenesis/TILLING is an effective method to achieve knockouts. However, as we understand more about plant molecular biology, there is an increasing list of traits where the identity of the target gene is known to a high degree of probability. In such cases, transgene-induced gene mutation can be used. Two major methods are RNAi and zinc-finger mutagenesis.

RNA interference (RNAi) is a powerful tool to suppress the expression of unwanted genes. RNAi can be triggered by generating transgenic lines expressing RNAs capable of forming a double-stranded hairpin (see Fig. 10.6e). Compared with previous antisense approaches, RNAi is usually much more effective at reducing levels of target gene transcripts and has other advantages as follows. RNAi-encoding transgenes are inherited in a much more stable way, and the degree of down-regulation of the target gene can be modulated by varying the strength of the transgene promoter. Also, unlike other technologies, RNAi can knock out all members of a multigene family. Finally, the DNA constructs are relatively easy to make and can be used in the latest generation of transgene vectors.

Practical RNAi technology only dates from about 2003, but it has already been demonstrated to be useful in generating variation for important traits in crop plants. For example, delayed senescence has been produced in wheat; increased lysine and tryptophan levels in maize grain proteins; gossypol reduction in cottonseed; and male sterility in rice, tobacco and tomato. In the future it is likely that RNAi methods will be used much more widely for crop improvement. However, it is worth pointing out that RNAi is not a radically novel technology in terms of creating genetic variation. It is simply an alternative way to generate knockout mutations that can in principle be generated in several other ways, albeit more crudely and expensively.

Zinc-finger mutagenesis is an even more recently refined method for the targeted knockout of plant genes. The technology is based on construction of a zinc-finger nuclease that recognizes a specific DNA sequence in a genome and induces a double-stranded break at this target locus. Zinc-finger nucleases are engineered proteins made up of zinc-finger-based modules that recognize specific DNA sequences. These DNA-binding modules are fused to an endonuclease domain. Once the zinc-finger nuclease binds to a particular stretch of DNA, the nuclease domain will introduce a double-stranded DNA break in two different places. This form of DNA cleavage results in an overhang that is difficult to repair, leading to a loss of function in the target gene. Alternatively, if a homologous gene is introduced at the same time as the zinc-finger nuclease, the result will be replacement of the target gene with another gene.

Although first reported in the mid-1990s, zinc-finger mutagenesis was initially just a research tool. However, several reports in the late 2000s have demonstrated that newer versions of the technology may soon be applicable for the efficient generation of targeted mutations in crop plants and/or replacement of poorly performing genes with improved versions. An advantage is that mutagenized plants can be backcrossed to their respective wild-type varieties to create genetic lines that contain the desired mutation but are devoid of any transgenic DNA sequences. A comparison of the various gene silencing/mutagenesis technologies is shown in Table 10.2.

Transgene insertion

The major transgene-insertion technologies have changed relatively little since the 1990s, although *Agrobacterium*-mediated transformation can now be used in more plant species and improved biolistics methods can used to transform plastid as well as nuclear genomes. Several additional transformation methods have been developed, such as floral dipping, electroporation and silicone carbide fibres, but these have not been used extensively in crop breeding.

Agrobacterium transformation

Agrobacterium-mediated transformation is still the most common method for the stable insertion of transgenes into plants. Although initially limited to a narrow range of dicot species, it is now possible to use *Agrobacterium* in many monocots, including crops such as rice and wheat. Although *Agrobacterium* methods generally result in low transgene copy numbers, some of the DNA can become fragmented and inserted elsewhere in the genome. Sometimes this leads to gene silencing (see above), which can also happen if the transgene is inserted into an existing gene. Such side effects are difficult to predict and

Table 10.2. Comparison of different gene silencing/mutagenesis technologies.

Mutagenic agent	Effect on DNA structure or gene expression	Comments
Ethylmethane sulfonate (EMS)[‡]	Point mutations, CG to AT transitions	Large range of mutations including gain and loss of gene function
Di-epoxy butane (DEB)[‡]	Point mutations, 6–8 bp deletions	High efficiency, hundreds of mutations per genome Difficult to locate mutations in genome
Gamma-rays or X-rays[‡]	Chromosome breaks and rearrangements	Large insertions, deletions and rearrangements Mostly loss of function mutations Medium efficiency
Fast neutrons[‡]	Deletions, <1 kb	Difficult to locate mutations in genome
T-DNA[*]	DNA insertion	Insertion of specific DNA sequences Mostly loss of function mutations Low efficiency
Transposons[*]	DNA insertion/deletion	Easy identification of mutations Targets single genes
RNAi or zinc-finger constructs[*]	DNA insertion so that transcribed gene product causes gene silencing	Causes loss of function Very effective Requires knowledge of target gene function

[‡] non-transgenic method; [*] transgenic method

are best tackled by creating large populations of initial transformants so that breeders can screen out any unwanted material. Advances in DNA marker technology are enabling many undesirable events, such as transgene fragmentation, to be screened out from initial transformants without needing to wait several generations to observe possible phenotypic abnormalities.

Multiple gene insertion – 'gene stacking'

The manipulation of many desirable traits via transgenesis will require the insertion of several, and perhaps over a dozen, genes. With conventional insertion technology and using similar or identical regulatory sequences (promoters, terminators etc.) for each transgene, there is a risk of interference between the various transgenes leading to gene silencing. This has led to development of methods including co-transformation and use of polycistronic transgenes to enable multiple gene insertion into plants. It is possible to transform separate explants with different transgenes to produce a series of transgenic lines, each expressing a different transgene. These transgenic lines are then crossed to create progeny carrying multiple transgenes. Although this strategy has been used successfully, it is relatively lengthy and labour intensive and the various transgenes will be dispersed in various genomic locations where they may be subject to different position effects. A polycistronic transgene contains several tandemly linked genes under the control of a single promoter and is mostly used in plastid expression systems (see below).

Co-transformation involves the simultaneous addition of several transgenes, either as different T-DNAs in a single *Agrobacterium* strain or in different strains that are inoculated together on to explants. The co-introduced transgenes tend to integrate into the same chromosomal position, which makes it much easier to detect and remove unwanted position effects at an early stage. Co-transformation has been used commercially to create 'stacked traits', such as crops expressing both herbicide tolerance and insect resistance transgenes. It was also used to engineer the 'golden rice' varieties that accumulate ß-carotene in their grains (see Chapters 11 and 12). In the coming years an increasing proportion of transgenic crops is likely to contain 'stacked traits' created via co-transformation.

Methods of targeted transgene insertion have existed for many years in non-plant systems such as mammals and yeast. A major technique is homologous recombination, where a transgene is inserted into a predetermined location in a genome. Transgenesis via homologous recombination has been less successful in plants because targeted integration tends to occur at very low frequencies. For example, transgene insertion frequencies of <0.1% were obtained with homologous recombination in various plants. In most plants, extrachromosomal DNA integrates almost exclusively into random, non-homologous sites. However, research in model systems has shown that specific protein complexes, such as Rad50/Mre11/Xrs2, can sometimes facilitate targeted transgene insertion. A significant advance in achieving targeted gene insertion into rice plants was reported by a Japanese group in 2002, but much more work is needed before such techniques can be applied routinely to crops.

Plastid transformation

Although early transformation methods focused on the nuclear genome, the plastid genome has several advantages for efficient transgene expression. Plastid genomes are much smaller and simpler than nuclear genomes. Transgenes can be targeted to specific sites by homologous recombination, which avoids problems encountered with random insertion into nuclear genomes as discussed above. Also, thanks to their prokaryotic nature, plastid genomes can be used to house polycistronic transgenes. This means that a complete set of genes, for example encoding an entire biosynthetic pathway, can be introduced as a single transcriptional unit. Plastid transformation has been achieved in many crops, including tobacco, soybean, potato, tomato, cotton, *Brassica* spp., lettuce, petunia and *Lesquerella fendleri*.

The main method for introducing transgenes into plastids is biolistics. Although only one or a few plastids may be transformed initially in a given cell, by doing several rounds of selection during the regeneration process it is possible to end up with cells in which all the plastids are transformed. A leaf mesophyll cell might contain up to 100 chloroplasts while cells in starch-accumulating tubers might contain up to 100 amyloplasts. Each of these plastids will contain as many as 100 copies of the plastid genome. Therefore, if a single transgene is incorporated into each plastid genome, a given cell might have 10,000 copies of the transgene. All of these transgenes can be expressed at high levels without the gene-silencing problems sometimes seen in nuclear genomes. Using plastid expression, recombinant proteins can accumulate to as much as 45% of total soluble protein. Another advantage of plastid transformation is that the gene products (enzymes) operate in a separate compartment that is apart from the main cytosolic compartment. This can be important where the transgenic metabolism might interfere with ongoing cytosolic metabolism.

Inducible and transient expression systems

Almost all first-generation transgenic crops contain transgenes that are regulated by highly active constitutive promoters such as 35S or ubiquitin-1. Many newer transgenic varieties that express output traits, such as modified grain quality, use more highly regulated tissue-specific promoters. However, there is also a wide range of inducible and/or transient gene promoters that enable transgenes to remain inactive until their expression is required.

Inducible expression systems

There are two basic forms of inducible transgene system. The first is induced by an endogenous signal from the transgenic plant itself, or by an associated event such as a particular stress. The second form is induced by an exogenous chemical signal applied by an operator. Some inducible promoters can respond to both endogenous and exogenous signals. For example, wound-inducible promoters will drive transgene expression when a plant suffers mechanical damage; but, because many such promoters are activated by jasmonic acid released after wounding, they can also be activated by spraying unwounded plants with exogenous jasmonic acid. Several promoters have been isolated that are induced by specific pathogens, senescence, temperature or drought. In a practical example, a soybean heat-shock promoter has been used to express foreign proteins in the moss *Physcomitrella patens*.

Gene induction systems using exogenous chemicals have a particular advantage if there are issues about transgene products being in the open environment, e.g. for some pharmaceutical compounds or toxins. In such cases, induction of transgene expression in a crop can be restricted in time and space to enable it to be highly controlled and monitored. It is even possible, as has been done in tobacco, to harvest the crop, remove it to a secure processing facility, and then apply the exogenous signal by spraying the harvested leaves. This enables the transgene product to accumulate and to be isolated and processed in isolation and safety. Transgene-inducing compounds include ethanol vapour, acetaldehyde, copper, hormone analogues, various detergents, steroids and the insecticide methoxyfenozide. The 'biocontainment' of certain transgenic plants is discussed in Chapters 11 and 12.

Transient expression systems

In some circumstances it may be useful for a plant to express transgenes that are not integrated into any of its genomes. For example, transgenes can be delivered into plant cells as parts of disabled viruses or as isolated plasmids. By using powerful promoters, such as 35S, such transgenes can be expressed at high levels for several days or even weeks. However, because they are not integrated into a stable genomic environment, these foreign DNA segments are normally eventually removed from cells and the plants revert to being non-transgenic. Such transient expression systems might be useful to enable a plant to cope with a sporadic pest or disease threat. Alternatively, they can be used for the rapid accumulation of a novel product without the risk of 'escape' of the transgene via pollination or via transmission to subsequent generations. The transgene can be introduced into intact plants in the field or within contained conditions, via viral vectors. Although several forms of transient expression system are being tested in model plants and crops, their cost and complexity might mean that they are used for high-value products rather than commodity crops.

10.6   Selection Technologies

The various transgenic and non-transgenic genome enhancement technologies discussed above are providing increasingly efficient methods for the creation of new genetic variation in plants. However, even the most sophisticated of these methods still produces a range of variants that must be screened further in order to select the most suitable progeny, whether for use in research or for crop breeding. The efficient and rapid selection of plant variants has greatly benefited from DNA-marker technologies and these are increasingly being used in crop breeding programmes.

Marker-assisted selection (MAS)

Several types of marker can be used to assist the selection of favourable traits in plant breeding. Morphological markers, such as pigmentation, leaf shape or dwarfism, are easy to observe but are few in number and only rarely affect agronomic traits. The major biochemical markers are isozymes, which are variant forms of enzymes that can be distinguished by electrophoresis. Specific isozymes can be associated with several simple and complex traits and have been used to select such traits in crops like tomato. However, the utility of isozymes is limited by their small number (fewer than 100, even in major crop species) and their restricted ability to detect polymorphisms. By far the most useful class of markers are those based on DNA sequences. Such markers are now being applied to almost every aspect of plant and animal breeding, and in medicine and basic research.

DNA-based MAS can save time and money in crop breeding programmes as follows. In order to select most characters of interest, it is normally necessary to grow up and analyse each new generation of the crop before it is possible to perform phenotypic selection of appropriate plants. Many traits, such as disease resistance or salt tolerance, cannot be measured until plants have been grown, often to full maturity, and then tested in the field. However, MAS enables breeders to use the presence of specific DNA markers in newly germinated seedlings to select those plants that are likely to express the required characters. Therefore breeders no longer have to wait for plants to reach maturity, which can take many months and involve expensive maintenance of the growing plants. Molecular markers have been developed for most of the major commercial crops, including several tree species. In addition to their increasingly prominent role in genetic improvement of crops, molecular markers are useful for many other applications such as characterizing crop genetic resources, management of gene banks and disease diagnosis.

A DNA-based molecular marker is used to identify a segment of genomic DNA within which allelic variation in sequence has allowed its location to be genetically mapped. In breeding programmes, such markers are chosen because of their close proximity to a gene of interest so that the marker and target gene are inherited together. This enables breeders to use the marker as a relatively straightforward way to screen very large populations for the presence of a target gene without needing to perform complex phenotypic tests. Hence, MAS can be used to track the presence of useful characters in large segregating populations in crop-breeding programmes. Using molecular markers, breeders can screen many more plants at a very early stage and save several years of laborious work in the development of a new crop variety. In the case of wheat breeding, for example, it has been estimated that MAS may result in an overall cost saving of 40% relative to conventional phenotypic selection, in addition to improved genetic gains.

Several types of DNA-based marker have been developed for basic research and plant breeding and these are constantly being refined to increase their utility and decrease costs (see Fig. 10.7). Early markers included RFLPs (restriction fragment length polymorphisms), AFLPs (amplified fragment length polymorphisms) and RAPDs (random amplified polymorphic DNA). More recently, MAS has used much cheaper and more informative PCR-generated markers, such as microsatellites and SNPs. Microsatellites are short sequences of repetitive DNA, such as poly(TG), that can be amplified using PCR: they are highly polymorphic with respect to the numbers of the repeat units between individuals in a population. Even in relatively inbred crop species, microsatellites are polymorphic, enabling individuals to be genotyped separately. Detailed genetic maps based on microsatellites are available for most major crop species. Some of the most useful markers are 'single nucleotide polymorphisms' (SNPs). SNPs occur very frequently in genomes, e.g. once every 60–120 bp in maize, and are used widely for both research and breeding.

The use of MAS in crop breeding was initially restricted to a few economically important temperate crops that are bred and marketed by major private-sector firms, but the list of MAS-enabled crops is expanding. As well as annual crops such as cereals and legumes, MAS has been useful in perennial crops, including subsistence and cash crops in developing countries. Examples include oil palm, coconut, coffee, tea, cocoa and tropical fruit trees such as bananas and mangoes. Several public-sector initiatives and public–private partnerships have resulted in cheaper and easier MAS breeding systems. MAS technologies have also benefited from more efficient screening methods including PCR, DNA/DNA hybridization and DNA sequencing. Most MAS technologies in crops now use PCR-based methods, such as sequence-tagged microsatellites and

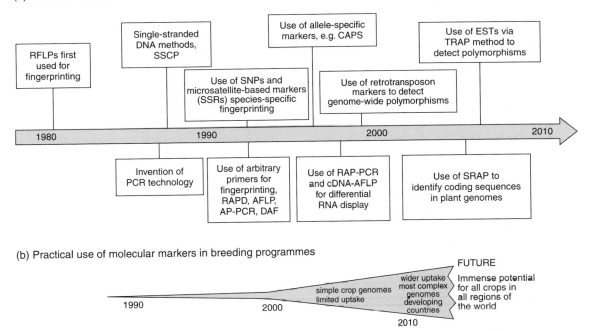

(a) Molecular markers as research tools

(b) Practical use of molecular markers in breeding programmes

Fig. 10.7. Development of molecular marker techniques, 1980–2010. (a) Timeline for invention of key marker technologies. (b) Timeline for practical use of key marker technologies in crop improvement. Abbreviations: AFLP, amplified fragment length polymorphism; AP-PCR, arbitrarily primed PCR; CAPS, cleaved amplified polymorphic sequence; cDNA-AFLP, cDNA amplified fragment length polymorphism; DAF, DNA amplification fingerprinting; EST, expressed sequence tag: RAPD, random amplified polymorphic DNA; RAP-PCR, RNA fingerprinting by arbitrarily primed PCR; RFLP, restriction fragment length polymorphism; SNP, single nucleotide polymorphism; SRAP, sequence-related amplified polymorphism; SSCP, single strand conformation polymorphism; SSR, simple sequence repeats; TRAP, target region amplification polymorphism. Adapted from Agarwal *et al.* (2008).

SNPs. By using DNA markers in conjunction with other new breeding technologies like clonal propagation, it should be possible to make rapid strides in the creation and cultivation of greatly improved varieties of many developing country crops (see Box 10.4).

10.7 Germplasm Resources

Plant germplasm resources, most commonly seeds, are key raw materials for accessing novel and useful genetic variation. As we have seen in previous chapters, crop production has benefited greatly from the global dissemination of seeds and other propagules. However, because most crops are now grown away from their centres of origin and are relatively inbred, local breeders may not have ready access either to traditional landraces of a given crop or to wild relatives that could be vital

sources of genetic variation. This led to the establishment of national, regional and more recently global germplasm centres. Unfortunately, local seed banks have been vulnerable to warfare, neglect and/or political manipulation, which makes it essential to regard plant germplasm as a global resource that can be shared by all people. In 2004, the International Treaty on Plant Genetic Resources for Food and Agriculture (ITPGRFA) provided an internationally agreed framework for the use of all plant genetic resources for food and agriculture.

Germplasm conservation and sharing have also benefited from initiatives such as the Global Crop Biodiversity Trust, and also from the use of new biotechnologies. Many biotechnologies are being employed for germplasm management in the widespread network of public-sector seed banks and resource centres across the world. For example, in

Box 10.4. MAS, genomics-assisted breeding and developing countries

In the future, marker-assisted selection, or MAS, could be part of a genomics-assisted breeding strategy. This would involve the use of bioinformatics-supported genomic and metabolomic resources within breeding programmes. For example, the immediate wild ancestor of rice, *Oryza rufipongon*, is a genetically diverse species containing alleles that confer agronomically useful transgressive variation when crossed with elite cultivars of *O. sativa*. However, there is currently no way of predicting where to look for such wild alleles. The integration of whole-genome mapping and marker analyses, coupled with QTL cloning and EcoTILLING, would enable genes from such wild relatives to be identified and incorporated into crop cultivars. Advances in sequencing and screening technologies since 2000, and especially their lower cost and increasing user-friendliness, make it feasible to consider the extension of such high-tech approaches to even relatively minor or non-commercial crops in the future.

Despite improvements over the past decade, a major challenge in extending MAS technology to developing countries is the cost and technical sophistication of the initial investment. For each crop, mapping populations must be created, genomic markers assembled and genetic maps compiled. Whether MAS technology is worthwhile for a particular crop depends on its genomic organization, the availability of technical infrastructure and expertise, and capital for set-up costs. Such calculations are especially important when developing countries are deciding whether to invest scarce resources in such technologies.

In some developing countries, MAS is beginning to produce significant results in the relatively few crop breeding programmes in which it has been deployed and future prospects are promising. One example is the development using MAS of a new pearl millet hybrid with resistance to downy mildew disease, first approved for release in India in 2005 and sown on 500,000 ha by 2009. Other examples include new rice varieties with bacterial blight resistance in India and flooding tolerance in the Philippines. Use of newer marker systems such as SNPs has decreased genotyping costs by 10-fold while enabling data throughput to increase 10-fold. This has made it feasible to produce SNP markers for developing-country crops such as cassava and cowpea. During the 2010s, it is likely that even relatively minor, non-commercial crops will benefit from the huge potential gains available from MAS technology.

---

addition to relatively established technologies such as cryopreservation, artificial seed production, somatic embryogenesis and other forms of *in vitro* cell or tissue culture, newer tools such as molecular markers, genomics and bioinformatics are having an increasing impact on germplasm management. Thanks to modern preservation methods, it is no longer necessary to keep plants as growing specimens, and they are often preserved instead as seeds, cuttings, or even DNA. One of the most recent initiatives is the Svalbard Global Seed Vault, on the remote Arctic island of Spitsbergen (Svalbard), which was announced in 2006 as a seed repository designed to maintain the security of its germplasm collections even in the event of a global disaster such as a nuclear war.

10.8 Summary Notes

- Modern biotechnologies are increasingly reliant on non-biological hardware (e.g. automated DNA sequencers, metabolite analysers and non-invasive phenotyping) and software (e.g. data analysis and information handling) systems.

- Core technologies such as genomics, proteomics and QTL analysis are greatly extending the range and precision of trait manipulation by breeders.

- The extension of variation, or 'genome enhancement', can be achieved by a range of non-transgenic technologies such as TILLING and wide hybrids.

- Transgenic technologies are being improved to minimize undesirable outcomes such as gene silencing and the transfer of unwanted DNA, to target transgenes to specific genomic locations, and to provide more precise forms of transgene regulation.

- Marker-assisted selection is increasingly being applied to improve efficiency and the range of traits that can be manipulated in crop breeding programmes.

- The collection, management and distribution of plant germplasm resources on regional and global scales are an important aspect of crop improvement.

Further Reading

Research and crop improvement technologies

Cook, D.R. and Varshney, R.K. (2010) From genome studies to agricultural biotechnology: closing the gap between basic plant science and applied agriculture. *Current Opinion in Plant Biology* 13, 115–118.

Edwards, D. and Batley, J. (2010) Plant genome sequencing: applications for crop improvement. *Plant Biotechnology Journal* 8, 2–9.

Kanehisa, M. and Bork, P. (2003) Bioinformatics in the post-sequence era. *Nature Genetics Supplement* 33, 305–310.

Mardis, E. (2008) The impact of next-generation sequencing technology on genetics. *Trends in Genetics* 24, 133–141.

Paterson, A.H., Freeling, M., Tang, H. and Wang, X. (2010) Insights from the comparison of plant genome sequences. *Annual Review of Plant Biology* 61, 349–372.

Shendure, J. and Ji, H. (2008) Next-generation DNA sequencing. *Nature Biotechnology* 26, 1135–1145.

Selection technologies

Agarwal, M., Shrivastava, N. and Padh, H. (2008) Advances in molecular marker techniques and their applications in plant sciences. *Plant Cell Reports* 27, 617–631.

Bernardo, R. (2008) Molecular markers and selection for complex traits in plants: learning from the past 20 years. *Crop Science* 48, 1649–1664.

Guimarães, E.P., Ruane, J., Scherf, B., Sonnino, A. and Dargie, J. (eds) (2007) Molecular marker-assisted selection. Current status and future perspectives in crops, livestock, forestry and fish. FAO, Rome. Available at: www.fao.org/docrep/010/a1120e/a1120e00.htm.

Xu, Y. and Crouch, J.H. (2008) Marker-assisted selection in plant breeding: from publications to practice. *Crop Science* 48, 391–407.

Yin, X., Stam, P., Kropff, M.J. and Schapendonk, H.C.M. (2003) Crop modeling, QTL mapping, and their complementary role in plant breeding. *Agronomy Journal* 95, 90–98.

Genome enhancement I: non-transgenic technologies

Dirks, R., Van Dun, K., De Snoo, C.B., Van Den Berg, M., Lelivelt, C.L.C., Voermans, W., Woudenberg, L., De Wit, J.P.C, Reinink, K., Schut, J.W., Van Der Zeeuw, E., Vogelaar, A., Freymark, G., Gutteling, E.W., Keppel, M.N., Van Drongelen, P., Kieny, M., Ellul, P., Touraev, A., Ma, H., De Jong, H. and Wijnker, E. (2009) Reverse breeding: a novel breeding approach based on engineered meiosis. *Plant Biotechnology Journal* 7, 837–845.

Harjes, C.E., Rocheford, T.R., Bai, L., Brutnell, T.P., Kandianis, C.B., Sowinski, S.G., Stapleton, A.E., Vallabhaneni, R., Williams, M., Wurtzel, E.T., Yan, J. and Buckler, E.S. (2010) Natural genetic variation in lycopene epsilon cyclase tapped for maize biofortification. *Science* 319, 330–333.

Rafalski, J.A. (2010) Association genetics in crop improvement. *Current Opinion in Plant Biology* 13, 174–180.

Ribaut, J.M., deVicente, M.C. and Delannay, X. (2010) Molecular breeding in developing countries: challenges and perspectives. *Current Opinion in Plant Biology* 13, 213–218.

Shu, Q.Y. (ed.) (2009) Induced Plant Mutations in the Genomics Era. FAO, Rome. Available at: www.fao.org/docrep/012/i0956e/i0956e00.htm.

Slater, A., Scott, N.W. and Fowler, M.R. (2008) *Plant Biotechnology: Genetic Manipulation of Plants*, 2nd edn. Oxford University Press, UK.

Tuberosa, R. and Salvi, S. (2006) Genomics-based approaches to improve drought tolerance of crops. *Trends in Plant Science* 11, 405–412.

Xu, Y. (2010) *Molecular Plant Breeding*. CAB International, Wallingford, UK.

Genome enhancement II: transgenic technologies

Halpin, C. (2005) Gene stacking in transgenic plants – the challenge for 21st century plant biotechnology. *Plant Biotechnology Journal* 3, 141–155.

James, C. (2011) Global status of commercialized biotech/GM crops: 2010. *ISAAA Brief 42*, ISAAA, Ithaca, New York.

Latham, J.R., Wilson, A.K. and Steinbrecher, R.A. (2006) The mutational consequences of plant transformation. *Journal of Biomedicine and Biotechnology* 2006, 1–7.

Maliga, P. (2004) Plastid transformation in higher plants. *Annual Review of Plant Biology* 55, 289–313.

Shukla, V.K., Doyon, Y., Miller, J.C., DeKelver, R.C., Moehle, E.A., Worden, S.E., Mitchell, J.C., Arnold, N.L., Gopalan, S., Meng, X., Choi, V.M., Rock, J.M., Wu, Y.-Y., Katibah, G.E., Zhifang, G., McCaskill, D., Simpson, M.A., Blakeslee, B., Greenwalt, S.A., Butler, H.J., Hinkley, S.J., Zhang, L., Rebar, E.J., Gregory, P.D. and Urnov, F.D. (2009) Precise genome modification in the crop species *Zea mays* using zinc-finger nucleases. *Nature* 459, 437–441.

Wu, G., Truksa, M., Datla, N., Vrinten, P., Bauer, J., Zank, T., Cirpus, P., Heinz, E. and Qiu, X. (2005) Stepwise engineering to produce high yields of very long-chain polyunsaturated fatty acids in plants. *Nature Biotechnology* 23, 1013–1017.

11 Social Context of Plant Biotechnologies

11.1 Chapter Overview

The purpose of this chapter is to examine the wider social and economic context of plant biotechnologies. Practical applications of science in agriculture often have social and economic implications that lead to controversy. Indeed, new technologies in general are often controversial. In order to understand such controversies, we must consider the institutional, economic, legal, commercial and political factors that interact with and influence the scientific areas that underpin new technologies. One of the most striking features of modern plant biotechnology is the way in which one particular set of methods, namely transgenesis or GM technology, became so controversial. After 1999 commercial cultivation of GM crops and the import of foods or other products derived from GM crops were banned or restricted in many countries. By the early 2010s, these restrictions had been eased in most of the world, but in Europe GM crops and products remained highly curtailed. We will look at how the commercial landscape shaped the initial portfolio of first-generation GM crops, how this affected consumer response to the technology, and how the technology and its context are gradually changing as it spreads across the world.

11.2 Science, Technology and Controversy

Most new technologies attract little critical attention and soon become part of daily life with people using new products without being aware of how they are made or function. In a few cases, however, a new technology might become so controversial that its adoption is delayed or it is even rejected, despite its possible usefulness. As discussed previously, there are many forms of plant biotechnology in addition to the use of genetic engineering to create GM plants. However, GM technologies attracted a great deal of attention soon after they were first publicized and were still controversial in the 2010s after two decades of widespread use in agriculture.

There was opposition to all forms of transgenic technology, including the use of microbial, animal and plant systems, from the 1970s. During the 1990s, however, the objections became much more focused on plants (and to a lesser extent on animals). The first transgenic plants were created in 1983 and released commercially in 1993. At first, the opposition to transgenic or GM crops was based mainly on the principle that genes should not be inserted into higher organisms either on ethical grounds or due to the perceived risks that potentially dangerous new life forms might be created. However, once first-generation transgenic crops were released on a large scale in the mid-late 1990s, much of the opposition became as much about the *provenance* of the technology as its perceived risks. In particular, campaign groups stressed that almost all transgenic crop varieties were patented and owned by a few large multinational companies.

This led to concerns that global food supplies could be controlled by a few companies for their own interests. Moreover, consumers saw little or no benefit from first-generation transgenic crops. These had modified input traits designed to facilitate crop management rather than to produce higher quality or cheaper food. Public disquiet was exacerbated in 1999 by claims that transgenic potatoes might have adverse effects on laboratory rats, and that transgenic maize might kill monarch butterflies. Although many claims about GM technology are no longer taken seriously by most of the scientific community, public disquiet still remains and since 1999 most transgenic crop products have been effectively excluded from European markets. To a great extent, this situation was due to pressure from anti-GM campaigning groups, but their success was only made possible by consumer uncertainty about the relative risks and benefits of GM technology and concerns about its provenance.

Vaccines and rainbow biotechnologies

GM crops are not unusual in being controversial. As far back as the early 18th century, the development by Jenner of a vaccine against smallpox led to a widespread public outcry in Britain. As shown in Fig. 11.1, people feared that injection with cow-derived vaccines might cause them to become cow-like and it took decades before public health authorities could set up national vaccination programmes. Indeed, despite its proven benefits, vaccination can still be a controversial issue. For example, in 1998, a poorly researched but widely publicized study about a supposed link between the MMR (measles, mumps and rubella) vaccine and autism was published in the *Lancet*. The resulting controversy led many to withdraw from public vaccination programmes. This in turn led to increased incidence of childhood measles and several avoidable deaths. Although the scientific community eventually refuted the original study and MMR vaccination rates gradually recovered, lingering doubts about the vaccine still remain with some sections of the public.

In many cases, initial hostility to a technology is gradually replaced by acceptance as it becomes more familiar and where its advantages are readily apparent to consumers. In other cases, acceptance might take much longer and might require fundamental changes to the technology or how it is used, i.e. its 'context' as discussed above. White biotechnology (the use of microorganisms for production of useful compounds) has been around for over a century and has rarely been controversial (see Chapter 1). In the 1990s, however, transgenic microbes were used for the first time and the technology was extended to manufacture high-value medical products such as recombinant proteins or

The Cow-Pock __ or __ the Wonderful Effects of the New Inoculation! __ vide the Publications of Y Anti Vaccine Society

Fig. 11.1. An early example of fears caused by a new biological technique. Following the widespread introduction of vaccination in the 19th century, groups such as the 'Anti-Vaccine Society' campaigned against it, for example by claiming that the 'unnatural' injection of cowpox virus might cause growth of bovine tissues in people.

peptides of human origin. Initially, this new form of white biotechnology caused controversy, especially in Europe. For example, Greenpeace initially campaigned against the use of all transgenic organisms, whether for medicine or agriculture. However, during the 1990s the risks of using contaminated human blood products led patient groups, such as haemophiliacs, to demand that recombinant products from transgenic microbes should be made available as a safer alternative.

The obvious advantages of recombinant medications and the fact that they were literally life-saving products soon overcame the rather vague arguments against their use. By the mid-1990s, almost all anti-GM groups had ceased objecting to recombinant medications and other products of transgenic microbes, such as food-processing enzymes. By the 2000s, use of transgenic microbes in white biotechnology had become routine and non-controversial. However, the use of similar transgenic technologies in animals, i.e. so-called red biotechnology, still remains controversial in some quarters today. Therefore, while there is virtually no opposition to using transgenic *E. coli* to manufacture human insulin, some people object to using transgenic sheep to synthesize human insulin in their milk. As we will see below, there is even more resistance to the use of GM plants (green biotechnology) for the manufacture of medically related compounds in biopharming.

In 2010, the farming and human consumption of GM salmon (an example of blue biotechnology) was approved in the USA for the first time. The initial announcement generated some public controversy (with predictable headlines about 'frankenfish') and concerns about their escape into the open sea. However, it is planned that the GM fish will be sterile females held in land-based tanks. Given this initial controversy, it will be interesting to see if these GM fish products are a commercial success and if blue biotechnology becomes less controversial in the future. In part, objections to the use of GM animal and plants are based on the issue of containment. With over a century of experience, we know that microbes can be securely contained in large-scale sealed systems as used in white biotechnology. Moreover, transgenic microbes can be engineered to ensure that, in the highly unlikely event of an escape, they cannot survive on their own. Larger animals such as sheep are slightly more difficult to contain securely; farmed fish pose even higher risks; and plants grown in the open environment are the most difficult of all to contain. Strategies for the biocontainment of transgenic plants used in bio-pharming are explored in Chapter 12.

Animal cloning

GM technologies are not the only controversial forms of biotechnology. The cloning of animals (but not plants) has also caused considerable public concern. The cloning of plants has been practised for thousands of years and human clones (i.e. identical twins) have always been with us. However, the ethics and rationale behind modern animal cloning have been challenged. In contrast to plants, it is relatively difficult to produce clonal animals from somatic cells in the lab. The first higher animal clone was Dolly the sheep, produced in 1996 from an adult mammary gland cell. The announcement of Dolly caused much controversy with speculation about the creation of hordes of genetically identical humans or monstrous new animals. Some of this speculation was similar to earlier scare stories about GM plants, with headlines such as 'Frankenfoods' and 'superweeds'. In response, many governments restricted the use of cloned animals and banned research aimed at creating cloned humans.

While some objections to animal cloning are based on ethical concerns, others centre on the technology itself. As with aspects of transgenesis, early versions of animal cloning were relatively crude and inefficient. For example, the creation of Dolly took 277 attempts of which only 27 resulted in embryos. These embryos were implanted into 13 surrogate mothers but only one pregnancy went to full term. In contrast to the success rate of only 0.4% for Dolly the sheep, rates of 10–20% have been obtained more recently for cloned cattle. However, cloned embryos still suffer from a high incidence of abnormalities and often result in abortion. At the molecular level, epigenetic changes in animal clones lead to incorrect programming of embryo development. Some abnormalities can persist into adulthood and result in conditions such as obesity or a compromised immune system. On the plus side, it seems that almost all clonal abnormalities disappear in subsequent sexual generations, which implies that the original abnormalities were caused by epigenetic factors alone.

Given these limitations to existing versions of animal cloning technology, its use is often questioned, even by people who have no ethical objections to cloning *per se*. Although cloning is not necessary for animal breeding it can be a useful tool. For example, it can assist the conservation of threatened livestock breeds or wild animals; the rapid multiplication of superior livestock genotypes; and production of research lines for experimental studies. In the future, cloning may also be used routinely to produce adult animals from cells that have been genetically engineered. Even with its limitations, cloning via nuclear transplantation is more efficient than current methods of producing transgenic animals that rely on pronuclear injection of DNA.

Food irradiation, nuclear power and mobile phones

Sometimes, potentially useful new technologies, such as food irradiation, are curtailed or delayed for many decades and even eventually abandoned. In the 1980s, it was proposed that some foods could be treated with high-energy radiation to kill microbial contaminants and extend shelf life, both in retail outlets and during storage at home. Much of the opposition to food irradiation focused on two aspects. First was the suspicion that food companies would use it to disguise bad food or store time-sensitive foods for much longer in order to boost profits. Second, there were public fears about the term 'irradiation' and possible side effects on the food. Many people might have accepted a process that allowed food to be stored without spoilage but the term 'irradiation' was associated with nuclear radiation and radioactivity.

Successful campaigns were launched in the 1990s by pressure groups seeking to prevent food irradiation. These developments coincided with food safety scares such as the BSE, *E. coli*, and salmonella outbreaks that severely dented public confidence in food safety standards. As a result, uptake of irradiation technology was restricted, especially in Europe, with consumers remaining wary and the industry hesitant to make the necessary investments. By 2011, only one licence for irradiation of a few herbs and spices had been granted in Europe. Although a wider range of irradiated foods is available in the USA, the technology has not lived up to its original promise and may never be used in much of the world.

Nuclear power is probably one of the most controversial technologies of the past 50 years. This is mainly due to perceived safety risks of the reactors themselves, plus the unresolved problems in disposing of long-lived radioactive waste. Interestingly, while remaining wary of its drawbacks, some environmentalists now embrace nuclear power because of its low carbon footprint compared with fossil-based power generation. This is an important factor in discussing plant biotechnologies because it highlights one of the ways in which people often judge new technologies, which is by balancing perceived risks against benefits. For example, the publication of some limited and disputed evidence about the risk of radiation from prolonged use of mobile phones by young children has not deterred the mass adoption of this technology by billions of people. In this case most of the population has decided that the very real benefits of mobile phones greatly outweigh the rather uncertain risks that were highlighted by a few inconclusive and disputed scientific studies.

Non-transgenic crop biotechnologies

As noted previously, plant biotechnology comes in many different forms and, while transgenesis has been controversial, most other crop-related biotechnologies have attracted much less attention. In order to understand why this is the case, it is useful to compare mutagenesis with transgenesis. This example is particularly interesting in the light of the many objections to food irradiation (see above) compared with the lack of objections to crop mutagenesis, even though both technologies can involve the exposure of tissues to high-energy γ-radiation produced by nuclear reactors.

As we saw in Chapter 9, induced mutagenesis was first used to manipulate plant genomes in the 1920s and has been used in breeding programmes around the world. However, the vast majority of the population is unaware that powerful DNA-altering chemicals or nuclear radiation have been used to create thousands of new crop genotypes in a much more random and uncontrolled manner than transgenesis. So why has mutagenesis remained such a low-key and uncontroversial form of biotechnology, while transgenesis has a much higher and more controversial profile? Much of the answer to this question is related to the differing contexts of the two biotechnologies. Some of this context relates to when the technology was developed. Crop mutagenesis was first used on a large scale during the 1960s when there was much less public interest in or scrutiny of science and technology. The work attracted hardly any publicity from the media. Also, products from crop varieties created by irradiation or chemical mutagenesis did not require distinctive labelling. In contrast, plant transgenesis was deliberately publicized as a novel and potentially world-changing technology in the 1980s and 1990s, when there were much higher levels of public interest and concern about science and technology.

Another big difference between mutagenesis and transgenesis lies in their provenance. Mutagenesis was mostly developed and applied by public sector scientists as a public good rather than for commercial use. Its development within public research and breeding institutes meant that it was mainly focused on long-term crop improvement rather than on short-term profit making. This is not to say that private companies have not used mutagenesis in their own breeding programmes: many commercial varieties have been developed using mutagenesis. However, plants developed by mutagenesis still cannot be patented in the same way as GM varieties. Because mutagenesis (unlike genetic engineering) only manipulates existing genes and has poor patent protection, many companies have understandably tended to use transgenesis instead. With some important exceptions, mutagenesis has been mainly used to improve developing-country crops for poorer farmers rather than for the commercial crops of interest to agbiotech companies.

These factors mean that the social context of mutagenesis technology is quite different from transgenesis. However, the scientific context of the two technologies is rather similar. Both are highly intrusive and 'artificial' methods aimed at altering plant genomes in order to create new genetic variation, some of which might be useful for research or crop improvement. Of the two technologies, mutagenesis is the more crude and is much more likely to result in undesirable alterations to the target genome. Indeed, it would be very easy to mount a scare campaign against radiation or chemical mutagenesis using emotive terms like 'nuclear radiation', 'carcinogenic chemicals', 'unnatural', 'γ-rays', 'cobalt-60', 'DNA damage', 'abnormality' and so on. It is unlikely that such a campaign will be mounted against mutagenesis precisely because of its wider social, as opposed to its scientific, context.

The comparative risks associated with various plant-breeding technologies, as assessed by the National Research Council in the USA, are shown in Fig. 11.2. This shows that mutagenesis actually carries the highest risk of the 12 methods compared. Transgenic tools are less risky, especially if genes are transferred from closely related species, while conventional selection within closely related popula-

tions is the least risky method. This example demonstrates how the wider social context can sometimes be an important factor in determining public reaction to new technologies. First-generation transgenic crops were developed in a context that was mostly private sector, for-profit, and initially focused on richer countries. This has contributed to the more controversial public profile of GM crops compared with mutagenized varieties. One way in which GM technology might become more widely acceptable in the future would be if its social context were to become more like that of mutagenesis.

Maize hybrid technology is another non-transgenic system that was eventually widely adopted, although in this case the context could not have been more different from mutagenesis. Maize hybrids were initially developed by public-sector scientists in the USA but were applied commercially by private-sector seed companies (see Chapter 9). As with many new technologies the acceptance of maize hybrids took a long time. During the 1920s and 1930s, many farmers resisted hybrid maize because of its high cost and because they would no longer be able to save seed, which they would need to buy from seed companies. It took several decades of patient marketing of hybrids by companies, the demonstration that higher seed costs would be far outweighed by the extra profit from higher yields, and the shock of the Great Depression and Dustbowl before US farmers eventually adopted maize hybrids *en masse* in the 1940s.

Non-biological technologies

Many other agricultural technologies were initially controversial and in some cases they provoked violent reactions. This has been especially true for mechanization, which often threatens jobs of existing farm workers. In 19th-century Britain there were many cases of farm labourers smashing new harvesting machinery and objections to mechanization still occur in modern times. For example, in the 1960s and 1970s US tomato pickers tried to resist new mechanized cropping systems developed at the University of California, Davis. In this case the mechanical technologies were developed alongside biological changes, such as delayed ripening and thicker skins, that made tomato fruits more robust and less likely to be damaged when harvested by machine. As noted in Chapter 9, some of this work eventually led to the first commercially released GM crop, the slow-ripening FlavrSavr™ tomato.

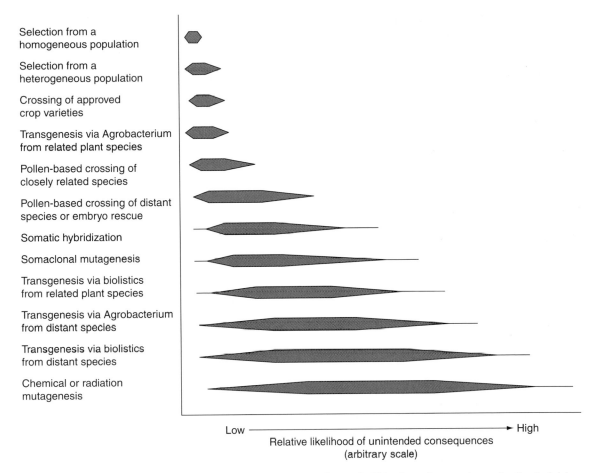

Selection from a homogeneous population

Selection from a heterogeneous population

Crossing of approved crop varieties

Transgenesis via Agrobacterium from related plant species

Pollen-based crossing of closely related species

Pollen-based crossing of distant species or embryo rescue

Somatic hybridization

Somaclonal mutagenesis

Transgenesis via biolistics from related plant species

Transgenesis via Agrobacterium from distant species

Transgenesis via biolistics from distant species

Chemical or radiation mutagenesis

Low ⟶ High

Relative likelihood of unintended consequences
(arbitrary scale)

Fig. 11.2. Comparative risks associated with various plant-breeding tools. This shows the spectrum of estimated risks of unforeseen and undesirable consequences arising from the use of various plant-breeding technologies. Selection from existing populations gives the lowest risks while mutagenesis-based breeding gives the highest risks. The various transgenic breeding methods produce a range of risks lying between these two extremes. Adapted from NRC (2004).

There are similar controversies about mechanization today, for example in sub-Saharan Africa, where crop yields have lagged behind other regions of the world. In part, this yield lag has been due to the slow pace of mechanization in Africa. Although much of this region is not suited to mechanized farming, UN reports suggest that certain forms of mechanization could greatly benefit food production in Africa, although some traditional agronomic practices might need to be amended as a result.

Chemical technologies, such as use of fertilizers and biocides in modern intensive farming, are regularly criticized on environmental and safety grounds and, to a great extent, the organic farming movement began as a reaction against the use of chemical inputs. Although organic farming often involves high standards of crop and animal husbandry, average yields are significantly lower and organic foods are more expensive than those from conventional farming. As discussed in Chapter 9, modern levels of food production depend on the judicious use of chemical inputs. These inputs can be misused and one of the most frequent examples is the overuse of fertilizers (leading to runoff into watercourses) and biocides (possibly affecting non-target organisms). Some of these issues can be tackled by better use of the technology itself, e.g. using imaging systems so that inputs are only applied where they are needed, rather than to the entire crop. Alternatively, new technologies such as biological methods (e.g. biofertilizers or biopesticides – see Chapter 12) can be used to deliver the inputs upon which most crops depend for optimum yield.

Wider concerns about modern agriculture

We can conclude from this section that the controversy about GM crops is not an isolated case nor is it particularly unusual in terms of new technologies. In addition, the controversy about GM crops reflects a wider sense of unease about many aspects of modern agriculture. Some factors underlying these concerns can include the following:

- greater public awareness, scrutiny and questioning of most new technologies;
- increased concern about food safety, especially after *E. coli*, BSE, foot and mouth, salmonella and avian influenza outbreaks;
- concerns about the environmental impact of many technologies;
- questions about the long term sustainability of intensive agriculture;
- increasing globalization of commerce, including agribusiness;
- the growth of giant transnational agbiotech corporations;
- laws allowing companies to patent genetically engineered organisms; and
- ethical concerns about the mixing of genes from different organisms.

In the remaining sections of this chapter, we will look at some of these wider issues as they relate to plant biotechnology.

11.3 Institutional and Commercial Context of Plant Biotechnology

Plant breeding and biotechnology are unusual, but not unique, in relying on a mixture of public, for-good and private, for-profit developers. A similar mixture is found in areas such as nuclear power and some large-scale infrastructure developments such as national transportation networks. In historical terms, crop improvement has always involved mixtures of public and private groups. Pre-scientific empirical breeding was mostly carried out by individual farmers but larger schemes such as land drainage, irrigation or the introduction of new crops often involved state intervention. With the advent of scientific breeding in the late 19th century much of the innovation in plant breeding and biotechnology

shifted from the private sector to the public sector as national research centres developed new crop varieties to feed growing populations. However, with increasing privatization and a focus on molecular technologies the pendulum swung back towards the private sector after the 1980s, although in the 21st century there may be a more balanced mixture between the public and private sectors.

Plant breeding before the 1980s

Modern scientific methods of plant manipulation began to be used in the 18th century. This work was initially funded by a small number of relatively wealthy individuals with an interest in the exploitation of plant products. This was the era of 'imperial botany' (see Chapter 8) where new crops were transplanted around the world and more efficient management systems, such as crop rotations, were being devised. Although there was some limited state involvement in these enterprises, such as the establishment of national botanical gardens for economic purposes, most crop improvement was carried out by and for the private sector. This situation began to change in the 19th century when the importance of maintaining food supplies for the growing populations of newly industrializing countries became apparent. Within Europe the main focus was on improving crop yields in the limited arable land that was available. In contrast, vast new areas of cropland were being opened up in the rich soils of the American Midwest.

Many governments established publicly funded research institutes aimed at all aspects of crop development. One of the pioneers of public sector crop improvement was the USA, where the Land Grant Universities and the USDA were established as early as 1862. Many of these institutions remain at the forefront of plant biotechnology R&D today. National crop research centres carried out a mixture of basic and applied work aimed at crop improvement. Some of these centres also worked with extension services to train farmers in scientific agriculture and to disseminate new crop varieties developed by researchers. Many technologies, including tissue culture, mutagenesis, quantitative breeding methods and hybrid production, were developed in these research centres and universities. Most European countries set up similar networks, although none of them approached the scale of the US effort.

As shown in Table 11.1, the public sector dominated much of 20th-century crop improvement

Table 11.1. Scientific and commercial landmarks in agronomy and crop breeding.

Date	Innovators	Innovation	Sector
1650s	Robert Child *et al.*, E	Early experiments on the effect of nitrates on cereals	Pri
1655	Jamaica	British conquest begins global sugar and Atlantic slave trade	Pri
1694	Camerarius, G	First report of sexual reproduction in plants	Pri
1701	Jethro Tull, E	Invention of the seed drill	Pri
1718	Thomas Fairchild, E	First man-made interspecific hybrid	Pri
1730s	Charles Townshend, E	Popularized four-course crop rotations	Pri
1758	Thomas Knight, E	Use of hybridization for practical plant improvement	Pri
1761	Josef Kölreuter, G	First practical hybrid crop variety	Pri
1825	John Lorain, USA	Described the possibility of growing maize as a hybrid crop	Pri
1840s	Justus Liebig, G	Use of inorganic fertilizers to improve soil fertility	**Pub**
1850	Various, France	Experiments with crop protection agents	Pri
1866	Gregor Mendel, AH	Published principles of inheritance	**Pub**
1900	DeVries *et al.*	Independent rediscovery of Mendel's work on plant heredity	**Pub**
1903	Wilhelm Johannsen, D	'Pure line' theory for selection of true-breeding cultivars	**Pub**
1904	William Bateson, UK	Discovered genetic linkage and established concept of genetic maps to describe the order of linked genes,	**Pub**
1905–1908	Bateson and Punnett, UK	Showed that some genes can modify the action of others	**Pub**
1908–1910	Shull and East, USA	Research lays foundations for commercial maize hybrids	**Pub**
1910	Rowland Biffen, UK	Used Mendelian approach in wheat breeding	**Pub**
1918	Ronald Fisher, UK	Laid the foundations of quantitative genetics	**Pub**
1919	D.F. Jones, USA	'Double cross' maize hybrids allow commercial production	Pri
1925–1938	Nikolai Vavilov, USSR	Discovered centres of origin and diversity of ancient crops Established plant germplasm collection and maintenance	**Pub**
1928	Stadler, USA	Use of X-rays to induce mutations in maize	**Pub**
1930s	Eur and North America	Development of plant cell and tissue culture *in vitro*	**Pub**
1934–1935	Nilsson-Ehle and Gustaffson, Sweden	Started mutation experiments in plants and reported erectoid mutants in barley with compact head type and stiffer straw	Pri
1937	Blakeslee and Avery, USA	Chromosome doubling for wide crosses and somatic hybrids	**Pub**
1940	Harlan, USA	Use of bulk breeding in commercial seed production	Pri
1940–1950	Eur and North America	Switch from animal to mechanical power on farms	Pri
1944	Avery *et al.*, USA	Discovery of DNA as hereditary material in most organisms	**Pub**
1950s	UN FAO and IAEA	Irradiation mutagenesis available to developing countries	**Pub**
1950s	Worldwide	Production of haploid crop varieties via microspore culture	**Pub**
1953	Watson and Crick, UK	Proposed double-stranded antiparallel helix model for DNA	**Pub**
1959	Eur and North America	Widespread use of the first systemic fungicides	Pri
1960s	Eur and North America	New screening technologies for quality characters	**Pub**
1964	Guha and Maheshwari, I	Developed doubled haploid technologies	**Pub**
1966–1971	India and Pakistan	High-yield dwarf wheat varieties initiated the 'Green Revolution', tripling food production in three decades	**Pub**
1971	Worldwide	Establishment of CGIAR to coordinate agricultural research and development in developing countries	**Pub**
1970s	Worldwide	Huge rise in fertilizer and biocide use increases crop yields	Pri
1960–1990	Asia	Development of 'miracle rice', a semi-dwarf, fertilizer-responsive form that led to a doubling of global rice yields	**Pub**
1980s	Worldwide	Commercial use of mass-propagation of clonal explants	Pri
1983	Europe and USA*	First reports of transgenic plant cells	***Both***
1987	USA*	Transgenic Bt technology for insect resistance	Pri
1992	USA*	First transgenic crop, Flavr Savr® tomato, from Calgene Inc.	Pri
1992	USA*	Transgenic virus-resistant papaya grown in Hawaii	***Both***
1996	USA*	Commercial release of transgenic herbicide-tolerant and insect-resistant soybean, maize, cotton and rapeseed	Pri

Continued

Table 11.1. Continued.

Date	Innovators	Innovation	Sector
1990s	Eur and North America	Development of DNA-based molecular markers	**Pub**
1990s	Eur and North America	Automated phenotype screening via GC-MS and NMR	Pri
2000–2008	West Africa	Widespread release of new African rice, NERICA	**Pub**
2002	USA*	First transgenic biopharmed crops	Pri
2002	East Africa	High yield, drought-tolerant hybrid sorghum release	**Pub**
2003	Worldwide	Extension of MAS technology to a wider portfolio of crops	**Pub**
2004	China	World Food Prize for hybrid rice varieties in China	**Pub**
2004	Worldwide	Sequencing of the first crop genome, rice	**Pub**
2005	USA	First use of TILLING in a crop species	**Pub**
2005	Worldwide	>3000 mutation-bred crop varieties in >59 countries	**Pub**
2005	South America	Emergence of region as major food and feed exporters	*Both*
2006	USA	First approval of cell culture-based vaccine biopharming	Pri
2007	India	50% increase in cotton production with 207 Bt varieties	Pri
2007	USA	Commercial release of multiple 'stacked' GM traits	Pri
2009	Worldwide	Sequencing of maize and oil palm genomes	*Both*
2010	Brazil	First GM crops bred by a developing country	*Both*
2012	South-east Asia	Release of original GM 'golden rice' to farmers	**Pub**
2015	South-east Asia	Possible release of improved GM 'golden rice' to farmers	**Pub**

AH, Austria-Hungary; D, Denmark: E, England; Eur, Europe; G, Germany; I, India.
*indicates use of GM crop varieties.

until the 1980s. Mendelian genetics were applied to crop breeding and biological questions behind crop performance were addressed. Public-sector scientists developed the maize hybrid systems later commercialized so successfully by the private sector. The same is true for many other crop varieties that were then adopted and refined by seed companies. Many aspects of early transgenic technology were developed in public-sector labs and later adapted for commercial use by agbiotech companies. Finally, public-sector scientists developed the technologies behind Green Revolution crops of the 1960s and 1970s.

By the 1970s, the public sector was the dominant force in global plant breeding. For example, in 1978, the public sector in the USA produced the overwhelming majority of commercial varieties of important crops such as soybeans (89%), wheat (86%), barley (95%), oats (86%) and rice (92%). In most developed countries, this situation changed during and after the 1980s as agricultural institutes were privatized, downsized or closed down. Government policies increasingly discouraged applied research because this was regarded as the job of the private sector. As discussed below, during this period the private sector greatly increased its investment in plant breeding, with a major focus on transgenic biotechnology. In contrast, many public-sector plant researchers chose to work on model systems and to pursue fundamental science instead of more applied crop-related studies.

Rise of the private sector

Until the mid-20th century, the private sector took little part in the more basic aspects of crop improvement, which were expensive, required skilled specialists and were already being done in public research centres and universities. Instead most crop-related companies focused on more profitable non-biological areas such as agrochemicals and machinery. For each dollar spent on seed, a farmer spent tens or hundreds of dollars on equipment and agrochemicals. A few companies involved in the retail seed market carried out some breeding work. This was especially true for hybrid maize where, because farmers had to repurchase seed each year, seed companies had a guaranteed income and an incentive to bring out new varieties on a regular basis. However, most seed companies were relatively small in comparison with the huge agrochemical or farm-equipment companies. This situation began to change in the 1960s–1980s as the farm equipment and agrochemicals markets became increasingly saturated, and the potential of biological technologies became more apparent.

During the 1980s and 1990s, many seed companies were taken over by agrochemical companies. This was partly an attempt to diversify away from their dependence on petrochemical products in the wake of the oil shocks of the mid-1970s. Because they already did lab-based R&D on agrochemicals, these companies also had the financial and scientific resources necessary to establish their own up-to-date plant science research operations with particular expertise in biochemistry and molecular biology. This expertise was especially relevant to the emerging field of recombinant DNA manipulation being explored in university labs. During the 1980s, plant transformation with foreign genes was successfully applied to crop plants and the private sector began to realize the immense scope and potential power of this new technology to enhance genetic variation.

By the mid-1980s, private-sector spending on plant breeding in the USA reached an annual $200 million, from a mere $20 million in the late 1960s. As shown in Fig. 11.3, this trend then continued with spending approaching $600 million by the late 1990s, at a time when public-sector funding was static or declining. One of the problems with transgenesis is the high cost of producing new GM crop varieties. The only way for developers to recoup this investment would be to have a legal monopoly on sales of GM seeds. This was achieved when GM organisms were granted patent protection in 1985, making it possible for transgenic plants to be patented in a similar way to inventions such as new forms of machinery. Now that new GM crop varieties could be treated like mechanical inventions, companies could retain ownership of their seed even after selling it to farmers so that the latter could be prevented from saving such seed or selling it to others. As discussed in Box 11.1, the particular way that patenting has been applied to transgenic plants is still controversial.

During the 1980s and early 1990s, the combination of groundbreaking scientific progress in developing transgenesis as a practical technology for crop improvement, plus the establishment of strong legal mechanism to protect its ownership, resulted in the rapid growth of private-sector companies specializing in plant genetic engineering. At first, many of these companies were small ventures set up by scientists originally from the public sector. Some of them were very scientifically innovative and in 1993 the small biotech company Calgene produced the first GM crop. However, the high cost of the initial R&D, the even higher costs in moving from lab to marketplace, plus the lengthy regulatory procedures applied to GM varieties, soon made it apparent that most GM crop development would only be feasible for larger companies.

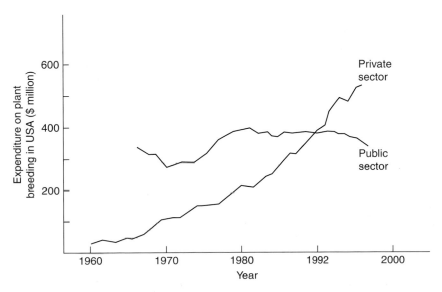

Fig. 11.3. Public- and private-sector expenditure on plant breeding in the USA. While US public-sector spending on plant breeding stagnated, private-sector spending rose 14-fold from 1960 to 1996. Data from Byerlee and Dubin (2010).

Box 11.1. Patenting and plant biotechnology

A patent is a form of legal protection that gives an inventor the exclusive right to exploit an invention, typically for 10–20 years. The system is designed to encourage innovation by rewarding individuals or companies who invent and/or invest in new technologies or processes by enabling them to profit from their efforts. Before the 1980s living organisms could not be patented, but that changed with the Diamond versus Chakrabarty case in the USA where the Supreme Court ruled that a patent could be granted on a GM bacterium (a *Pseudomonas* strain able to break down crude oil). Patent protection was soon extended to transgenic plants and animals, enabling a company to retain a high degree of control over a GM crop variety plus any progeny that carry the patented transgene(s).

Patents can also be granted for the invention of new devices or processes. This can include the methods used to produce a transgenic plant, e.g. the transformation technique or the selection protocol. One of the issues of concern to some in the agbiotech field is that several of these key enabling technologies are owned by a few large companies. This can inhibit the wider development of transgenic crops and is a particular disincentive to their deployment in developing countries. Another problem in some areas of agbiotech is the issue of 'broad claims' where a patent claim might be extended so that it covers not just a particular organism or process but a much broader group. For example, the invention of a technique to transform one wheat genotype might be extended to cover all wheat genotypes, all members of the Triticeae, all species of cereal, all monocots, or even all higher plants. Another example might involve the invention of an antibiotic resistance marker where the claim might be extended to cover all antibiotic markers, or even all selection systems for transgenic plants. In some cases, the granting of such broad claims has acted as a disincentive for other companies to develop new innovations in plant biotechnology.

This problem was highlighted for public-sector researchers in 2000 by the 'golden rice' episode. Golden rice is a transgenic provitamin A-enriched variety developed by a Swiss/German university group to benefit the poor in developing countries. The researchers discovered that, during the development of their transgenic plants, they had inadvertently used techniques protected by no fewer than 70 patents, originally held by 31 different organizations. This shows the immense complexity of agbiotech patenting. In this particular case, the companies and other institutions were eventually persuaded to partially waive their patent rights for the use of the golden rice. However, this permission was limited to use of the rice in the poorer developing countries where there would be no commercial market anyway. The existence of hundreds of broad patent claims like the ones described above can deter smaller companies or public-sector researchers from trying to apply GM technology either for public good or for a new commercial purpose.

Looking to the future, there are signs of improvement in the area of agbiotech patenting. There are several initiatives to develop 'open source' biotechnologies that are freely available to developing countries. Some of the major agbiotech companies have agreed to donate aspects of their patented technologies to selected countries, e.g. in sub-Saharan Africa. And, finally, some developing countries are developing their own proprietary biotechnologies rather than licensing them from multinationals.

During the mid–late 1990s, there was a spate of acquisitions and mergers with most agbiotech start-up companies being taken over by large multinational agrochemical companies. For example, Calgene and many other biotech and seed companies were acquired by one of the emerging giants of the agbiotech world, Monsanto. Other large agrochemical multinationals, such as Dow Chemical, Du Pont, BASF and Syngenta, acquired most of the remaining smaller independent agbiotech companies. As shown in Table 11.2, by 2006 the 'big six' companies controlled 66% of R&D spending. This concentration was even more marked in the commercial market for GM crop varieties, where well over 90% of seed sales came from a single company, Monsanto. By the 2010s, only a few small agbiotech companies remained and these were mostly either service providers or working in small niche, high-risk areas such as biopharming (see Chapter 12), where the multinationals were less active.

In this section we have seen how the institutional and commercial context of plant biotechnology has changed and how this had a particular effect on GM technologies. Although these were initially the products of public-sector institutions, the changing scientific, legal and commercial circumstances of

Table 11.2. Private-sector firms and R&D expenditures by type of activity in 2006.

Market sector	Number of companies	Agricultural R&D ($ billion)	Total R&D spending (%)
'Big six' agrochemical/agbiotech companies*	6	2.03 + 1.57 (chemicals + seed/biotech)	66
Other agrochemical companies	122	0.62	11
Other seed companies	82	0.63	12
Other agbiotech companies	45	0.17	3
Fertilizer companies	No data	0.45	8
Total	255	5.47	100

*The 'Big six' are BASF, Bayer, Syngenta, Du Pont, Dow and Monsanto. Adapted from Piesse and Thirtle (2010).

the 1980s meant that commercial GM crops were developed within a very different context compared with earlier forms of plant-breeding technology.

11.4 Scientific Context of Plant Biotechnology

As discussed in Chapter 9, the crops and traits altered in first-generation GM crops were determined by a combination of scientific and commercial factors. The science dictated that only simple monogenic traits where genes were already available could be manipulated. High investment costs limited the application to a few major crop species for which seed sales could be controlled. And the ownership of proprietary herbicides by a few agrochemical/agbiotech companies made the development of herbicide-tolerant traits a commercially attractive proposition. For these reasons, the vast majority of first-generation GM crops were modified to express either or both of two relatively simple input traits, namely herbicide tolerance and/or insect resistance. This is also why, since 1996, large-scale commercial transgenic agriculture has been limited to four crops, namely maize, soybean, rapeseed and cotton. As late as 2011, these four crops expressing one or more simple input traits still made up over 99% of the global area of transgenic crops.

Improving first-generation GM technologies

As discussed above, first-generation GM crops are very restricted in their traits and in the underlying technologies involved in their production. The focus on rapid commercial returns and the scientific challenges of manipulating complex traits meant that this narrow group of traits and technologies dominated the first two decades of commercial GM crop development. In contrast to many other modern technologies, GM crops have shown a relatively slow pace of scientific innovation. Another criticism of some first-generation GM crops is the slow response to public concerns about issues such as the use of antibiotic resistance markers, transgene spread in the environment and consequences of random transgene insertion into genomes. Such concerns have been voiced since the 1980s but had still not been addressed in many GM varieties being grown during the 2010s.

Many scientists rightly point to the lack of rigorous evidence behind most of these public concerns. For example, there is no evidence for claims that antibiotic resistance genes present in some GM food plants could be functionally inserted into gut bacteria to generate antibiotic resistant strains that go on to become pathogenic. However, the lack of public trust in the industry makes it important to address such concerns wherever possible, however misplaced they may be. This is also good commercial practice as in the maxim, 'the customer is king'. Moreover, there are relatively straightforward methods to replace or remove such markers in GM plants and to target genes to specific genomic locations (see Chapter 10). While these may add to production costs, their use would show a responsiveness to consumer concerns that has been lacking in some parts of the agbiotech industry. There are also scientific solutions to risks of unwanted transgene spread as discussed in the section on biocontainment in Chapter 12.

Scientific concerns and controversies

One of the features of GM technology that contributed significantly to public concerns was the perception that peer-reviewed scientific studies had revealed risks to health and the environment. This resulted in a decisive shift in public opinion against GM crops and foods in Europe in 1998–1999. Anti-GM movements had been active in Germany in the early 1990s and gradually spread across Europe, becoming particularly acute in the UK. The perception of scientific scepticism about GM crops was largely fuelled by the release of information, mostly via the mass media, from two experimental studies. These were by Pusztai *et al.* on the effects of GM potatoes on the dietary system of rats, and by Losey *et al.* on the effects of GM maize pollen on monarch butterflies. In the first case, GM food intake was associated with abnormalities in experimental animals and, by extension, people. In the second case, the environment, in the form of monarch butterflies, seemed to be at risk from GM technology. To this day, both studies continue to be cited as serious arguments against GM crops by environmentalists, politicians and even by some scientists.

Both of these studies attracted international media attention and both were published in scientific journals, which added to their credibility. However, both studies were soon shown to be incomplete and unable to justify their claims about the dangers of GM technology. The monarch butterfly study, originally published in *Nature*, was refuted by a series of follow-up studies in several US labs (including that of the original authors) but was never withdrawn by the journal. The GM potato study, originally published in *Lancet*, was immediately criticized widely for serious flaws and was never confirmed elsewhere, but again it was not withdrawn by the journal and so it remains to be cited by those objecting to GM technology.

Public concerns about the GM potato study in the UK were exacerbated by the treatment of Arpad Pusztai, the head of the team that performed the research. He was effectively dismissed from his job, accused of scientific fraud and investigated by an Audit Commission, a working group of the UK Royal Society, and a Parliamentary Committee in London. This rather heavy-handed treatment made Pusztai into a martyr for the anti-GM movement. It also reinforced existing public scepticism about the conduct of some scientists, businesses and politicians in the wake of several other food scares including mad-cow disease, *E. coli* and salmonella outbreaks.

As with the MMR vaccine episode in the late 1990s (see above), a few flawed scientific studies about the GM technology were highly publicized by the media without any balancing mention of the hundreds of other reputable studies that give rise to contrary conclusions. Unfortunately, even in scientific circles the GM debate is often characterized in simplistic pro- and anti- terms. In reality, many independent scientists have been critical of some aspects of the commercial use of GM technology such as its slow response to consumer concerns and concentration in a few large companies. However, the vast majority of plant biologists would also agree that the technology itself is not fundamentally flawed and that it has great potential to serve alongside other crop improvement technologies in the future.

11.5 GM Crops: Consumer Reaction and Global Impact

While some aspects of modern agriculture, such as intensification and the increasing use of chemical inputs, have elicited adverse comments, crop biotechnologies such as mutagenesis, cloning, tissue culture and hybrids barely registered with most of the public. In contrast, the introduction of GM crops was highly controversial in many parts of the world. As shown in Table 11.3, the most intense reaction to GM crops was seen in Europe, but even here several food products made from transgenic crops were openly on sale until 1999. The 'tipping point' where the initial acceptance of GM crops turned into outright rejection in Europe was the Pusztai affair in 1998–1999, as discussed above.

Why were GM crops rejected in Europe?

Despite some campaigning by anti-GM groups in the early 1990s, when foods derived from GM crops first appeared in shops in the mid-1990s they were generally accepted by consumers. In North America, GM products were unlabelled so most shoppers were unaware of them. In the UK the first GM food product, a tomato paste produced by Zeneca, was voluntarily labelled by the company and was a modest commercial success. However, public disquiet increased when unlabelled GM soybeans were exported from the USA to Europe, where they could potentially be found in 60% of supermarket food items. In 1998, the UK media increasingly focused on GM foods after Pusztai's claims about abnormalities in rats were shown on national television (see above). In mid-1999, anti-GM groups such as Greenpeace picketed supermarkets, demanding withdrawal of all GM foods. By autumn all the major UK food retailers had effectively banned GM items and this soon spread to the rest of Europe.

Table 11.3. Timeline of the GM crop controversy.

Date	Area	Event	Responses
1983	Europe, USA	First reports of transgenic plant cells	Muted
1992	USA	First GM crop, Flavr Savr® tomato, released by Calgene Inc.	Modest consumer success but failed due to poor breeding lines
1996	USA	Release of GM herbicide-tolerant and insect-resistant soybean, maize, cotton and rapeseed varieties	Popular with farmers, little public response
1996	Europe	First attempts to export unlabelled GM soybeans from USA to Europe	Public protests, demand for food labelling
1997	UK	Release of tomato paste made from transgenic varieties	Modest consumer acceptance
1998	UK	Pusztai affair: claims that GM potatoes may cause abnormalities in rats	International concern, supermarket boycotts, strong criticism by scientists
1999	UK	Ewen and Pusztai publish rat work in *Lancet*	Renewed international concern
1999	USA	Monarch butterfly affair: Losely *et al.* claim GM maize causes increased death rate in monarchs	International concern, especially in Europe
1999	UK, Europe	Supermarkets ban sale of foods containing ingredients derived from GM crops	International concern, especially in Europe
2000	USA	StarLink affair: GM maize not approved for human use found widely in food products	International concern, trade boycotts, $500M cost to company
2001	USA	Chapela affair: claims that GM maize from USA is contaminating indigenous Mexican landraces	International concern, work criticized by some scientists
2001	USA	Follow-up studies refute claims that GM maize is toxic to monarch butterflies	Muted
2002	USA	Prodigene affair: vaccine-producing GM maize found in soybean crop	International concern, company fined and subsequently collapsed
2003	Europe	EU establishes threshold of 0.9% above which foods must be labelled as GM	Welcomed by consumer groups
2006	Europe	Surveys show increased opposition to GM foods	Muted
2007	USA	Commercial release of multiple 'stacked' GM traits	Muted
2010	Brazil	Locally developed GM soy approved for use	Muted
2010	Europe	Amflora™ potatoes are first EU approved GM crop for 10 years	Muted
2011	Europe	*Eurobarometer* survey shows declining opposition to GM crops, especially in developing countries	Muted

To ordinary consumers, the abrupt banning of GM foods in 1999 confirmed many of their fears about the technology. Experience of previous food scares, and especially mad-cow disease, had decreased public trust in food companies, politicians and even independent scientists. The public rejection of GM foods has been characterized as an ill-informed reaction fuelled by sensationalist campaigns by anti-GM lobbyists and the media. While there is an element of truth in this point of view, it is also important to remember that no first-generation GM crops contained traits that made them attractive to the average shopper. The agbiotech companies had traditionally regarded farmers as their primary customers and herbicide-tolerant and insect-resistant crops were designed to benefit farmers by decreasing costs and improving profitability.

However, by the 1990s, food scares and concerns about industrialized farming led to increased public scrutiny of how food was produced, processed and sold. This meant that food shoppers were now also the customers of agbiotech companies. However, unlike farmers, shoppers derived no benefit from input trait modified first-generation GM crops, which were no cheaper, better tasting or more nutritious than non-GM alternatives. In fact, according to the rules of 'substantial equivalence' consumers were assured these GM crops were essentially the same as the non-GM alternatives. For the average European consumer therefore, GM food did not offer any benefit to balance the risks being reported in the media. Also, there were many alternative non-GM foods so rejection of GM technology had little impact on consumer choice. As discussed previously, this can be compared with mobile phones where alleged risks from radiation during constant use were balanced by the immense benefits of the technology for which there was no comparable alternative.

Has the impact of GM technology been exaggerated?

Many of those objecting to GM technology say that it is unnecessary and has not been used to address key traits such as yield, drought or salinity. Indeed, as discussed above, after two decades of commercial use GM technology had only been used to manipulate a few traits in a small number of crops. However, one area where considerable impact has been claimed is that the amount of GM crops planted was rising steadily each year. According to industry sources, by the early 2010s the global area of crops carrying GM traits had almost reached 150 Mha and was still rising. This is 10% of the global arable land area, which at first sight is an impressive achievement. But such statistics can be rather misleading unless taken in the context of the overall plant breeding process. This is because crop varieties labelled as 'GM' additionally carry many other trait combinations that are the result of other forms of breeding technology. In many cases it is the latter traits that are the most useful in the crop rather than the GM traits.

For example, several agbiotech companies have used non-GM methods such as mutagenesis, wide crossing and MAS to develop new oilseed varieties with higher levels of useful fatty acids such as oleic acid. These varieties are marketed for the improved nutritional or industrial qualities resulting from their altered fatty acid profiles. However, the varieties are also classified as transgenic because in one of the final stages of the breeding process they were crossed with older GM herbicide-tolerant and/or insect-resistant varieties. In another example, between 2009 and 2011, novel traits such as drought tolerance were produced in maize via non-GM methods. However, in many cases the subsequent incorporation of a transgenic trait such as herbicide tolerance (with its stronger patent protection) caused the new varieties to be labelled as GM although their new traits were unrelated to GM technology.

In the future this could lead to almost any new crop variety being classified as GM simply because a transgene from an older variety had been crossed in at some stage of the breeding programme. Clearly, this can lead to a misleading idea of the true impact of this or any other breeding technology. Hence, by following the same logic applied to GM varieties it could be claimed that there were several hundred million hectares of 'mutation bred' crops and over a billion hectares of 'wide crossed' crops simply because these crops contained one or more traits that had been altered via mutagenesis or wide crossing. While it is true that the impact of GM technology has been very significant in the four crops where it has been widely applied, its impact has been relatively modest in terms of the bigger picture of global agriculture and the dozens of complex traits involved in crop performance. However, GM technology is still developing and may well have a much greater impact in future decades, as we will now discuss.

11.6 Future Perspectives

As discussed in Chapters 10 and 12, after over a decade of relative stagnation, new transgenic traits are gradually being produced. Many second-generation GM varieties carry traits, such as enhanced nutritional content, that are more relevant to consumers. Such varieties may be more acceptable to the relatively GM-sceptical European public. In addition, an increasing number of GM varieties have been developed in countries such as Brazil and China. During the next decade we can expect many more transgenic crops with improved protein, lipid, starch and vitamin profiles, and processing or storage properties. Many more crop species will be transformed and the technology will be used much more in developing countries, often for public-good applications as well as for commercial gain. As the utility and provenance of transgenic crops is extended, and possible risks are addressed (e.g. by eliminating antibiotic resistance genes), it seems likely that the technology will gradually become more acceptable in a similar manner to white and red biotechnologies.

Whereas first-generation GM technology was widely seen as narrowly based, irrelevant to consumers and mainly the property of corporate agribusiness, second-generation technologies will have a broader scope, addressing environmental and consumer needs, and a truly global reach in terms of ownership and application. Meanwhile, it is important that GM technology is not discussed as a polarized topic where people are either pro or anti. In reality the issues are far more complex and the technology continues to evolve in ways that are changing its context, provenance and scientific basis. The GM technology of the 2010s and 2020s will be very different from that of the 1990s and this should be recognized in public discussion of its merits and drawbacks. To conclude, GM technology is already well established in large parts of the world and, while its impact in terms of crop breeding has been exaggerated, it has great potential to serve as one of many items in the toolkit of 21st-century plant breeders.

11.7 Summary Notes

- Some new technologies become controversial due to factors such as context, provenance and direct usefulness to consumers.
- While GM technologies have been controversial, other more 'risky' technologies such as mutagenesis have elicited little public response. In part this is due to the differing provenance and context of the two technologies.
- Earlier crop-breeding technologies were developed in public-sector labs with relatively weak forms of ownership, but GM technologies were primarily developed in the private sector and benefited from strong patent protection.
- Several aspects of first-generation GM technology, such as random gene insertion and use of antibiotic resistance markers, have been controversial but can be readily avoided.
- Public concerns about GM foods were heightened by selective media reports of a few scientific studies that were later disproved, by their lack of perceived consumer benefit, and by

a distrust in government and industry following earlier well-founded food scares.
- The impact of first-generation GM technology on crop breeding and food production has often been exaggerated but new forms of the technology have great promise for the future.

Further Reading

Science, technology and controversy

Avise, J.C. (2004) *The Hope, Hype and Reality of Genetic Engineering*. Oxford University Press, UK.

Halford (ed.) (2006) *Plant Biotechnology: Current and Future Applications of GM Crops*. Wiley, New York, USA.

Kind, A. and Schnieke, A. (2008) Animal pharming, two decades on. *Transgenic Research* 17, 1025–1033.

Pringle, P. (2003) *Food, Inc. Mendel to Monsanto – the Promises and Perils of the Biotech Harvest*. Simon and Schuster, New York.

Wells, D.N. (2005) Animal cloning: problems and prospects. *Revue scientifique et technique, International Office of Epizootics* 24, 251–264.

World Bank (2007) *Agricultural Mechanization: Issues and Options*. World Bank, Washington, DC.

Institutional, scientific and social contexts of plant biotechnology

Byerlee, D. and Dubin, H.J. (2010) Crop improvement in the CGIAR as a global success story of open access and international collaboration. *International Journal of the Commons* 4, 1.

Ewen, S.W.B. and Pusztai, A. (1999) Effect of diets containing genetically modified potatoes expressing *Galanthus nivalis* lectin on rat small intestine. *Lancet* 354, 1353–1354.

Losey, J.E., Rayor, L.S. and Carter, M.E. (1999) Transgenic pollen harms Monarch larvae. *Nature* 399, 214.

Melo-Martín, I. and Meghani, Z. (2008) Beyond risk. A more realistic risk–benefit analysis of agricultural biotechnologies. *EMBO Reports* 9, 302–306.

Murphy, D.J. (2007) *Plant Breeding and Biotechnology: Societal Context and the Future of Agriculture*. Cambridge University Press, UK.

National Research Council (NRC) (2004) *Safety of Genetically Engineered Foods*. National Academies Press, Washington, DC.

Piesse, J. and Thirtle, C. (2010) Agricultural R&D, technology and productivity. *Philosophical Transactions of the Royal Society B* 365, 3035–3047.

Rommens, C.M. (2010) Barriers and paths to market for genetically engineered crops. *Plant Biotechnology Journal* 8, 101–111.

Shelton, A.M. and Sears, M.K. (2001) The Monarch butterfly controversy: scientific interpretations of a phenomenon. *Plant Journal* 27, 483–488.

12 Future Challenges for Plant Biotechnology

12.1 Chapter Overview

This final chapter looks at how plant biotechnology can help address some of the most important challenges now confronting humanity. These include food production, resource depletion, environmental degradation and climate change. We will survey some of the major traits involved in yield, quality and environmental response that will help to address these challenges. In particular, many staple crops in developing countries perform well below their maximum potential with much scope for yield improvement. Other crops are so deficient in certain nutrients that the improvement of nutritional quality is now a key objective for many breeders. We will look at several strategies aimed at the 'biofortification' of crops with nutrients such as vitamins, iron and essential oils and amino acids. Finally, two newer areas where biotechnology is playing a key role are biofuels and biopharming. We will survey new crops that could provide 'second-generation' biofuels without affecting food production. Biopharming is the production of high-value molecules in transgenic plants. We will briefly examine some plant production systems under development and their 'biocontainment' to minimize the environmental risks of this technology.

12.2 Population, Land Use and Food Production

In many senses the challenges of population growth, land use and food production are as old as agriculture itself. However, during the early 21st century these challenges have come together with factors such as climate change and globalization in a way that requires new approaches. Meeting these challenges will require the full range of plant biotechnology tools, such as MAS, 'omics, transgenesis and TILLING. But what exactly is the magnitude of the challenges posed by population, land use and food production, and how might plant biotechnology be used to address them?

Population growth is mainly in Asia and Africa

In Chapter 9 we saw that the global human population in 2000 was about 6.0 billion. This was 3 billion more than in the pre-Green Revolution era of 1960. By 2011 the population was close to 7 billion and is projected to reach over 9 billion by 2050. The need to find sufficient food for an extra 2–3 billion people over the next few decades is comparable to the challenge facing Green Revolution breeders of the 1960s and 1970s. Beyond 2050 it is projected that population growth will level off to reach a stable plateau of 10–12 billion people by the 22nd century. As long as people can be fed in sustainable ways, it might be possible to maintain such a population on a long-term basis. However, this task is made more complex because the population increase is not distributed evenly across the world and greater efforts to maximize food production will be required in some regions than others.

In general, developing-country populations are increasing much faster than the older industrialized regions and China (see Fig. 12.1). Between 2010 and 2050, the population of North America is projected to increase by about 25% while declines of 7% in China and 9% in Europe (including the former Soviet Union) are predicted. In contrast, population increases are predicted of 38% for India, 45% for Brazil and 100% for sub-Saharan Africa. In 2050, almost half of the additional global population will live in Africa, while most of the remainder split between southern Asia and South America. Therefore, although food production should be increased everywhere, it will be especially important to focus on improving the yield and quality of African crops. It is normally better to produce food locally in

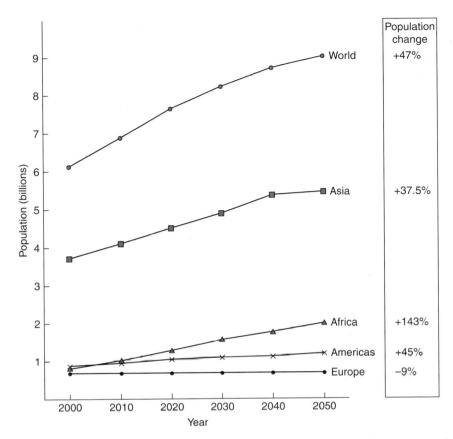

Fig. 12.1. Population increase, 2000–2050. Projections of total population size for the major continents. Data from UN Scenario of IIASA projections.

terms of cost, environmental impact and food security. During the late 20th century, enough food was produced to feed everybody in the world but due to its unequal distribution periodic famines still occurred. For example, the Ethiopian famine of the 1980s was due to a combination of drought and poor food distribution. Surplus food stocks were available but were not always made available or were too expensive to be afforded by many people in the affected areas.

Additional land for crops is available but limited

The pattern of global land use is shown in Fig. 12.2. Only about 11.5% of the total area is currently used for arable crops. Although most of this arable area is used to grow food crops, there are increasing pressures to cultivate more crops for animal feed, such as fodder cereals and soybeans, to supply growing demands for animal products in expanding economies such as China and India. More arable land is also required for fibre crops such as cotton and flax.

In Chapter 9, we saw that much of the increase in food production during the 19th and 20th centuries was made possible by the massive conversion of virgin land to support arable farming, but the conversion of land to arable use slowed down considerably during the late 20th century. It is often stated that there is little potential for further expansion of arable land and that increasing crop yields is the only option for feeding growing populations. As shown in Fig. 12.2b, almost 90% of the earth's surface is not used for crop cultivation, but most of this land is unsuitable for arable use. This is true for much of the world, and especially for the traditional regions of high crop production, such as Europe, North America and China. However, as discussed in Box 12.1, the assumption that no

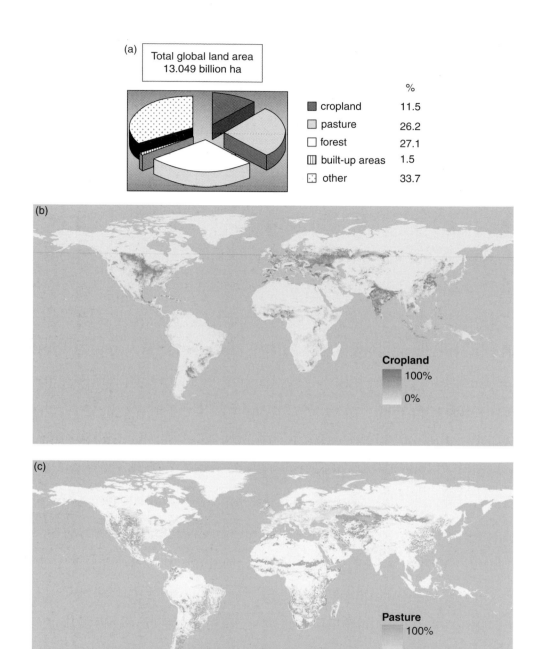

Fig. 12.2. Global land use. (a) Global land use in 1996, showing percentages of land covered by crops (11.5%), pasture (26.2%), forest (27.1%), built-up areas (1.5%) and others (33.7%), respectively. Much of the cropland is still underutilized and some pasture could be converted to arable use if there were a need to increase food production in the future. Note: data do not include Antarctica (Sources: FAO and World Resources Institute, 1998). (b) Global distribution of arable land. (c) Global distribution of pastureland.

more land is available for farming has been challenged by developments in regions such as the Brazilian cerrado, where significant expansion of arable farming began in the early 2000s.

Large areas of land could also be converted to arable use in other parts of the world. As shown in Fig. 12.2c, about 26% of the land surface is currently used for pasture. Some of this land might be convertible to arable use for producing food and/or biofuel crops. In many regions of Asia, Africa and South America, non-agricultural land is now being converted to cropland. However, in sensitive habitats such as peatland or tropical forests such conversions could have adverse environmental consequences. Therefore, although the FAO estimates that an additional 1 billion ha may be available for arable use in the next decade, the realistic figure may be much lower if consideration is given to environmental and sustainability issues. However, by using modern land-management and crop-breeding techniques, it is still likely that several hundred million hectares of new arable land could be sustainably developed. This amount of extra land could be sufficient to feed an additional 1 billion people.

12.3 Improving Crop Yield and Quality

Increasing crop yield

To what extent will it be necessary to use biotechnology to increase crop yields in the future? In the previous section we saw that it is conservatively estimated that, over the next decade or so, as much as 400–500 Mha of new land could be available for crop cultivation. This extra land could have a significant impact in raising crop production without requiring increased yields. But, even if all the extra land were used to grow food crops, it would only be sufficient to feed about 30–50% of the projected increase in human population. In reality, some of the land will also be used for non-food crops, such as grains for animal feed, fibre crops for textiles and biomass crops for biofuels. Other factors such as climate change and altered incidence of pests and diseases could also reduce crop productivity and require a compensatory increase in overall yield just to maintain current levels of production. Therefore we will still need to use modern breeding technologies to increase the yield of food crops for the foreseeable future.

Maximizing crop yield is a key aim in agriculture, but it is especially important in developing countries where population growth is highest and yields of indigenous crops are still low by modern standards. By increasing yield, more people can be fed from the same area of land. Higher yields also mean that less land is required for crop production, relieving pressure to develop pristine and often environmentally sensitive habitats such as rainforests or species-rich wetlands. We have already seen that, prior to the introduction of scientific breeding techniques in the early 20th century, average grain yields rarely exceeded 2 t/ha, even in the most favourable environments. In developed countries, the application of Mendelian genetics was an important step forward in realizing yield gains, but great progress also came from biotechnologies, such as maize hybrids (Fig. 9.7). A similar combination of breeding and management technologies could greatly increase yields in those developing-country crops that still perform well below their physiological limits.

As shown in Fig. 12.3 and Table 12.1, yield and production of the big-three cereal crops (maize, rice and wheat) are at historically high levels. Despite concerns in the early 2000s about breeders reaching a 'yield ceiling', yields of these key crops have continued to rise, largely thanks to modern biotechnologies. Use of novel approaches such as redesigning plant architecture (see below), plus the effects of warming in some regions, may allow further yield gains in the future. Crop yield is dependent on many factors, both genetic and environmental. Genetic factors that play a role in regulating yield include plant growth rate, seed/fruit/tuber size and composition, and plant architecture, as well as genes involved in environmental response such as pest/disease resistance or drought tolerance. Breeders are steadily improving their understanding of the genetics of these yield traits in existing crops in order to manipulate these often-complex characters.

Two approaches to yield improvements of particular promise are the manipulation of seed development and of plant architecture. For example, gibberellins are important regulators of plant height. Therefore, mutations or gene deletions that affect known gibberellin biosynthetic enzymes or signal transduction pathways can be predicted to result in dwarf phenotypes as seen in modern cereals. This may enable breeders to produce higher-yielding dwarf varieties of a much wider range of crops in the future, possibly including tree crops such as oil palm. Alternatively, manipulation of plant architecture could maximize yield-bearing structures, such as seeds and fruits, and reduce non-productive structures, such as unnecessary branches, thick seed coats or tall, slender stems. An example is the *tb1* gene, where a single mutation altered the morphology and cob size of the weedy teosinte plant, leading directly to the high yielding forms of domesticated maize grown today (see Fig. 7.2). Manipulation of *tb1*-like genes in other species might enable breeders to select novel plant architectures with considerably improved yields.

In grain crops, yields can be increased by developing larger seeds with higher levels of useful edible products (e.g. starch or oil). Many genes affect seed size, some of which are quite unexpected, such as ABC (ATP-binding cassette) transporters, which are important determinants of seed size in tomato and perhaps in many other plants. Seed size is a complex trait that has evolved under different selective pressures in different habitats. For example, plants producing large numbers of small seeds find it easier to disperse due to the abundance of seeds and higher likelihood to escape from predation. These are traits that are useful in wild plants but not crops. In contrast, plants with smaller numbers of large seeds often have higher tolerance to drought, herbivory, shading and nutrient-deficient soils. These large-seed traits are ideal in crops where water and nutrient availability and crowding are often limiting factors on plant performance. Large seed size is therefore highly desirable in most crops, not only because it can increase yield directly, but also because such plants are more resilient in their response to stress.

Improving crop quality

Farmers have been selecting quality traits such as improved taste or nutritional content for many millennia. In principle, seeds can now be selected or engineered to contain any specified proportion of macronutrients (starch, protein and oil) and/or micronutrients (vitamins and minerals).

Macronutrients

All crops supply one or more of the major macronutrients, starch, protein and oil, but the amount and composition of these compounds is not always

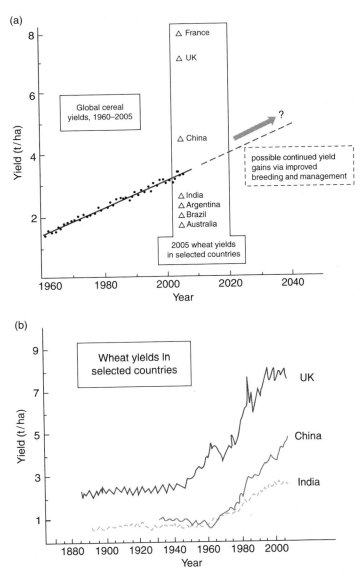

Fig. 12.3. Long-term trends in cereal yields. (a) Globally averaged trends from 1960 and projected to 2040 still show a steady increase in yields with potential for further rises with improved breeding and/or management (Source: FAOSTAT, http://faostat.fao.org). (b) Wheat yields in selected countries from 1890 to 2010 show the dramatic effects of the Green Revolution after 1960 with a plateau in UK yields after 1990 but continued rises in China well into the 2000s. Data updated from Fischer *et al.* (2009).

ideal in each crop. For example, many seed proteins are deficient in some essential amino acids while some edible oils have relatively high levels of saturated fatty acids. The type of starch, protein or oil in seeds and fruits can be manipulated by both transgenic and non-transgenic methods. In the future, more precise manipulations may be possible

to produce so-called 'designer crops'. For example, as discussed in Chapter 4, the overall amount or nutritional value of seed or tuber starch, protein or oil has been improved in several crops, and in a few cases commercial varieties are now available. Other varieties in the pipeline include: oilseed crops with enhanced monounsaturated and very

Future Challenges for Plant Biotechnology

Table 12.1. Global food production of selected major crops.

Crop	Area		Output		Yield (t/ha)
	Mha	%	Mt	%	
Wheat	241	29.4	606	20.9	2.8
Maize	158	19.3	792	27.3	5.0
Rice	156	19.0	660	22.8	4.2
Soybean	90	11.0	221	7.6	2.4
Sorghum	47	5.7	63	2.2	1.4
Rapeseed	31	3.8	51	1.8	1.6
Dry beans	27	3.3	18	0.6	0.7
Sugarcane	23	2.8	107	3.7	4.7
Sunflower	22	2.7	27	0.9	1.2
Potato	19	2.3	309	10.7	16.7
Sugarbeet	5	0.6	44	1.5	8.4
Total*	819*	99.9	2898	100.0	

*Note that this is the total area of the crops shown in the Table: the total area for *all* crops in 2011 was about 1500 Mha. Data from FAO.

long-chain polyunsaturated fatty acid contents; grain crops with lysine-enriched storage proteins; and potatoes with enhanced starch or protein compositions.

Micronutrients and 'biofortification'

Dietary micronutrients include many vitamins and inorganic minerals that are essential for normal human development but required in much smaller quantities than the macronutrients. Biofortification is the manipulation of crops or other food sources to provide improved levels of micronutrients that may be lacking in the present diet of people or livestock. Almost all crop staples are nutritionally deficient in some respect. For example: most seed proteins are deficient in one or more essential amino acids; white rice contains almost no pro-vitamin A; and many food crops are poor sources of essential minerals such as iron and zinc. Low levels of vitamins or nutrients in a food crop may not be problematic if it is one of many items in a varied diet where other foods may supply the missing vitamins or nutrients. However, for poorer populations where a single crop like rice or cassava might make up the bulk of their diet, such deficiencies can lead to malnutrition, sickness and even death.

As people become richer, the variety of foods in their diet increases and their health improves dramatically. However, even relatively affluent and well-fed (or overfed) populations may suffer from nutrient deficiencies. For example, it is estimated that only 10% of adolescent girls and 30% of boys

in the USA currently achieve the recommended intake of dietary calcium. Therefore it is important that breeders should try to improve the nutritional quality of crops wherever this is feasible. In many developing countries, staples such as rice and cassava might deliver sufficient calories to prevent starvation, but their lack of certain nutrients means that populations suffer from malnutrition. The prevalence of such undernourishment is estimated at 33% in sub-Saharan Africa, 23% in South Asia, 12% in East Asia and 10% in Latin America. Three dietary micronutrients, namely iron, zinc and vitamin A, are recognized by the World Health Organization (WHO) as particularly deficient in these regions and all three are priority targets for crop manipulation as follows.

Vitamin A

The orange-coloured pigment β-carotene found in many plant tissues is also the major dietary precursor of vitamin A group compounds such as retinol and retinal that play essential roles in processes including vision, immune response, skin and bone growth and embryo development. Due to low dietary intake of vitamin A, an estimated 127 million pre-school children are affected by vitamin A deficiency, an early symptom of which is night blindness that progresses to total blindness in up to 0.5 million children each year. A major reason for low dietary vitamin A intake is the dearth of β-carotene in staple crops such as rice, tropical maize, sweet potato and cassava.

Several breeding approaches have been taken to improving the vitamin A content of these crops but these have been made more difficult by the complexity of the genetic trait. As shown in Fig. 12.4, biosynthesis of β-carotene from geranyl geranyl pyrophosphate involves at least three enzymes, all of which may be lacking in a particular crop. The Vitamin A for Partnership Africa (VITAA) is focused on vitamin A in the sweet potato, which is the fifth most important global crop on a fresh weight basis and is especially important in Africa where traditional white varieties have little vitamin A. New orange-fleshed varieties with high vitamin A levels have now been developed using wide crosses. One future challenge is to propagate enough planting material (normally bundles of vine cuttings) to meet the high levels of demand. Other targets are to improve postharvest handling and food-preparation methods to ensure retention of β-carotene levels in foods during processing.

Although traditional cassava varieties are white, higher β-carotene forms are available and could potentially be used for breeding, but in practice this is made difficult because the crop is normally propagated vegetatively and its high heterozygosity prevents varietal recovery via backcrossing (see Chapter 9). This makes the transgenic approach attractive for cassava improvement. As outlined in Box 12.2, vitamin A enhancement in rice and maize has used modern breeding tools such as transgenesis and association genetics.

Minerals

In Asia, people subsisting mainly on cereal crops grown in mineral-deficient soils are at risk of deficiencies in such minerals as iron, zinc, calcium, copper and selenium. The WHO has identified iron and zinc as the most important minerals in terms of crop-related dietary deficiency. More than 1 billion

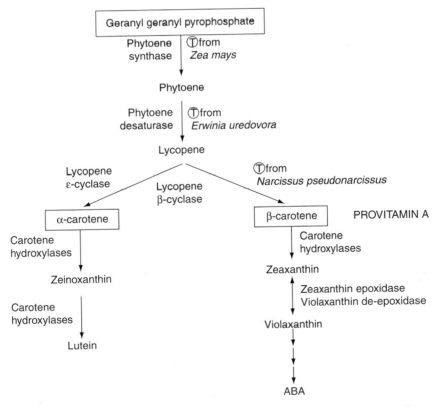

Fig. 12.4. Biosynthesis of ß-carotene (provitamin A) in transgenic 'golden rice'. Three transgenes ⊤ from maize, rice and daffodil were added to rice in order to convert geranoyl geranoyl pyrophosphate to ß-carotene. Because of the presence of ß-carotene the resulting rice grains were yellow rather than white and are popularly known as 'golden rice'.

Box 12.2. Biotechnological approaches to improving vitamin A in crops

Modern biotechnology can provide new tools to manipulate complex traits involved in crop quality as shown by the following two examples, one transgenic and one non-transgenic.

Golden rice

The best-known transgenic approach to vitamin A biofortification is 'golden rice', which was initially developed in the 1990s by a Swiss/German public-sector group. This new rice variety has yellow, rather than white, grains due to accumulation of the pigment β-carotene or provitamin A. Compared with these early varieties, later versions of golden rice have a reported 23-fold increase in provitamin A levels. Development of lab versions of golden rice was just the start of a lengthy process of backcrossing into local varieties and field tests that took more than a decade. In 2005–2007, the original golden rice trait was crossed into the popular IR64 variety at IRRI in the Philippines and outdoor field trials of 20 potential breeding lines started in 2008. Field trials of the improved golden rice variety show fivefold more provitamin A than the original lines.

The newer improved golden rice lines might be ready for release to farmers in developing countries by 2015. A further challenge will be to ensure that newly expressed provitamin A can withstand processing, storage and cooking, while remaining bioavailable after consumption. The journey of golden rice from the laboratory to the cooking pot took well over a decade and there is concern that the lengthy delays and regulatory hurdles faced by the developers of this pro-poor trait could act as a disincentive to future attempts to use transgenesis to improve food quality. One possible way forward would be for poorer countries to use open-source GM technologies or to develop their own methods as discussed in Box 12.1.

Tropical maize

Much of the maize grown as a subsistence crop in the African tropics is made up of white varieties with very low levels of β-carotene. As a result, millions of children in the region suffer from vitamin A-related blindness. Vitamin A deficiency is also a leading cause of early childhood death and a major risk factor for pregnant women in Africa. Accumulation of β-carotene is a complex trait in maize but new methods such as association mapping can now be used to identify genes that regulate such traits. Initially, conventional QTL mapping was used to confirm that two genes, *lcyE* and *crtRB1*, played a large role in regulating β-carotene levels in maize kernels. Expression profiling showed that *lcyE* is more active in the endosperm relative to the embryo and its expression was highly correlated with carotenoid levels, explaining 70–76% of the genetic variance in maize accessions.

Inexpensive PCR-based markers were then developed to differentiate all alleles of the two genes for use in MAS. The favourable alleles for both genes are rare in most maize germplasm and are especially rare in tropically adapted maize cultivars. By using several donor lines of maize and simple PCR-based marker systems, it was possible to introgress the most favourable *crtRB1* and *lcyE* alleles into tropical germplasm in a few years of breeding. This strategy has already generated new breeding lines of tropical maize with β-carotene concentrations as high as 15 μg/g. An advantage of using non-transgenic breeding technologies is the much reduced level of regulatory and IPR barriers involved in getting improved crops to farmers when compared with GM varieties such as golden rice.

people, mostly female, suffer from iron-deficiency anaemia leading to impaired childhood growth, cognitive development, resistance to infection, work capacity, productivity, and health in pregnancy. Annually, >0.8 million deaths are attributed to iron deficiency. Although dietary supplements of iron can be provided these are relatively expensive and evidence shows that coverage of at-risk populations is generally poor. For example, a study in India showed that only 30% of pregnant women and 10% of adolescent girls were using iron-folate supplements provided by the government.

In contrast, it is estimated that biofortification of iron-deficient crops, such as rice, would be both more effective and much cheaper in the long term compared with the use of supplements. Unfortunately, iron storage in rice is a complex trait that might involve as many as 43 genes belonging to five protein families. In a trial study, rice has been genetically engineered to have higher levels of iron, coupled with a reduced amount of phytates, which can reduce the uptake of iron in the gut. The iron-rich transgenic rice was produced by inserting three groups of transgenes: a ferritin gene from the bean

Phaseolus vulgaris doubled the overall iron content of the grain; a phytase gene from the fungus *Aspergillus fumigatus* decreased phytate levels; and an endogenous cysteine-rich metallothionein was added to stimulate dietary uptake of iron. So far this work, which was carried out by the same Swiss-based university group that produced the original 'golden rice', has not been developed into a crop-breeding programme and transgenic iron-rich rice is still a rather distant prospect.

It may also be possible to improve zinc content using transgenic methods. Zinc is essential for a healthy immune system and its deficiency in children is associated with poor growth, reduced motor and cognitive development, and increased susceptibility to infectious diseases. Most transgenic studies on iron and zinc are being carried out by university groups in industrial countries. In contrast, plant breeders in developing countries have focused almost exclusively on non-transgenic strategies. Genetic variation has been found for traits, such as high iron and zinc contents and low phytate levels, in major cereal crops such as rice. In particular, breeders at IRRI in the Philippines have found that some existing aromatic varieties of rice contain double the normal amount of iron, as well as greatly enhanced levels of zinc.

Crop-breeding institutes such as IRRI are concentrating on this approach to 'biofortification' of staple crops as part of the CGIAR Global Challenge Program on Biofortification. The target is to use non-transgenic breeding strategies to produce improved grains that are enriched in iron, zinc, vitamin A, selenium and/or iodine, depending on the dietary needs of various target populations in different parts of the world. Breeders at IRRI have already shown that new rice varieties can be produced with fivefold higher amounts of iron in the post-milled grains. The main advantage of plant-breeding approaches for mineral enhancement is that investment is only required at the research and development stage, after which the nutritionally enhanced crops provide micronutrients at no additional cost. Furthermore, mineral-rich plants tend to be more vigorous and more tolerant of biotic stress, which means that yields are likely to improve with mineral content.

12.4 Improving the Environmental Responsiveness of Plants

The importance of the environment in determining plant growth and development was discussed in Chapter 6. In this section we will examine how biotechnology might help mitigate the impact of environmental stresses resulting from climate change and industrialization. Changes in global temperature, atmospheric CO_2 and ozone, and altered rainfall patterns will greatly alter the biotic and abiotic stresses experienced by crops. By learning more about these processes it may be possible to develop new trait combinations to maintain or increase crop yields despite such challenges. Although the exact nature of future climate change in each region of the world cannot be predicted, it is clear that many crops will require modification of stress-tolerance traits. Key abiotic traits are responses to drought, salt and heat/cold stresses. Biotic targets are less predictable as the threats are more numerous and tend to be crop specific, but the development of new surveillance and rapid response methods will be especially useful here.

Climate change

Due to changes in climate patterns, the environment in many crop-growing regions is likely to alter considerably. While it may be possible to mitigate some of these changes in the long term by reducing greenhouse gas emissions, climate scientists predict that significant changes will still occur during the remainder of the 21st century. By 2050, it is predicted that atmospheric CO_2 concentrations will reach 550 ppm and ozone will reach 60 ppb (Fig. 12.6). Although there will be considerable regional differences, average global temperatures may be about 2°C higher than in 2000 and rainfall will be reduced in some key crop-growing regions. Each of these climate-related factors is likely to have some effect on plant productivity, although the impact of these and other climate-related factors on agriculture remains uncertain. This uncertainty makes it difficult to develop specific crop traits that will enable us to cope with climatically related threats, as discussed in the case of CO_2 in Box 12.3.

Although elevated CO_2 levels could increase crop yields substantially, especially in C3 crops like wheat, many of these gains could be cancelled out by other factors such as drought. However, higher CO_2 levels may give breeders additional opportunities to increase yields under the new conditions. Higher ozone levels will decrease crop yield by at least 5%, but it may be

Box 12.3. Will crop yields increase with higher atmospheric CO_2 concentrations?

Between 1960 and 2010, atmospheric CO_2 concentrations increased from 315 to almost 390 ppm. This was primarily due to the combustion of fossil fuels for energy and transport. Over the coming decades, continuing release of CO_2 may result in concentrations as high as 550 ppm. The effects of these levels of CO_2 on crop yields are difficult to predict because several different physiological processes may be affected in different ways in different crops.

Photosynthesis

The major effect will be on rates of photosynthesis in C3 plants where CO_2 levels are normally rate limiting (see Chapter 2). This means that the photosynthetic rates in C3 crops such as wheat and soybean could increase by as much as 40% if CO_2 levels reach 550 ppm, although increases in grain yield would be somewhat lower. However, this effect would be much reduced in C4 plants because their internal CO_2 levels are already three- to sixfold higher than atmospheric concentrations. Therefore, yields of C4 crops such as maize, millet, sorghum and sugarcane may increase only marginally at higher atmospheric CO_2 concentrations.

Other processes

Rising atmospheric CO_2 concentrations will affect other physiological processes that can have a significant effect on crop yields. Some pests, such as aphids and weevils, respond positively to elevated CO_2. The quality of the crop may be adversely affected as shown in the case of wheat where higher CO_2 levels result in reduced protein deposition during grain development. This means that, although grain yields may increase in wheat, flour quality will deteriorate. As well as encouraging crop growth, higher CO_2 levels will stimulate the growth of other plants, especially weedy C3 species. This may encourage new, more vigorous weeds able to adapt faster than crops to elevated CO_2 conditions. Finally, increased CO_2 levels will tend to reduce water consumption by all crops, but any consequent yield gains may be cancelled out by the effect of increased temperatures on evaporation rates.

Future predictions

Given the uncertainty about the extent of future CO_2 increases and the many different ways in which CO_2 can affect plant processes, it is impossible to predict precisely how crop yields will respond on a global basis. In a series of papers from the Royal Society published in 2010, it was concluded that overall crop yields would benefit from higher CO_2 levels. However, this yield increase will be offset to some extent by other aspects of climate change, such as reduced rainfall. Ironically, if as a result of reduced global emissions CO_2 levels are reduced or stabilized well below 550 ppm, crop yields may fall even more rapidly than they would if we maintained slightly higher levels of CO_2.

Implications for biotechnology

This analysis of the impact of higher CO_2 levels on crop yields does not give us any clear targets for key traits that could be manipulated to cope with the new conditions. The uncertainty about the impact of CO_2 and other environmental factors reflects our lack of knowledge about the physiological effects of these complex and interactive processes on plant growth and development. The general message is that we need more research into the impact of the environment on a wide range of agronomic traits in all the major crop groups.

---

possible in some crops to select varieties that are less prone to ozone damage. Better drought tolerance and response to higher temperatures are clearly major traits that should have high priority in crop breeding programmes. Providing all available tools of plant breeding can be used, overall crop yields could be increased by 50% by 2050, despite the impacts of climate change. However, this will only be possible by the manipulation not just of yield itself but also the indirect effects of climate change and especially the stress-related traits of crops.

Abiotic stress

Globally, abiotic stresses already severely affect over 170 Mha of farmland. They are a particular threat to crop yields in regions impacted by climate change or increasing soil salinization. Drought and salinization are already the most common causes of sporadic famine in arid and semi-arid regions, and are one of the most significant threats to agricultural productivity. For example, as much as one-third of global arable land may become degraded by salinization over the next 25 years.

Although abiotic stress is often regarded as an external (i.e. environmental) factor in crop performance, there is also a great deal of untapped genetic variation in the responses to such stresses in the major crop groups. In particular, genetic diversity within crop groups, whether in the form of wild relatives, conserved landraces or other genetic resources, can be a powerful source of useful variation for abiotic stress tolerance.

Drought and aridity are the major water-related stresses for most crops, but flooding can also be a serious problem that may be exacerbated by climate change. In some crops, such as rice, flooding tolerance appears to be a relatively simple genetic trait. This has enabled breeders to use hybrid technology to transfer a submergence tolerance gene *Sub1* from the Indian rice 'FR13A' to the IR64 rice variety 'Submarino 1' that was distributed to Asian farmers in 2010. Another component of abiotic stress tolerance in crops that has been much neglected by researchers and breeders is the rhizosphere. While the structural and inorganic components of the rhizosphere have been well studied, very little work has been done on biological communities, such as rhizosphere bacteria that can both promote plant growth and reduce the impact of stresses such as drought, salinity and poor soil nutrition.

It has been claimed that there is significant potential for transgenesis in modifying stress-related traits. However, our limited knowledge of stress-associated metabolism in plants makes this difficult to achieve in practice. Another problem is the synergistic effect of different stresses on crop performance. It is often the combination of such stresses that reduces crop yield in the field, rather than the effect of a single category of stress. While molecular biologists have tended to focus on single stresses applied in highly controlled environments, recent studies have shown that simultaneous application of several stresses gives rise to unique responses that are not predicted by extrapolating from effects of stresses given individually. The simultaneous presence of multiple stresses is the norm in open environments, so the success of molecular approaches to their relief in crops will probably require a broader approach by researchers and breeders in the future (see Chapter 6).

Salinity

Globally, salt and nutrient stresses affect over 100 Mha of farmland. Efforts to select salt-tolerant crop varieties, while partially successful, have been hampered by the complexity of the trait and the number of genes involved. Salt tolerance in most crops in the field is a complex multigene trait that has evolved differently in several plant groups. However, there have been some successes in developing salt tolerance in model plants in the lab. For example, transgenic tobacco engineered to accumulate elevated levels of mannitol was able to withstand high salinity. Lab and small-scale field studies have shown that transgenic plants accumulating betaine or trehalose have enhanced salt tolerance. Finally, rapeseed plants expressing an *Arabidopsis* vacuolar transport protein tolerated as much as 250 mM sodium chloride (about half the concentration of seawater and enough to kill most crops) without significant impact on seed yield or composition.

Despite these encouraging reports, it is not clear whether such simple modifications will have a sustained effect on crop yields in more complex real-world cropping systems, where osmotic stress is often linked with a combination of additional factors such as periodic aridity, mineral/salt build-up and/or erosion. A more broadly based approach being pursued in the public sector is the CIMMYT 'Genomics towards gene discovery' project. This is aimed at understanding complex traits, such as salt or drought stress, via genomics, using crops where extensive genetic maps are already available. These crops include Andean roots and tubers, barley, cassava, chickpea, coconut, cowpea, finger millet, forages, groundnut, lentil, maize, *Musa*, pearl millet, beans, pigeon pea, potato, rice, sorghum, soybean, sweet potato, wheat and yam. An important aspect of the project is to investigate interactive effects of stress responses with the aim of developing multiple stress tolerance in crops.

Drought tolerance

Drought tolerance, like salt tolerance, appears to be regulated by complex sets of traits that may have evolved separately in different plants depending on the dynamics (i.e. timing and intensity) of water shortage in each case. In the near future, it is likely that aridity will increase in several parts of the world, with the FAO estimating that, by 2025, 1.8 billion people will be living in regions of water scarcity. This will be caused by factors such as localized lower rainfall due to climate change, or the diversion of upstream water supplies from rivers, e.g. for dams or irrigation. In the case of rice alone, over 70 Mha are already affected by drought stress. Given the predicted increase in long-term aridity, it is surprising that until relatively recently there have been relatively few well-resourced attempts to produce drought-tolerant crops, even by publicly funded organizations.

Basic research using reverse genetics and other genomic approaches is beginning to give clues about some aspects of drought tolerance mechanisms. For example, it was reported that the *erecta* gene, involved in transpiration efficiency, might regulate some of the genetic variation for drought tolerance in *Arabidopsis*. As with salinity, advanced non-transgenic breeding methods can be used to improve the performance of existing drought-tolerant crops in arid regions. One of the most important of these crops is pearl millet, which is grown on over 40 Mha in Africa. The similarity in gene order, or synteny, between the pearl millet genome and that of the other major cereals means that, once their loci are identified, drought-tolerance traits could be potentially introduced into local varieties via MAS. At present, most of the transgenic approaches to drought tolerance are under way in the private sector. They include collaborative ventures between several major multinational agbiotech companies aimed at commercial crops such as maize. The resulting varieties are likely to carry traits such as enhanced root growth for maize grown under high-input conditions.

Biotic stress

The climatic factors discussed above will greatly affect patterns of crop pests and diseases. While we have a fairly good idea of how to improve crop yield or quality, we cannot predict the pests that might be in a particular region in 20 years' time. Therefore, the best strategy to cope with biotic threats will be to develop as many biotechnology tools as possible so that we are better able to detect and then deal with them promptly. This will be especially important for surveillance and diagnostics where molecular tools can be used with great effect. It will also be important to learn more about the major groups of crop pathogens and how they can be better controlled with biological as well as chemical tools. In Table 12.2, it can be seen that, in the absence of conventional crop protection agents (i.e. biocides, see Chapter 9), losses to biotic threats would be approximately doubled in the three major cereal crops. This illustrates the potential seriousness of new biotic threats resulting from climate change. Biotechnology can help with the detection of such threats and can also provide new bio-based approaches to crop protection that may reduce the need for chemical biocides in the future.

The severe impact of pathogens and the limitations of chemical control have stimulated efforts to genetically engineer resistance in crops. In the case of viral or insect resistance, where single-gene resistance traits are common, several approaches have been successfully developed (see Chapter 10). But this has been more difficult to address for many fungal and bacterial pathogens where resistance traits are often more complex, although there have been some successes here as well. For example, the *Xa21* bacterial blight resistance gene has been transferred to five rice varieties. Antifungal agents such as phytoalexins or chitinases have also been expressed in plants, and a gene encoding an antifungal defensin protein from lucerne was transferred to potatoes. The transgenic potatoes became resistant to the soil-borne oomycete *Verticillium dahliae*. However, the resistance was relatively

Table 12.2. Yield losses due to biotic threats for three major crops. Actual crop losses are compared with estimated losses in the absence of physical, biological or chemical protection for wheat, rice and maize.

Biotic threat	Actual losses			Potential losses without crop protection agents		
	Wheat	Rice	Maize	Wheat	Rice	Maize
Weeds	7.7	10.2	10.5	23.0	37.1	40.3
Animal pests	7.9	15.1	9.6	8.7	24.7	15.9
Pathogens (fungi, bacteria, viruses)	12.6	12.2	11.2	18.1	15.2	12.3
Total	28.2	37.4	31.2	49.8	77.0	68.5

Data from Oerke (2006).

Chapter 12

specific and the potatoes were still susceptible to the late-blight pathogen *Phytophthora infestans*.

Newly emerging threats

New crop pests and diseases are constantly emerging and, with global transportation and trade, they can spread rapidly across the world. Climate change is likely to worsen the situation as it will create new opportunities for such threats to arise. Modern biotechnologies can be used as part of surveillance and breeding programmes to first detect and then combat such threats. For example, one of the most serious crop diseases to emerge in recent years is a highly virulent strain of the wheat black stem rust, *Puccinia graminis*. The new rust, called Ug99, first emerged in Uganda in 1998–1999, spread around eastern Africa in the early 2000s and may spread further to the major wheat-growing areas of the Indian subcontinent.

The new Ug99 strain has overcome most of the rust resistance genes bred into wheat over the past 50 years since the beginnings of the Green Revolution. Over 1 billion people live in potentially affected areas and annual wheat production of 120 Mt is threatened. The serious threats to food security posed by Ug99 and other emerging crop pathogens can only be satisfactorily addressed by international efforts using all available technologies. The Ug99 threat is now being tackled by the Borlaug Global Rust Initiative, which is a multinational programme including CIMMYT, ICARDA, Gates Foundation, FAO and USDA. It is likely that other new threats will emerge in the coming years, which makes it particularly important both to maintain and increase crop monitoring around the world and to prioritize research on improved approaches to combating biotic threats.

Disease diagnostics

Biotechnologies can contribute to combating newly emerging threats in the areas of surveillance/detection and in breeding for resistance. It is likely that initial detection of the Ug99 wheat rust outbreak was delayed due to a reduction in the disease monitoring after a period of 40 years without outbreaks. In the future, improved molecular kits such as microarray-based systems might enable surveillance to be carried out more extensively and cost effectively. By their very nature, new threats are an unknown quantity, but the more we understand about the relationships between crops and pest/disease organisms in general, the better placed we will be to mount rapid and effective responses.

As discussed in Box 12.4, the rapid identification of new pathogens, and especially their genome sequences, will facilitate development of control strategies based on previous experience with related disease organisms. Such measures have already been of immense benefit in the case of new human or animal pathogens, such as the coronavirus that causes SARS (severe acute respiratory syndrome) or the virulent influenza A-type viruses that can cause human-lethal versions of avian or swine flu. For example, within a few days of the April 2009 outbreak of influenza A (H1N1) in Mexico, the genome sequence of the new variant of the virus was publicly available on the Internet, enabling the rapid development and global distribution of vaccines.

Bioprotection

Bioprotection is the use of biologically based crop protection systems against threats such as pests and diseases. One example is so-called 'biological control', which can be defined as: 'the use of living organisms to reduce the population density or adverse impact of pest organisms'. As an alternative to biological control, the crop itself can be bred to express resistance to pests, either by using endogenous genetic variation or genes introduced by transgenesis. Another form of bioprotection, and one of the commonest methods found in developing countries, is the use of microbial agents as biopesticides (see Fig. 12.5). Often these agents have the additional benefit of replacing chemical pesticides that might be unaffordable and/or environmentally undesirable. As a result, the use of microbial pesticides such as Bt toxins (see Chapter 9) and biocontrol agents such as pheromones, growth regulators and hormones is growing steadily.

A related strategy is integrated pest management (IPM). For example, in Malaysia, Bt sprays are used to control insect pests of oil palm such as the bagworm group and the rhinoceros beetle (*Oryctes rhinoceros*) and large-scale Bt production facilities have been set up. In combination, strains of the insecticidal fungus *Metarhizium* and the virus *Oryctes* are used to control the rhinoceros beetle. Using 'white biotechnology' methods of scale-up, the *Metarhizium* Technology Centre in Australia has produced nearly 0.5 t of pure *Metarhizium*

Box 12.4. Disease diagnostics

Biotechnology offers important tools to diagnose plant diseases of all kinds. These tools are of particular value when the identification of the causal agent is difficult (e.g. many viral diseases exhibit similar symptoms), or when detailed knowledge of the pathogen is necessary to develop proper management measures. As well as DNA-based methods, immuno-diagnosis, including enzyme-linked immunosorbent assay (ELISA) and monoclonal antibodies, is commercially applied in many countries. Diagnostic techniques are also routinely used for quarantine systems and for production of seeds and other propagation materials. The importance of prompt diagnosis is shown by the initial failure to detect the virulent Ug99 pathogen of wheat (see main text) as it spread through eastern Africa. By 2010, this pathogen had reached some of the major wheat-growing regions of Asia and threatened to become a global disease. It is clear that similar disease outbreaks could affect other staple crops with potentially disastrous results.

Rice is an even more important staple than wheat, feeding over 3 billion people. Despite many breeding advances, between 10% and 30% of the rice crop is still lost to the rice blast fungus, *Magnaporthe oryzae*. This highly adaptable pathogen is constantly mutating to produce more virulent strains. New strains that are able to overcome resistance genes that have been bred into many commercial rice varieties.

Molecular marker techniques are being used to monitor *M. oryzae* populations and to detect such genetic changes. Examples include RAPD and RFLP markers and the use of PCR directed at DNA transposons in *M. oryzae*. These methods can quickly establish accurate DNA fingerprints in large numbers of samples so that extensive regions can be routinely monitored. A newly developed fingerprinting method called retrotransposon-microsatellite amplified polymorphism (REMAP) has recently been used to target repetitive DNA elements as part of this surveillance effort.

In the future, the availability of high-throughput, low-cost, next-generation genome sequencing technologies (see Chapter 10) should facilitate the rapid and accurate discrimination between benign strains and new pathogenic strains of this and other crop disease agents. Such technologies will be especially useful in enabling the presence of several different pathogens to be determined simultaneously. In particular, the emerging field of metagenomics, where the genomes of large groups of species can be sequenced, can be used to study and manage the extensive microbial communities that live on or interact with rice and other staple crops. Data derived from such a 'community biology' approach will enable us to follow the dynamics, movement and interactions of such pathogen assemblies on crop plants.

---

fungal spores for future crop treatments. In India, Bt sprays produced at village level by fermentation technology have been used for crop protection. Yet another example of biocontrol is the 'sterile insect technique'. This involves release of large numbers of factory-reared sterilized male insects into target areas. By mating with females in the native pest population, the sterile insects can drastically reduce or even eradicate pest populations.

Fungi are being increasingly used as highly specific pest management agents that can often replace chemical pesticides. One example is the desert locust, a sporadic pest that can severely affect food production over wide areas of northern Africa. Between 2003 and 2005, conventional control via chemical sprays required 42 million l of mainly organophosphate pesticides sprayed across 13 Mha. While there were no reported instances of serious animal or human health problems, safety measures were expensive and there was significant environmental damage. As an alternative, FAO and other partners developed combined bio- and chemical control strategies. These involve *Metarhizium* fungi, which are pathogens of locusts and grasshoppers, plus the biocontrol agent phenyacetonitrile, which is a naturally occurring hormone that affects swarming behaviour in locusts. One isolate of *Metarhizium anisopliae* is the proprietary agent Green Muscle®, which is produced commercially by a South African company and used in the field with considerable success. Biocontrol methods can be used alongside conventional chemical control as cheaper, more effective and less environmentally damaging approaches to crop management.

12.5 Improving the Sustainability of Agriculture

Modern agriculture is an energy-intensive process with a large environmental footprint in terms of energy consumption, emissions and ecological effects.

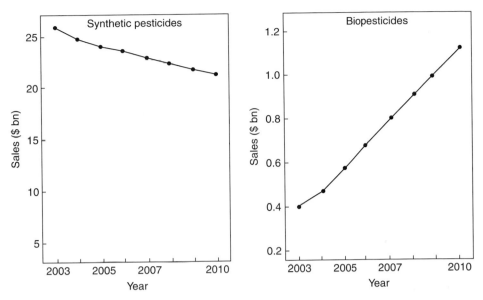

Fig. 12.5. Sales of synthetic pesticides and biopesticides, 2003–2010. As a result of more efficient application methods and more stringent controls, sales of synthetic pesticides declined by about 1.5% per annum after 2003. At the same time the biopesticide sales increased almost sixfold. These trends are likely to continue as new biopesticides and transgenic crops expressing biocides are developed in conjunction with the application of integrated pest management.

Energy is required to construct and run the many machines that have largely replaced human and animal power, and also to synthesize chemical inputs such as fertilizers and biocides. Almost all of this energy is derived from non-renewable fossil sources such as gas, oil and coal. Other non-renewable inputs include potash fertilizer, which is extracted from mineral deposits. Farming and associated land-use change accounts for over 30% of global greenhouse gas emissions. Examples of land-use change include the felling or burning of tropical forests and exposure of tropical peatlands to air as they are converted to farmland and generate huge amounts of atmospheric CO_2.

Many other pollutants are generated from agricultural systems, including runoff into watercourses of excess fertilizer and animal slurry. Agricultural land already occupies 38% of the world's surface and this is likely to increase as populations increase in both numbers and affluence. Historically, agriculture has had profound effects on biodiversity with the loss of wild habitats to farming being a major cause of plant and animal extinctions. As agriculture continues to expand in extent and intensity, its environmental footprint will increase further and these adverse effects will multiply and become more difficult to halt or reverse.

The particular form of energy-intensive agriculture practised in the early 21st century may not be sustainable in the long term. This applies especially to the use of non-renewable resources such as fossil fuel or conventional mineral potash deposits. There are many biotechnology-based approaches that can be used to improve the sustainability of modern agriculture without drastic effects on its proven ability to produce sufficient food for an expanding world population. A few examples that we will examine below include strategies for reducing some chemical inputs and the generation of renewable fuels to partially replace fossil fuels.

Reducing chemical inputs

In some cases, agrochemicals can have undesirable side effects, such as runoff of surplus fertilizer or the toxicity of biocides to non-target species. The development of new biological methods to deliver fertilizers and biocides therefore has the potential to improve the energy and environmental footprint of agriculture without jeopardizing crop yields. As shown in Fig. 12.5, while sales of synthetic pesticides declined from 2003 to 2010, those of biopesticides increased almost sixfold, although the latter still only make up about 5% of global pesticide sales.

Although the widespread use of agrochemicals has greatly increased crop yields, such chemicals are often relatively energy-intensive to manufacture, transport and apply in the field. The production of synthetic ammonia for crop fertilizers currently consumes about 5% of global natural gas production, or just under 2% of world energy output. Natural gas is overwhelmingly used for the industrial synthesis of ammonia, but other energy sources, together with a hydrogen source, can also be used for the production of nitrogen compounds suitable for fertilizers. The cost of natural gas makes up about 90% of the cost of producing ammonia. The increase in price of natural gases over the past decade, along with other factors such as increasing demand, has contributed to increases in fertilizer prices and stimulated the search for biological alternatives.

Potassium and phosphorus for fertilizers come from mines and such resources are limited. More effective fertilizer utilization practices may decrease present usage from mines. Improved knowledge of crop-production practices can potentially decrease fertilizer usage of P and K without reducing the critical need to improve and increase crop yields. Biofertilizers include renewable biological sources of all categories of plant fertilizers such as nitrogen, potassium and phosphorus compounds. They also include the use of nitrogen-fixing bacteria and/or mycorrhizal fungi to improve plant performance. Recent studies have shown that plant-growth-promoting rhizobacteria can enhance nutrient uptake by crops and also induce systemic tolerance to other abiotic stresses such as drought and salinity. As with biopesticides, use of bionutrition strategies carries the benefit of reducing input costs for farmers as well as preventing nitrate and phosphate accumulation in soils and runoff into sensitive watercourses.

These strategies can be used as alternatives to chemical supplements that also enhance the nutritional status of crops. Biofertilizers can provide farmers with an alternative to chemical inputs as well as improving the quality of their soils. For example, rhizobial inoculants can replace chemical fertilizers for the production of soybean, groundnut and mung bean crops. In Kenya, a *Rhizobium* inoculant, known as BIOFIX, has been developed for sorghum crops and elsewhere in Africa biofertilizers are being developed for cowpea, groundnut and rice. Another *Rhizobium*-based biofertilizer for common bean has been commercialized in Mexico since 2003 under the commercial name of Rhizofer. The biofertilizer is sold either on its own or together with spores of the mycorrhizal fungus *Glomus intraradices* to help the plant acquire soil nutrients and to solubilize phosphates. The use of this biofertilizer can significantly reduce costs compared with chemical fertilization, as well as improving soil biodiversity and biological activity.

The nutritional status of the soil can also be enhanced by using fungal inoculants to accelerate the breakdown of organic fertilizer. In the Philippines, inoculation of rice straw with the fungus *Trichoderma* reduced composting time to as little as 21–45 days. Production units have now been set up for the fungal agent and rice and sugarcane farmers adopting this technology use significantly less chemical fertilizer, have higher crop yields and higher net incomes. Here, the main advantage of replacing chemical fertilizer with organic fertilizer was its effect on soil nutrient content as well as soil tilth and texture, making it superior to chemical fertilizers. In addition, organic fertilizers can be used in conjunction with local reactive phosphate rocks and elemental sulfur to stimulate the release of phosphorus from phosphate rocks for plant nutrition and biological nitrogen fixation. These examples from developing countries are now being studied for possible adoption in more intensive farming systems, including the USA and Europe.

12.6 Biofuels

The dramatic increase in carbon emissions from fossil fuels and the resulting increase in atmospheric CO_2 levels are shown in Fig. 12.6. As well as being implicated in climate change, these emissions are unsustainable in the long term because reserves of fossil fuel are limited and non-renewable. Estimates of reserves vary but accessible oil and gas supplies may be severely depleted by the end of the century (Fig. 12.6c). Well before this happens, fuel will become prohibitively expensive for many uses. One strategy to replace fossil fuels is to use plant-derived oils or starches as renewable alternatives. However, these 'first-generation' biofuels are controversial because they come from existing food crops such as maize, sugarcane, oil palm and rapeseed and could exacerbate food shortages. This has led to research into new second- and third-generation biofuel crops, such as lignocellulosic biomass to produce

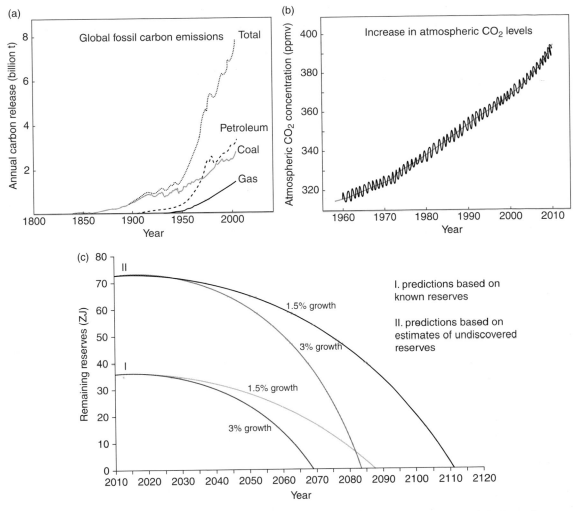

Fig. 12.6. Fossil fuel emissions, 1800–2005, and atmospheric CO_2 concentrations, 1960–2010. (a) Carbon release from the combustion of fossil fuels increased steadily during the first half of the 20th century, but accelerated dramatically after 1950 at a rate that shows few signs of slowing down through the 2010s. (b) Due to the increased combustion of fossil fuels, atmospheric CO_2 concentrations rose by about 25% between 1960 and 2010. (c) Predicted rates of fossil fuel depletion based on already-known reserves (lower lines) and estimated global reserves including undiscovered fields (upper lines). Data from Stephens *et al.* (2010).

bioethanol and microalgae for conversion to biodiesel. As shown in Fig. 12.7, global energy demand is projected to increase steadily in the coming decades, mostly due to rising populations and industrial activity in countries such as China, India and Brazil. Renewable fuels such as biofuels will account for a small but steadily increasing proportion of this demand (see Fig. 12.8), so it is important that higher biofuel production does not result in decreased food production.

First-generation biofuels

During the 2000s many governments encouraged the use of first-generation biofuels as a way to decrease net CO_2 emissions. Between 2006 and 2010 the amount of bioethanol produced from starch crops increased from 53 to 95 Mt while biodiesel from oil crops increased from 7 to 20 Mt. The justification was that carbon emitted by fuels derived from crop photosynthesis would be refixed

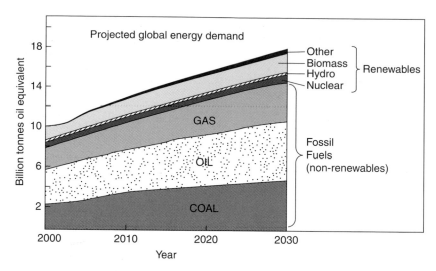

Fig. 12.7. Projected rate of global energy demand, 2000–2030. For much of the 21st century, global energy demand will continue to be met largely from non-renewable fossil fuels, although the proportion of energy from plant-based biomass and other renewables will increase as fossil sources become scarcer and more expensive. Data from World Energy Outlook (2007).

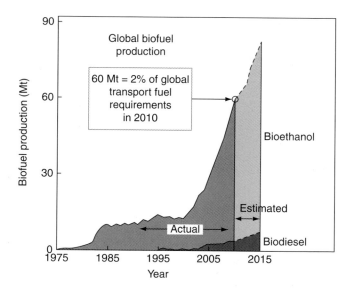

Fig. 12.8. Global production of biofuels, 1975–2015. Biofuel production, mostly as bioethanol and often obtained from food crops, increased dramatically after 2000 and is projected to continue to rise as non-crop alternatives, such as microalgae and lignocellulosic crops, are developed. However, the 50 Mt of biofuel produced in 2010 fulfilled a mere 2% of global transport fuel requirements. Data from Licht and Berg (2006).

the next time the crop was grown, resulting in a much lower net release of CO_2 compared with non-renewable fossil fuels. However, this assumption has been disputed because crops also require fossil energy for their inputs and cultivation. While some biofuel crops such as tropical oil palm or sugarcane may have a favourable carbon balance, the environmental credentials of crops like maize or sugarbeet are less certain. For example, the financial and energy costs of planting, weeding, spraying

Chapter 12

(with energy-requiring chemical inputs), harvesting, transporting and processing a biofuel crop may offset much of the gain in using a carbon-renewable fuel versus the fossil-derived alternative.

A second criticism of first-generation biofuels is that they come from crops normally grown for human food and/or animal feed. As noted above, there is increasing pressure on arable land to feed the growing world population, which can make it difficult to justify using land to grow biofuel crops to power vehicles for relatively well-off people in Europe and the Americas. So can plant biotechnology help develop new sources of renewable biofuels that do not impact on food production? Potential solutions include the development of new biofuel crops that grow on land that is unsuitable for food production. An alternative tropical oil crop being developed in countries such as India is *Jatropha curcas*, whose oil-rich nuts can potentially yield 1.5–2.5 t/ha biodiesel. Unfortunately, while *Jatropha* has been successfully grown on land that is unsuitable for food crops, its oil yields and other performance traits have been disappointing and will require considerably more R&D in the future.

Second-generation lignocellulosic crops

Lignocellulosic or biomass crops such as *Miscanthus*, poplar, switchgrass and willow are grown as sources of woody or fibrous materials. Many of these crops will grow on land that is unsuitable for arable farming and therefore do not compete with food crops. At present the woody stems of conventional biomass crops are normally burned in small-scale power plants to produce electricity for local consumption, but this is a relatively inefficient process. A better alternative is to convert lignocellulosic biomass to liquid fuels such as bioethanol via microbial fermentation.

Many companies are investing in bioengineering suitable biomass crops and the microbes required to process them to bioethanol. For example, transgenic and other breeding technologies may be used to alter the cellulose/lignin content of the plant so that it can be more easily digested and converted to ethanol. Transgenic approaches are also being used to engineer yeasts and other microbes into more efficient systems to ferment the various types of lignocellulosic biomass produced by different crops. While there are many challenges in scaling up lab-based processes to commercially feasible dimensions, new investment in these technologies makes it likely that there will be significant progress in developing lignocellulosic feedstocks over the next decade. This technology could also be applied to process the fibrous residues of other crops, such as cereal stubble, into more effective biofuels.

Third-generation microalgal fuels

Some species of microalgae can accumulate as much as 50% of their mass as storage lipids similar to those found in oil crops (see Chapter 4). The lipid is normally stored as triacylglycerols accumulated either in cytosolic oil bodies or in plastidial bodies called plastoglobuli. These triacylglycerols can be easily converted to biodiesel via transesterification. Examples of lipogenic microalgae include many species of green algae, diatoms and cyanobacteria. Microalgae normally grow much faster than land plants and, providing high-oil varieties can be identified and cultured on a sufficiently large scale, they could generate large quantities of biodiesel. As an extra bonus, some diatoms and green algae grow well in brackish or salty water and can therefore be cultivated without needing to use scarce freshwater resources.

It has been estimated that, if existing microalgae could be grown on a large scale without yield losses, they could produce 10–50 times more biodiesel per hectare than conventional first-generation crops such as rapeseed or oil palm. If microalgae were grown in shallow ponds over an area of 0.2 Mha, 1 billion t of biodiesel could be produced each year. Therefore an area of 3.8 Mha of non-arable land could potentially generate sufficient biodiesel to replace all petroleum transport fuels currently used in the USA. This is only 2% of the current arable hectarage in the USA and a mere 0.4% of the total land area of the country. The potential of microalgae as biofuel production systems is explored further in Box 12.5.

12.7 Biopharming

The first transgenic plants expressing human proteins were produced in the 1980s but commercial development of biopharming has faced several technical challenges. These include low rates of recombinant protein accumulation, difficulties in extracting active proteins from plant tissues, and variations in post-translational modification patterns compared with humans. For example, in several cases, plant-produced proteins, including

Box 12.5. Engineering microalgae to produce biofuels

Although they will require large land areas for economic cultivation, many microalgae grow well in sunny, warm habitats such as shallow ponds. It may therefore be possible to set up ponds as microalgal bioreactors in arid and little-used areas such as the Colorado Desert of the western USA. Currently both open and closed bioreactor systems are being tested. However, many technical challenges will have to be overcome before microalgae can fulfil this vision of a cheap, renewable alternative to fossil fuels. In particular, it is proving difficult to maintain oil yields as microalgal cultures are scaled up and contamination by other non-lipogenic algae and pathogens is a serious problem.

As well as improved bioreactor design, several transgenic and mutagenesis approaches are being used to improve oil yields. One strategy is to reduce the size of photosynthetic light-harvesting antenna systems to avoid excess light capture. Another approach is to transfer the genes responsible for oil secretion from slow-growing species such as *Botryococcus braunii* to more productive microalgae such as *Chlamydomonas* spp. If microalgae can be engineered to secrete oil, this will collect on the surface and could be continuously harvested rather than the current less efficient method of periodically collecting algal cells and extracting the oil from them.

Further into the future, microalgae and bacteria may be harnessed to produce gaseous fuels, such as biogas and biohydrogen. For example, *Chlamydomonas* can produce hydrogen under anaerobic conditions in the light and this can be enhanced by transgenic manipulation of photosystem II activity. Conversion of various forms of biomass to hydrogen can also be achieved using fermentation and/or anaerobic digestion. Potential feedstocks range from crops to organic waste, including sewage. Microalgal biofuel production therefore has considerable promise but it will probably take several decades before it is economically feasible.

---

follicle-stimulating hormone, measles virus and acetylcholine esterase, which were therapeutically active, but not necessarily glycosylated in the correct way. Despite being therapeutically active, their different glycosylation patterns compared with their human counterparts means that they may be treated as new drugs requiring more rigorous, expensive and lengthy clinical trials than if they were human-identical. However, in other cases, e.g. an immunoglobulin produced in transgenic lucerne, the glycosylation pattern was exactly the same as in humans. In addition to these technical problems, biopharming has faced considerable regulatory hurdles and social concerns (see Chapter 11).

The future of plant biopharming is uncertain, but the development of rigorously contained indoor production systems, rather than using openly grown crop plants, may enhance their acceptance. Because plant-based systems are not susceptible to infection by human viruses, they have safety advantages over mammalian systems. For example, in 2009, production of glucocerebrosidase (used to treat Gaucher's disease in which this enzyme is inactive) in mammalian cell lines was halted due to viral contamination. Within weeks a plant-derived version of the enzyme from carrot cell cultures was fast-tracked for use instead. Soon afterwards, the pharmaceutical company Pfizer Inc. acquired rights

to develop and commercialize this product. This is one of the first major pharmaceutical companies to adopt biopharming and should this trend continue the technology might have a very bright future indeed.

Improved systems for biocontainment

Biopharming has the potential to produce a huge range of useful pharmaceuticals safely and in large quantities at relatively low prices. However, in order to satisfy existing European regulatory and environmental requirements for the manufacture of therapeutics, improved production platforms will be necessary. In Chapter 11, we saw some of the adverse consequences of using mainstream food crops as biopharming production platforms, especially when these were grown in the same regions as edible varieties of the same crops that were destined for animal or human food chains. The following three alternative production platforms would greatly reduce such risks and enhance future prospects for biopharming technologies.

Non-food crops

One of the best non-food systems is tobacco, *Nicotiana tabacum*, a widely grown but inedible crop that is increasingly frowned upon as a source of an addictive and life-threatening cocktail of drugs and toxins. In technical terms, tobacco is one of the most amenable plants for genetic engineering, and its leaves can express therapeutic proteins at high yields. Tobacco seeds have also been used to express several viral proteins. Another, even more secure, strategy is to grow non-food plants in rigorously controlled containment zones rather than in open fields. Examples include polytunnels (inexpensive, tubular plastic versions of glasshouses) or in tightly sealed glasshouses from which even tiny pollen grains cannot escape. The biosecurity of such crops can be improved still further by use of inducible transgenes or by growing them as tissue or cell cultures. For example, hairy-root cultures of tobacco have been used to express murine monoclonal antibodies, human alkaline phosphatase, green fluorescent protein and ricin B subunit.

Non-crop plants

The small model plant *Arabidopsis* has been used to overexpress the VP2 protein from the chicken bursal disease virus and several bacterial-derived proteins. The two model legume species *Medicago truncatula* and *Lotus japonicus* have also shown promise as production systems for proteins such as phytase and a major house dust mite allergen. Hydroponic culture has been used to grow transgenic duckweed, *Lemna gibba* and *Lemna minor*. This is now being scaled up for commercial production of recombinant α-interferon (used for treatment of hepatitis C). A particular advantage of duckweed is that recombinant proteins are secreted from the plant into the culture medium from where they can be readily purified, hence avoiding problems associated with extraction from plant tissues. Also, because clonal *Lemna* plants are vegetatively propagated and do not produce pollen or seeds, the production system has a high degree of inbuilt biocontainment. In Europe, several plant-derived therapeutics have been approved for topical use (i.e. on the skin) and several products are in clinical trials, including α-interferon (Locteron®) from *Lemna minor*.

Lower (i.e. non-vascular) plants and algae have also been used as expression systems. For example, transgenic varieties of the moss *Physcomitrella patens* are able to secrete recombinant human growth factor and human serum albumin directly into the culture medium. The N-glycosylation patterns in *P. patens* are similar to higher plants and may allow for expression of many animal and human glycoproteins. Several multicellular algal species have also been transformed, and the common brown seaweed, kelp or *Laminaria japonica*, has been assessed as a potential marine bioreactor.

Cell/tissue cultures

Unicellular species, such as the green microalga *Chlamydomonas reinhardtii*, have been used to express a variety of human therapeutic proteins. The halophilic microalga *Dunaliella salina*, which is already cultured as a source of β-carotene, has been transformed to overexpress the hepatitis B surface antigen. In the future, many other species of microalgae and cyanobacteria could be developed for contained, reactor-based pharmaceutical production, especially where external release of such agents is problematic. For example, transgenes have been inserted into at least four species of diatom, namely *Phaeodactylum tricornutum*, *Cyclotella cryptica*, *Cylindrotheca fusiformis* and *Navicula saprophila*.

Cell cultures of higher plants are also being developed for commercial biopharming and in 2006 Dow Agrosciences became the first company in the world to receive regulatory approval for a plant-made vaccine for animals via its Concert™ plant-cell-produced system. This method is being used for expression of a range of animal vaccines, including those against livestock pathogens such as avian influenza. Carrot cultures are also being used commercially to produce human glucocerebrosidase. Chloroplast transformation of plant cell cultures is a way to further improve product accumulation rates and enhance biosecurity. Numerous studies have been published on the use of tobacco cell cultures in biopharming and a human cytokine granulocyte-macrophage colony-stimulating factor has been expressed in a number of plant cell culture systems, including tomatoes. The advantages of such improved containment systems is shown by the fact that, although approval for field-grown biopharmed crops in Europe remains unlikely, several human therapeutics from indoor-grown *Lemna* and plant cell cultures are already in clinical trials.

12.8 Summary Notes

- Increasing global populations in the 21st century will require increased food production. Although it may be possible to convert large areas of unused land to crop production, higher crop yields will also be required.

- Biotechnologies can contribute to improving crop yield and nutritional quality, especially in developing countries where needs are greatest.
- Crop biotechnologies can help mitigate the biotic and abiotic environmental threats associated with climate change and human activities such as pollution.
- New bio-based fertilizers and biocides will decrease requirements for chemical inputs and may improve the environmental footprint of agriculture.
- Biotechnology will be essential to the development of new generations of more efficient and sustainable biofuels that can partially replace fossil fuels without affecting food production.
- Plant systems have immense potential as safe and inexpensive manufacture of high value compounds such as pharmaceuticals and veterinary products.

Further Reading

Population, land and food production

Cremaq, P. (2010) Brazilian agriculture. The miracle of the cerrado. *Economist* 26 August 2010.

Jaggard, K.W., Qi, A. and Ober, E.S. (2010) Possible changes to arable crop yields by 2050. *Philisophical Transactions of the Royal Society B* 365, 2835–2851.

United Nations (2010) *Technology and Innovation Report: Enhancing Food Security in Africa through Science, Technology and Innovation.* United Nations, New York.

Various (2011) *The Future of Food and Farming, Final Project Report.* The Government Office for Science, London. Available at: http://www.bis.gov.uk/assets/bispartners/foresight/docs/food-and-farming/11-546-future-of-food-and-farming-report.pdf

World Resources Institute (1998) *World Resources 1998–99: Environmental Change and Human Health.* World Resources Institute, Washington, DC. Available at: www.wri.org/publication/world-resources-1998-99-environmental-change-and-human-health.

Increasing crop yield and quality

Farre, G., Ramessar, K., Twyman, R.M., Capell, T. and Christou, P. (2010) The humanitarian impact of plant biotechnology: recent breakthroughs vs bottlenecks for adoption. *Current Opinion in Plant Biology* 13, 219–225.

Fischer, R.A., Byerlee, D. and Edmeades, G.O. (2009) *Can Technology Deliver on the Yield Challenge to 2050?* FAO, Rome, Italy.

Gómez-Galera, S., Rojas, E., Sudhakar, D., Zhu, C., Pelacho, A.M., Capell, T. and Christou, P. (2010) Critical evaluation of strategies for mineral fortification of staple food crops. *Transgenic Research* 19, 165–180.

Nassar, N. and Ortiz, R. (2010) Breeding cassava. *Scientific American* May 2010, 78–84.

Oerke, E.C. (2006) Crop losses to pests. *Journal of Agricultural Science* 144, 31–43.

Improving the environmental responsiveness of plants

Blum, A. (2005) Drought resistance, water-use efficiency, and yield potential – are they compatible, dissonant, or mutually exclusive? *Australian Journal of Agricultural Research* 56, 1159–1168.

Gornall, J., Betts, R., Burke, E., Clark, R., Camp, J., Willett, K. and Wiltshire, A. (2010) Implications of climate change for agricultural productivity in the early twenty-first century. *Philisophical Transactions of the Royal Society B* 365, 2973–2989.

Jaggard, K.W., Qi, A. and Ober, E.S. (2010) Possible changes to arable crop yields by 2050. *Philosophical Transactions of the Royal Society B* 365, 2835–2851.

Semenov, M.A. and Halford, N.G. (2009) Identifying target traits and molecular mechanisms for wheat breeding under a changing climate. *Journal of Experimental Botany* 60, 2791–2804.

Improving the sustainability of agriculture

Figueiredo, M.V.P., Burity, H.A., Martinez, C.R. and Chanway, C.P. (2008) Alleviation of drought stress in the common bean (*Phaseolus vulgaris* L.) by co-inoculation with *Paenibacillus polymyxa* and *Rhizobium tropici. Applied Soil Ecology* 40, 182–188.

Gyaneshwar, P., Kumar, G.N., Parekh, L.J. and Poole, P.S. (2002) Role of soil microorganisms in improving P nutrition of plants. *Plant Soil* 245, 83–93.

Piesse, J. and Thirtle, C. (2010) Agricultural R&D, technology and productivity. *Philosophical Transactions of the Royal Society B* 365, 3035–3047.

Yang, J., Kloepper, J.W. and Ryu, C.M. (2009) Rhizosphere bacteria help plants tolerate abiotic stress. *Trends in Plant Science* 14, 1–4.

Zhang, H., Kim, M.S., Sun, Y., Dowd, S.E., Shi, H. and Paré, P.W. (2008) Soil bacteria confer plant salt tolerance by tissue-specific regulation of the sodium transporter HKT1. *Molecular Plant Microbe Interactions* 21, 737–744.

Biofuels and biopharming

Davies, M. (2010) Commercialization of whole-plant systems for biomanufacturing of protein products: evolution and prospects. *Plant Biotechnology Journal* 8, 845–861.

Hu, Q., Sommerfeld, M., Jarvis, E., Ghirardi, M., Posewitz, M., Seibert, M. and Darzins, A. (2008) Microalgal triacylglycerols as feedstocks for biofuel production: perspectives and advances. *Plant Journal* 54, 621–639.

Karg, S.R. and Kallio, P.T. (2009) The production of biopharmaceuticals in plant systems. *Biotechnology Advances* 27, 879–894.

Kayser, O. and Quax, W. (eds) (2006) *Medicinal Plant Biotechnology: Basic Research to Industrial Apps.* Wiley, New York, USA.

Licht, F.O. and Berg, C. (2006) Presentation at World Biofuels, Seville.

Murphy, D.J. (2007) Improved containment strategies in biopharming. *Plant Biotechnology Journal* 5, 555–569.

Sexton, S., Zilberman, D., Rajagopal, D. and Hochman, G. (2009) The role of biotechnology in a sustainable biofuel future. *AgBioForum* 12, 130–140.

Stephens, E., Ross, I.L., Mussgnug, J.H., Wagner, L.D., Borowitzka, M.A., Posten, C., Kruse, O. and Hankamer, B. (2010) Future prospects of microalgal biofuel production systems. *Trends in Plant Science* 15, 554–564.

World Energy Outlook (2007) International Energy Agency, Paris. Available at: http://www.worldenergyoutlook.org/2007.asp

Glossary and Abbreviations

ABCE model: an update of the previous ABC model of floral development, also known as the quartet model, whereby angiosperm flower pattern is determined by four classes of homeotic genes.

Abiotic stresses: non-living, environmental factors that may be harmful to growth or development of an organism: examples include drought, salinity and mineral deficiency (see Biotic stresses).

Abscisic acid, ABA: plant hormone involved in regulation of developmental processes including leaf senescence, seed storage product accumulation, seed desiccation, bud and seed dormancy, and stomatal closure.

Agbiotech: agricultural biotechnology – the use of biotechnology tools in agriculture.

Agrobacterium tumefaciens: bacterial species able to infect plants and cause crown gall disease. Carries the Ti plasmid and is used as a vector for the introduction of transgenic DNA into plants.

Alkaloid: diverse group of compounds involved mainly in plant protection and consisting of multiple ring structures containing nitrogen and carbon atoms most of which are synthesized from amino acid precursors.

Allele: one or more variants of a gene.

Allelopath: chemical agent released by a plant to inhibit growth of neighbouring plants, sometimes including members of the same species.

Allometric growth: pattern of growth found in many animals in which the body grows such that its overall proportions are broadly maintained and final size and shape are relatively fixed (see Indeterminate growth).

Allopolyploidy: see Polyploid.

Ampicillin: semi-synthetic β-lactam antibiotic used as a resistance marker in plant transformation.

Amplified fragment length polymorphism (AFLP): DNA marker (see MAS).

Anabolism: metabolic synthesis of complex molecules, normally requiring energy.

Aneuploidy: presence of an incomplete set of chromosomes in a genome such that the total chromosome number is not an exact multiple of the haploid chromosome number.

Angiosperms: seed bearing, flowering plants making up the largest and most diverse extant group of higher plants. Divided into the monocots and dicots, each of which contain many species of major economic importance.

Antibiotic resistance marker: gene encoding resistance to an antibiotic and included in a transgene cassette to facilitate selection of transgenic cells.

Antisense DNA: strand of DNA in the complementary or antisense orientation to a target gene sequence. Expression of antisense DNA in a transgenic organism leads to formation of the equivalent antisense mRNA, which can form inactive duplexes with sense mRNA causing reduced expression of the target gene (see RNA interference).

Antithetic hypothesis: a proposal that, during plant evolution, the diploid sporophyte generation was interpolated into the life cycle of charophytes through an ever more extensive delay of meiosis after fertilization.

Apomixis: form of non-sexual plant reproduction without meiosis or fusion of gametes in otherwise normal reproductive structures. Progeny are genetic clones of the parent plant (see Vegetative propagation).

Apoplast: extracellular network of interlinked cell walls that extends throughout a plant and acts as a pathway for water and solute movement.

Arabidopsis thaliana: also known as thale cress, a small plant of the Brassicaceae used in plant research owing to its rapid growth rate and small genome size.

Artificial seed: plant embryos produced *in vitro* and encapsulated in a semi-solid coating such as gelatin.

Assimilates: see Assimilation.

Assimilation: conversion via photosynthesis of simple molecules such as minerals and CO_2 into more complex molecules and structures such as carbohydrates, proteins, cells and tissues. Assimilates such as sucrose and amino acids are transported from sources to sinks through the phloem.

Autopolyploidy: see Polyploid.

Auxin: class of plant hormone involved in regulation of developmental processes including cell elongation, apical dominance, adventitious root formation and ethylene synthesis.

Avirulence (AVR) gene: AVR gene products in pathogens (also known as 'race-specific elicitors') are recognized by the products of resistance (R) genes in plants. A precise match between an elicitor from the pathogen and the corresponding receptor encoded by an R gene induces a strong defence response.

Avirulent: strain of a pathogenic species that is unable to establish infection in a potential plant host.

Backcross: genetic cross involving a hybrid and one of its parents.

Basta™: see Glufosinate

Biocide: agent used to kill an unwanted organism. Examples include herbicides, pesticides, fungicides, bactericides and nematicides.

Biocontainment: biologically based strategy to prevent the release of products from a plant production platform, such as an outdoor-grown or indoor crop or tissue/cell culture system, into the wider environment.

Biofortification: the manipulation of crops or livestock to provide enhanced levels of dietary vitamins or minerals, such as calcium or iron.

Biofuels: renewable fuels derived from recently living organisms as alternatives to non-renewable fossil fuels. Existing biofuels include bioethanol from cereals and biodiesel from oil crops (see First-generation and Second-generation biofuels).

Bioinformatics: organization and analysis of complex biological data sets such as DNA sequences, protein structures or metabolic pathways. Data are analysed by specialized software such as algorithms.

Biopharming: the use of transgenic plants or plant-derived production systems to produce high-value products such as vaccines or antibodies.

Biotic stresses: organisms such as pathogens, pests or competitors that adversely affect other organisms (see Abiotic stresses).

Blue biotechnology: biotechnological manipulation of aquatic organisms such as fish.

Brassinosteroids: group of plant hormones containing a sterol ring structure.

Breeding: the deliberate manipulation of biological organisms to produce new varieties that are useful to humans.

Bryophytes: group of relatively small non-vascular, seedless plants descended from the earliest plant colonists of the land.

Bt: *Bacillus thuringiensis* – bacterium producing insecticidal protein toxins. The Bt toxin gene has been transgenically added to some crops to provide inbuilt insect protection, and live Bt sprays are used as insecticides by some organic farmers.

Callus culture: mass of undifferentiated cells that develops when tissue explants are cultivated *in vitro*. Callus can grow as a solid mass of tightly packed cells or as a more fluid suspension of individual cells or small cell clumps.

Caryopsis: dry fruit in which the pericarp is tightly fused to the seed, to which it remains attached at maturity. All cereal crops, e.g. rice, wheat, maize and barley, produce grains that are caryopses rather than true seeds.

Catabolism: metabolic breakdown of complex molecules, often with release of energy.

cDNA: complementary DNA, made by transcribing a segment of mRNA into DNA using the enzyme reverse transcriptase.

Centromere: specialized region of a chromosome to which spindle fibres attach during cell division to form a chromosome pair. Many repetitive DNA elements tend to occur at or close to centromeres.

Cereal: cultivated species of the grass family, Poaceae, with relatively large starch-rich grains or caryopses. Important crops include maize, rice, wheat and barley.

CGIAR: Consultative Group on International Agricultural Research – an alliance of countries, international and regional organizations, and private foundations supporting research centres working to improve crops for the public good.

Charophyte: relatively complex group of green algae from which land plants are probably descended.

Chimera: organism made up of two or more genetically distinct cell types.

Chromatin: DNA/protein complex consisting of loosely packed euchromatin and more tightly bound heterochromatin.

Chromatography: group of techniques used to separate compounds in a mixture.

Chromosome: one of several DNA-protein structures making up a genome. Prokaryotes and organellar genomes consist of many identical copies of a small circular or branched chromosome. Eukaryotic nuclear genomes typically contain between 4 and 80 pairs of large linear chromosomes per cell.

CIMMYT: Centro Internacional de Mejoramiento de Maiz y Trigo (International Maize and Wheat Improvement Center) – a CGIAR crop improvement centre.

Clones: genetically identical organisms, normally derived from asexual reproduction or via propagation by humans. Although clones are genetically identical, they may differ phenotypically due to epigenetic or other environmental influences.

Codon: group of three DNA or RNA bases that specify a particular amino acid.

Collenchyma: living cells with thickened walls made up of cellulose with an irregular secondary thickening.

Complementary DNA: see cDNA.

Constitutive: expression of a gene in most or all tissues and developmental stages of an organism. This applies especially to 'housekeeping' genes involved in such key processes as respiration and protein synthesis/turnover.

Containment: process of restricting access to and escape by biological materials, such as genes, cell extracts, viruses, whole cells, tissues and whole organisms, that may be of concern. Containment can be achieved by physical methods, such as sealed laboratories or production facilities. Alternatively, biocontainment may be used in addition to or instead of physical containment.

Copy number: number of copies of a transgene incorporated into the genome of a recipient organism. Multiple transgene copies may lead to reduced gene expression via mechanisms such as RNAi or co-suppression.

Corn: originally a term for any grain crop, particularly cereals such as wheat, barley or rye. More recently, in North America maize has become known as corn.

Co-suppression: down-regulation of expression caused by presence of additional copies of part or all of a gene.

Cotyledons: the first or seed leaves of seed plants; they are present in the seed before germination.

Among angiosperms, eudicots and magnoliids have two cotyledons and monocots have one. After germination they may remain in the seed or are carried above ground to become the first photosynthetic organs.

Cultivar: cultivated variety of a crop – such varieties have normally been selected by breeding and are adapted for a particular climatic region or agricultural use.

Cyanobacteria: ancient class of photosynthetic prokaryotes that use two chlorophyll-containing reaction centres to carry out oxygenic photosynthesis using water as a reductant. All plastids in algae and plants are descended from cyanobacteria.

Cytokinins: plant hormone class involved in regulation of cell division, leaf expansion, and retardation or inhibition of apical dominance and leaf senescence.

Dermal tissue: external tissue around a plant that prevents water loss and acts as a barrier to invasion by pests and diseases such as fungi, bacteria, viruses and insects. Dermal tissue includes the epidermis, stomatal cells and trichomes.

Desaturase: enzyme inserting a double or triple carbon–carbon bond into a molecule.

Dioecious: having unisexual flowers or cones. Individual plants of a dioecious species carry either male or female flowering parts or cones (see Monoecious).

Diploid: cell or organism with two normally non-identical but often highly similar copies of each chromosome in the genome (see Haploid and Polyploid).

Domestication: process of adapting plants or animals for human use.

Dominant trait: allele that is expressed as a phenotype; see Recessive trait.

ELISA: enzyme-linked immunosorbent assay – highly sensitive method to detect polypeptide(s) in biological extracts.

EMBRAPA: Empresa Brasileira de Pesquisa Agropecuária (Brazilian Agricultural Research Corporation) – major public sector crop R&D organization in Brazil.

Endosymbiosis: form of symbiosis where one organism lives within another. Mitochondria and plastids are derived from endosymbiotic prokaryotes.

Epigenetics: relating to the transmission of information resulting in an alteration of gene function from a cell or multicellular organism to its descendants without that information being encoded in the DNA sequence of the gene(s) in question. Epigenetic changes can be caused by differences in DNA methylation or by histone modifications that affect chromatin architecture.

EPSP synthase: 5-enolpyruvyl shikimate 3-phosphate synthase – a key enzyme in the biosynthesis of aromatic amino acids. The herbicide glyphosate, also called Roundup™, is an inhibitor of the plant form of EPSP synthase, with lethal consequences for any affected plant. Transgenic plants containing a bacterial version of EPSP synthase are tolerant to this herbicide.

Escherichia coli: bacterium commonly found in the gut and also much used in commercial biotechnology and in molecular biological research.

EST: expressed sequence tag – a small portion of a gene that can be used to help identify unknown genes and to map their positions within a genome.

Ethylene: gaseous plant hormone involved in regulation of developmental processes including fruit ripening, abscission, senescence and lateral cell expansion.

Euchromatin: see Chromatin

Exon: DNA sequence in a gene that encodes regions of the primary RNA transcript that are removed during processing to form the mature mRNA (see Intron).

Ex situ **conservation:** the maintenance of biological specimens away from their normal habitat, normally under closely controlled conditions, such as in an arboretum (trees), or botanical (plants) or zoological (animals) garden. The term also refers to the keeping of stocks, such as seeds, cuttings or other propagules in germplasm repositories.

FAO: Food and Agriculture Organization – United Nations agency, set up in 1945, 'to raise levels of nutrition, improve agricultural productivity, better the lives of rural populations and contribute to the growth of the world economy'.

First-generation biofuels: fuels derived from biological sources such as crops. Examples include bioethanol from cereal crops and biodiesel from oil crops.

Flavr Savr™: an early transgenic tomato variety developed by Calgene.

Fruit: structure developing from the ovary wall in flowering plants that normally contains one or more seeds and may be fleshy or dry. The vast majority of edible crop products arise from fruits, including the cereal caryopsis, legume seedpod, and fleshy fruits such as apples, citrus and banana.

Functionality: as applied to fatty acids, the way in which a saturated hydrocarbon chain is modified after *de novo* synthesis to create functional groups such as double or triple bonds and epoxy or hydroxyl residues.

Gamete: a normally haploid egg or sperm cell formed after the reductive divisions of meiosis in male or female sporogenous tissues; see also Zygote.

Gene: DNA sequence comprising a transcription unit that is the template for mRNA formation, plus flanking non-coding regions involved in its regulation.

Gene pyramiding: see Gene stacking.

Gene silencing: total or partial abolition of gene function that may be deliberate or may be an unwanted side effect of transgenesis.

Gene stacking: insertion via various breeding technologies of several traits or genes into a plant to create an improved phenotype.

Genetic mapping: use of genetic techniques to construct a genome map.

Genetic marker: gene or DNA sequence that exists in two or more readily distinguishable forms or alleles and whose inheritance can therefore be followed after a genetic cross, enabling the gene/sequence to be mapped.

Genetic use restriction technology (GURT): breeding tool to ensure that a plant variety cannot usefully be propagated by the grower. Male-sterile or non-true breeding hybrids were early types of GURT. Newer GURTs often use transgenic methods to produce phenotypes such as pollen or seed sterility.

Genome: the genetic complement of an organism, including functional genes and non-coding DNA. The principal genome of eukaryotes resides in the nucleus but much smaller genomes are present in mitochondria and plastids.

Genomics: description and analysis of genomes.

Genotype: genetic constitution of an organism; see also Phenotype.

Germplasm: the genetic material of an organism. The term often relates to the collection or conservation of seeds, cuttings, cell cultures or other germplasm resources, in repositories such as gene banks.

Germplasm conservation: the identification and protection of biological resources. Plants may be conserved as whole organisms, propagules (seeds, pollen or cuttings), or as isolated DNA. Conservation may be *in situ*, i.e. by protecting the intact plant together with its adjacent community in its natural habitat, or *ex situ*, i.e. in a botanical garden or seed bank.

Gibberellins (GA): class of plant hormones that promotes stem elongation in buds and dormancy in seeds.

Glufosinate: also known as phosphinothricin, a non-selective herbicide and inhibitor of glutamine synthase in plants, disrupting photosynthesis and causing death within a few days. Transgenic glufosinate-tolerant plants expressing a bacterial phosphinothricin acetyltransferase gene are known under the trade names of Basta™ and Liberty Link™.

Glyphosate: a non-selective herbicide also known under the trade name Roundup™. Glyphosate is an inhibitor of the shikimate pathway enzyme EPSP synthase, resulting in a lethal blockage of protein synthesis. Transgenic glyphosate-tolerant plants expressing a bacterial EPSP synthase are known as RoundupReady™.

GM: genetically modified or genetically manipulated – a cell or organism into which DNA (or occasionally RNA) has been transferred. The transferred DNA may be a copy of DNA from a different organism or may be copied from the same organism. Alternatively, the transferred DNA may be completely synthetic and hence of non-biological origin. An organism containing any of these categories of introduced gene is called transgenic.

GMO: Genetically modified organism (see GM).

Golden rice: one of several transgenic rice varieties that express at least three additional plant or bacterial enzymes enabling them to accumulate sufficient β-carotene (provitamin A) to make their grains yellow or golden in colour.

Grain: seed-like structures of grain crops that are their major harvested products, e.g. all cereals and many grain legumes such as soybean and lentils. In the strict botanical sense, a grain refers only to the caryopsis of a member of the Poaceae.

Gramineae: plant taxon including grasses and cereals; see Poaceae.

Green biotechnology: use of advanced biotechnological tools for the improvement of plants, algae or cyanobacteria (see Red and White biotechnology).

Green Revolution: period during the 1960s and 1970s when development of disease-resistant, semi-dwarf wheat and rice varieties and greater use of fertilizers led to a huge increase in crop yields, especially in developing countries in Asia.

Ground tissue: comprises the bulk of the primary plant body. Parenchyma, collenchyma and sclerenchyma cell types are common in the ground tissue.

GURT: see Genetic use restriction technology.

Gymnosperms: major group of seed bearing but non-flowering higher plants, most of which are coniferous trees, including pines, firs, spruces and cycads.

Haploid: a eukaryotic cell or organism with a single set of chromosomes.

Heat-shock proteins (HSP): class of proteins involved in responses to a wide range of abiotic stresses including but not limited to heat shock.

Herbivore: plant-eating animal.

Heterochromatin: see Chromatin.

Heterosis: also called hybrid vigour, occurs when hybrids of genetically distinct (but often inbred) parents are more vigorous than either parent.

Histones: small basic nuclear proteins that form the DNA binding structures called nucleosomes, which are the fundamental units of chromatin.

Homeotic genes: genes encoding proteins that determine organ identity, e.g. the body plan in animals or flower pattern in plants.

Homologous recombination: method of targeted insertion of transgenes into predetermined locations in a genome that is still under development for plants.

Horizontal gene transfer: naturally occurring transfer of DNA segments, including entire genes, from one species to another unrelated species. Many genes have been transferred between plants, animals, viruses and bacteria and this is an evolutionarily important mechanism for generating new genetic diversity.

Hormone: small organic molecules that regulate plant growth and developmental processes; they and their synthetic analogues have numerous commercial uses.

Host resistance: occurs when a single plant variety is resistant to a specific pathogen but other members of the same species may be susceptible to the pathogen.

Hybrid: progeny of a cross between genetically distinct parents; hybrids may be fertile or sterile. Hybrids are most commonly formed by sexual crossing between compatible organisms, but cell fusion and tissue culture now allow hybridization between more distantly related organisms. Intraspecific (same species) and interspecific (different species) hybrids are key resources used in plant breeding.

ICARDA: International Center for Agricultural Research in the Dry Areas (Syria) – a CGIAR crop improvement centre.

Inbreeding depression: a reduction in fitness and vigour of individuals as a result of increased homozygosity through continual inbreeding in a normally outbreeding population.

Indeterminate growth: flexible pattern of body growth where final size and shape can be markedly influenced by the environment (see Allometric growth).

Input trait: a genetic character that affects how the crop is grown without changing the nature of the harvested product. For example, herbicide tolerance and insect resistance are input traits that assist crop management, but they do not normally alter seed or fruit quality or other so-called output traits that are related to the product(s) obtained from the crop.

In situ conservation: the maintenance of a species or population in its normal biological habitat. In the case of crops it is especially useful in the preservation of traditional landraces and/or their wild relatives, many of which are under threat.

Intron: DNA sequence in a gene that encodes regions of the primary RNA transcript that are removed during processing to form the mature mRNA (see Exon).

Inverse PCR: method to locate or 'tag' a known DNA sequence, such as a transgene, in a genome by digesting total DNA using restriction enzymes with no sites in the known sequence. One of the DNA fragments will contain the entire insert plus its flanking regions.

IPR: intellectual property rights – legal entitlement to own and exploit an invention for a specified period, often 10–20 years. Applies to many transgenic organisms.

IRRI: International Rice Research Institute (Philippines) – a CGIAR centre focused on rice improvement.

Jasmonates: plant hormone group involved in regulation of developmental and stress-related processes including pathogen responses and seed development.

Landrace: a genetically diverse and dynamic population of a given crop produced by traditional breeding and often seen as useful sources of novel genetic variation.

Lectin: a plant protein that binds reversibly to mono- or oligosaccharides, often acting as toxins in animals where they cause agglutination of cells.

Liberty Link™: transgenic plants tolerant to the herbicide glufosinate that express a bacterial phosphinothricin acetyltransferase gene.

Lignin: a complex heterogeneous polymer based on phenylpropane units that is the major structural component of wood.

Lignocellulosic biomass: fibrous and woody plant material, such as stems, that can be harvested and used to generate ethanolic biofuels via fermentation.

Lipid: substance soluble in organic solvents, e.g. phospho- and glycolipids, sterols, waxes and triacylglycerols. In plants, lipids function as the matrix of biological membranes, as storage reserves and as hormone-like mediators.

Lodicule: small structure at the base of grass flowers that expresses typical whorl 2 genes. It is probably a former petal or corolla that became redundant once grasses developed wind pollination and has now acquired a new function.

LOD (least odds of divergence) score: a measure of the likelihood that a particular QTL phenotype is linked to a group of physical markers on a chromosome.

LTR: long terminal repeat – a common class of retrotransposon.

MADS-box genes: genes encoding MADS-box transcription factors. In plants, they include developmental regulatory genes comparable to homeotic genes in animals. The protein region encoded by the highly conserved MADS box is called the MADS domain and is part of the DNA-binding domain.

Management: applied to crops, it includes their cultivation, harvesting and processing for human use.

MAS: marker-assisted selection – use of biochemical or (more commonly) DNA-based markers linked to genes of interest that enable the presence of such genes to be readily tested for in large breeding populations.

Mb: megabase, i.e. 1 million bases, of DNA.

Meristem: localized region of mitotic cell division contributing to newly formed tissue and classified as apical, primary or secondary.

Metabolome: list of the metabolites detectable in a tissue, cell or organelle under a specific set of conditions (see Proteome).

Microalgae: microscopic photosynthetic eukaryotes that form much of the marine phytoplankton. Researchers are attempting to engineer some forms of lipid-rich microalgae into bioreactors potentially capable of producing vast quantities of renewable, carbon-neutral biofuels.

Micro-RNAs: see sRNAs.

Microsatellite DNA: DNA markers used in MAS.

Monoecious: having unisexual flowers or cones. Individual plants of a monoecious species will carry both male and female flowering parts or cones, normally maturing at different times (see Dioecious).

Monogenic trait: genetic trait controlled by a single gene.

Mutagenesis: the process of causing mutations. Naturally occurring mutagenesis, e.g. via solar UV radiation, is a relatively slow process. Induced mutagenesis, e.g. via X-rays, γ-rays, DNA-damaging chemicals or targeted nucleases, is much used in genetics research and by plant breeders.

Mutation: inherited alteration in the DNA sequence of a genome. Somatic mutations normally only affect the cell, tissue or organism concerned and are not passed on to progeny. Germ line mutations are passed on to progeny and, if they result in viable offspring, they can become fixed in subsequent generations.

Mycorrhiza: one of numerous symbiotic fungal species associated with the roots of most higher plants. Mycorrhizae receive assimilates from plants in return for soil-borne nutrients and water.

Natural: a subjective and sometimes misleading term often describing a substance, process, organism or landscape not created or markedly influenced by humans.

NERICA: new rice for Africa – interspecific hybrid of Asian and African rice.

Non-host resistance: occurs when an entire plant species is resistant to a specific pathogen via a complex multigenic mechanism.

Nucleosome: the fundamental units of chromatin that are made up of a series of octomeric histone complexes each binding about 146 bp of DNA.

'Omics: collective term referring to a series of technologies for the collection, analysis and exploitation of large biological data sets that include genomics, proteomics, transcriptomics and metabolomics.

Oomycete: fungus-like organisms that include many virulent plant pathogens. Unlike fungi, oomycetes have cellulose cell walls and may be derived from algae that lost their plastids after assuming a heterotrophic lifestyle.

ORF: open reading frame – a DNA segment containing codons in the same reading frame that is uninterrupted by stop codons and normally makes up the transcribed region of a gene.

Output trait: a genetic character that alters the quality of the crop product itself, e.g. by altering its starch, protein, vitamin or oil composition.

Ozone layer: stratospheric ozone layer (15–30 km high) that absorbs high-energy solar UV radiation that would otherwise be harmful to most terrestrial organisms. In contrast, ground level (tropospheric) ozone can be a dangerous pollutant damaging plants and reducing crop yields.

Parenchyma: living, relatively unspecialized, cells found in most plant tissues.

Pathogen: disease-causing organism – the most important plant pathogens are fungi, viruses and bacteria.

Pathogenesis-related proteins (PR proteins): proteins whose synthesis is induced by the presence of potential pathogens. Some are involved in plant defence responses, but the function of many others remains unclear.

PBR: Plant Breeders' Rights – a form of intellectual property protection in the European Union (via UPOV) designed specifically for new varieties of plants.

PCR: polymerase chain reaction – a technique for rapidly copying a particular piece of DNA in the test tube (rather than in living cells). PCR has made possible the detection of tiny amounts of specific DNA sequences in complex mixtures.

Pest: animal that damages a useful plant. Most pest species are insects that damage plants mechanically and/or by herbivory (see Pathogen).

Pharmaceutical: physiologically active compound used in medical applications. Can be synthesized chemically or extracted from a biological source.

Phenotype: physical manifestation of the combined effects of the genotype and the environment for a given organism. Phenotypic traits can include external appearance, composition and behaviour.

Phloem: plant-wide transport network through which assimilates move from source to sink tissues.

Phosphinothricin acetyltransferase: bacterial enzyme that inactivates the herbicide glufosinate, and confers glufosinate tolerance when expressed in transgenic plants as part of the Liberty Link™ package.

Plasmodesmata: membrane-lined channels linking adjacent cells through their cell walls that enable the regulated movement of molecules through the symplast.

Plastids: DNA-containing organelles, found in all cells from plant-like organisms from algae to angiosperms, that originate from a cyanobacterial endosymbiont.

Pleiotropic effect(s): multiple phenotypic effects of a single gene.

Poaceae: a major monocot family, commonly called the grasses, that includes major crops such as the cereals plus minor grain crops such as tef (see Gramineae).

Polycistronic transgene: a transgenic DNA sequence containing several tandemly linked genes under the control of a single promoter.

Polygenic trait: genetic character regulated by several (sometimes many) genes, also called a complex or quantitative trait.

Polymerase chain reaction: see PCR, Inverse PCR and RT-PCR.

Polymorphism: occurrence of numerous allelic variants of a DNA sequence motif or an entire gene that are used for identification of individuals in breeding programmes and in research.

Polyploid: an organism containing more than two sets of chromosomes that can be either auto- or allopolyploid, depending on whether it arose from whole-genome duplication or from hybridization between two diploid progenitors.

Promoter: DNA region immediately upstream (5′) of an open reading frame that signals the transcriptional start site to RNA polymerase.

Propagule: part of a plant used in propagating a new adult generation; common propagules include seeds, pollen and cuttings or other tissue fragments.

Proteome: the expressed protein complement of an organism, tissue, cell or subcellular region (such as an organelle) at a specified stage of development and/or under a particular set of environmental conditions (see Metabolome).

Proteomics: description and analysis of proteomes.

Protoplast: a plant cell from which the cell wall has been removed, normally via digestion by cellulolytic enzymes.

Pulses: annual legumes cultivated for their starchy and relatively protein-rich seeds. Important pulses include peas, lentils and the numerous types of beans.

Pusztai affair: in 1998, UK-based plant researcher Arpad Pusztai made a highly publicized and much-disputed claim that rats fed on rations containing transgenic potatoes developed tissue abnormalities. After this, supermarkets across Europe withdrew food products containing ingredients from GM crops.

PVPA: Plant Variety Protection Act – legislation enacted in 1970 by the US Congress that extended UPOV-like legal protection to plant germplasm.

Quantitative genetics: study of continuously varying traits (such as height or weight) and their underlying mechanisms, and its use as a tool for crop improvement.

Quantitative trait loci (QTL): DNA regions associated with a particular trait, such as plant height. While QTL are not necessarily genes themselves, they are closely linked to the genes that regulate the trait in question. QTL normally regulate so-called complex or quantitative traits that vary continuously over a wide range.

Rachis: structure holding cereal grains onto the stalk of the plant, which in wild plants normally becomes brittle as the ears mature enabling grains to break off from the main plant and be dispersed. Domesticated cereals have a non-brittle rachis so they retain grain on the stalk for easier harvesting by farmers.

R&D (research and development): the discovery and application stages of a new technology. Research/discovery involves fundamental/basic research, while the development/application stage occurs in commercial settings and typically costs ten times more than the basic research.

RAPD: random amplified polymorphic DNA – PCR-based method of DNA profiling involving amplification of sequences using random primers.

Reactive oxygen species (ROS): oxygen radicals and non-radical derivatives of oxygen that may be formed during normal metabolic processes such as respiration and photosynthesis, or as a consequence of many forms of stress. ROS can damage biomolecules leading to injury, premature senescence or death. Examples include the superoxide (O_2^-), hydroxyl (OH), peroxyl (ROO) and alkoxyl (RO^-) radicals, as well as the non-radical intermediates singlet oxygen ($^1O_2$), hydrogen peroxide (H_2O_2) and ozone (O_3).

Real-time PCR: method for quantitative determination of DNA; see PCR.

Recessive trait: an inherited trait that is phenotypically obvious only when two copies of the gene for that trait are present. The recessive condition is masked by the dominant trait when both are present.

Recombinant DNA: DNA sequence created artificially from two or more sources incorporated into a single recombinant molecule. It is often made in order to create a transgenic organism.

Red biotechnology: use of transgenic animals, such as sheep or cattle, to produce useful compounds such as pharmaceuticals. Can also apply to other uses of biotechnology in animal systems, such as creating fish by transgenesis or using MAS in livestock breeding (see Blue, Green and White biotechnology).

Redundancy: presence of surplus capacity in a system that makes it more robust since failure of a single component is much less likely to result in whole-system failure. Biological examples include multiple independent routes to synthesize or degrade key compounds, multiple copies of genes or multiple sensory pathways.

Regeneration: the induction of tissue and organ differentiation to produce an adult plant from relatively undifferentiated cell or tissue culture.

Regulatory gene: a gene that regulates the expression of several other genes. Many regulatory genes encode transcription factors responsible for controlling complex developmental processes, such as flowering, or entire metabolic pathways.

Repetitive DNA: DNA sequences in a genome that are present in many copies. Can be either tandem or interspersed repeats (see Transposon and Retrotransposon).

Restriction enzyme: DNA-cleaving enzymes found in many bacteria, also called restriction endonucleases.

Retrotransposons: the most abundant class of transposon in eukaryotes and very common in some plant genomes. Retrotransposons require formation of an RNA intermediate to move within a genome. They are useful in phylogenetic and gene mapping studies and as DNA markers for advanced crop breeding.

Retrovirus: member of a class of RNA viruses that transcribes its RNA into DNA, after infecting a host cell. Related viruses in plants are pararetroviruses, some of which are pathogens of important food crops. Plant genomes harbour retroviral-like sequences and some plant retroviruses are used to facilitate gene transfer.

Reverse transcriptase: polymerase that synthesizes DNA from an RNA template.

Reverse transcription PCR: type of PCR where a reverse transcriptase is used for the first step, so that RNA can be used as the starting material.

RFLP (restriction fragment length polymorphism): a variation in the DNA sequence of a genome detected by fragmenting the DNA with restriction enzymes and analysing fragment sizes via gel electrophoresis. RFLPs are used in MAS.

Rhizosphere: soil region in the vicinity of plant roots and its associated flora, such as plant-associated bacteria and mycorrhizae.

RNA interference (RNAi): the sequence-specific alteration of gene expression by a small RNA (sRNA) molecule of 21–24 nucleotides. Such RNA sequences can originate from a pathogen or may be transgenically inserted to prevent expression of a target gene via complementary base pairing to its mRNA transcript (see Antisense DNA).

Roundup™: see Glyphosate and EPSP synthase.

R (resistance) genes: genes encoding receptors that recognize the products of specific avirulence (AVR) genes in a particular pathogen.

RT-PCR: see Reverse transcription PCR.

Saccharomyces cerevisiae: bakers' yeast.

Saccharomyces pombe: fission yeast.

Salicylic acid: plant hormone involved in the regulation of stress-related processes from pathogen responses to seed development.

SARS: Severe acute respiratory syndrome.

Sclerenchyma: dead, lignified cells such as fibres or sclereids that are the principal mechanical tissue in short-lived plants and in many young tissues. Generally found in tissues that have, at least temporarily, completed growth.

Second-generation biofuels: renewable fuels, still mostly under development, that are derived from biological systems other than food crops. Such fuels include ethanol from lignocellulosic biomass and hydrogen from waste biomass.

Selection: the process of identifying and choosing favourable genetic variants for breeding or propagation.

Shotgun cloning: the practice of cutting at random a large DNA fragment to reduce it into various smaller pieces that can then be cloned.

Single-locus probe: a DNA or RNA sequence able to hybridize (i.e. form a DNA–DNA or DNA–RNA duplex) with DNA from a specific restriction fragment on a Southern blot. A sequence of labelled DNA or RNA that can be used to identify a region of DNA tandem repeats that occurs only once in the genome.

Single nucleotide polymorphism (SNP): single DNA base change carried by some individuals in a population. One of the commonest and most useful forms of DNA marker used in molecular genetic studies; see MAS.

Sink: tissue or organ, such as a developing seed, tuber or root, that is a net importer of photosynthetic assimilates, usually via the phloem.

Site-directed mutagenesis: deliberate alteration in a DNA sequence in order to alter the function of a specific gene.

Somaclonal variation: genetic changes induced during plant cell or tissue culture, often involving epigenetic mechanisms such as DNA methylation, that may lead to unwanted phenotypes in regenerated plants.

Source: tissue or organ, such as a mature leaf, that is a net exporter of photosynthetic assimilates, usually via the phloem.

Species: group of genetically similar organisms normally capable of breeding with one another, but not with members of other species. While broadly applicable to animals, this definition is less useful for plants where interspecific hybridization is widespread and plays a key role in diversification, evolution and crop breeding.

sRNA: short or small RNA, class of small or micro RNAs, typically 18–40 nucleotides in length that do not encode protein products. sRNAs are involved in the regulation of many cellular processes, including gene expression, and are increasingly used as tools in biotechnology, see also RNA interference.

Storage proteins: specialized group of nitrogen-rich plant proteins synthesized during seed development that accumulate in cotyledon or endosperm cells and may comprise <25% total seed weight in some legumes.

Substantial equivalence: legal concept that applies when a new crop variety or food product is essentially similar in morphology and composition to existing versions. For transgenic crops, it means that the phenotype of the new variety should be virtually indistinguishable from non-transgenic counterparts, except for the presence and expression of the transgene(s).

Symplast: continuous intracellular network within plants where cellular contents are connected by plasmodesmata.

Synteny: conservation of gene order in organisms derived from a common ancestor.

Systemic acquired resistance (SAR): analogous to acquired immunity in animal systems, SAR is a generalized increase in resistance to further pathogenic attack following exposure of a plant to a pathogen.

T-DNA: transferred DNA from the tumour-inducing (Ti) plasmid of some bacterial species such as *Agrobacterium tumefaciens*.

Teosinte: the original wild grass, native to Mexico, from which cultivated maize is derived; it is now classified as a member of the same species, *Zea mays*.

Terpenoid: class of compound made up of one or more isoprene units that includes carotenoids, many plant hormones and important herbicides such as pyrethrins.

TILLING: Targeting induced local lesions in genomes – the directed identification of random mutations controlling a wide range of plant characters. A more sophisticated DNA-based version of mutagenesis breeding, TILLING does not involve transgenesis.

Tissue culture: cultivation of plant cells and tissues *in vitro*.

Totipotency: given the right conditions all plant cells are totipotent, i.e. they are able to express the entire genetic potential of their parent plant.

Tracheophytes: synonym for the vascular plants such as ferns, gymnosperms and angiosperms, which have a well developed vascular system of xylem and phloem.

Trait stacking: see Gene stacking.

Transcription: process of synthesizing a complementary strand of mRNA from the coding strand of a DNA double helix. The primary mRNA transcript is normally processed before the mature mRNA travels out of the nucleus and binds to a ribosome for translation into a protein.

Transcription factor: DNA-binding protein, often involved in regulation of several genes. Mutations in transcription factor genes are common mechanisms for radical phenotypic change in organisms, including aspects of the transition from wild to domesticated plants.

Transformation: uptake of exogenous DNA, normally via transgenesis.

Transgene: exogenous gene transferred to a recipient organism via genetic engineering to create a transgenic organism.

Transgenesis: the process of creating a transgenic organism.

Transgenic: an organism into which exogenous segments(s) of DNA, containing one or more genes, have been transferred (see GM).

Transgressive segregation: production of plants in the F_2 generation that are superior to both parents for one or more characters.

Translation: process of synthesizing protein on ribosomes via an mRNA template where each amino acid is specified by three RNA bases, known as a codon.

Transplastomics: targeted insertion of transgene(s) into the plastid genome. Because plastids are normally only maternally inherited, such transgenes are less likely to be transferred to other plants via cross-pollination than are nuclear transgenes.

Transposons: sometimes called 'jumping genes', these DNA sequences can move within the genome and relocate elsewhere, sometimes affecting gene function. While DNA transposons can move directly, retrotransposons require formation of an RNA intermediate.

Triacylglycerol: ester of glycerol with three fatty acyl side chains.

UPOV: Union for the Protection of New Varieties – established in 1960 in Europe to extend legal ownership rights to plant germplasm.

USDA: United States Department of Agriculture – government department that carries out and funds much of the public-sector crop-related research in the USA.

Vaccine: agent that triggers a protective immune response to a pathogen.

Variation: existence of genetically and phenotypically different forms within a population that is the 'raw material' for the operation of natural selection.

Vascular plants: see Tracheophytes.

Vascular tissue: transports food, water, hormones and minerals within the plant. Vascular tissue includes xylem, phloem, parenchyma and cambium cells.

Vector: a carrier. Many flowering plants rely on insects or other animal vectors to disseminate their pollen and, in some cases, seeds. Vectors can also be harmful: many pathogens are introduced to plants by animal vectors.

Vegetative propagation: form of non-sexual plant reproduction without meiosis or fusion of gametes that arises from vegetative structures, such as roots, stems or cuttings thereof. Progeny are genetic clones of the parent plant (see Apomixis).

Vernalization: promotion of flowering development by a period of low temperature, as seen in winter varieties of barley and wheat.

Xenoprotein: foreign or exogenous protein, often from a different species, expressed in a plant or other organism using transgenesis.

Waxy: class of mutants affecting granule-bound starch synthase resulting in a highly branched form of starch with reduced levels of amylose and a waxy seed texture.

White biotechnology: industrial-scale use of microbial systems, such as yeast or *E. coli*, to produce useful compounds such as antibiotics, enzymes and high-value food products. Historically non-transgenic microbes were used but transgenic organisms are used increasingly in modern processes, e.g. for vaccines or blood clotting factors (see Green and Red biotechnology).

Wide crossing: genetic cross where one parent is from outside the immediate gene pool of the other, e.g. a wild relative crossed with a modern crop cultivar.

Wild relative: plant or animal species taxonomically related to crop or livestock species; a potential source of genes for breeding new crop or livestock varieties.

Xylem: tissue made up of a number of lignified cell types, namely vessels, tracheids, fibres and non-lignified xylem parenchyma.

Younger Dryas Interval: period of sudden and profound climatic change involving widespread cooling and drying, from 12,900 to 11,600 BP.

Zinc-finger mutagenesis: gene knockout strategy using an engineered zinc-finger nuclease that recognizes and introduces a double-stranded break into a target DNA sequence in a genome.

Zygote: fertilized egg, normally created by fusion of haploid male and female gametes to produce a diploid cell that develops into an embryo.

Index

Page numbers in **bold** refer to illustrations and tables.